Met. O. 985

METEOROLOGICAL OFFICE

Meteorological Glossary

Sixth edition

LONDON: HMSO

UDC 551.6(02): 551.501.1

HMSO publications are available from:

HMSO Publications Centre
(Mail and telephone orders only)
PO Box 276, London, SW8 5DT
Telephone orders 071-873 9090
General enquiries 071-873 0011
(queuing system in operation for both numbers)

HMSO Bookshops
49 High Holborn, London, WC1V 6HB   071-873 0011 (Counter service only)
258 Broad Street, Birmingham, B1 2HE   021-643 3740
Southey House, 33 Wine Street, Bristol, BS1 2BQ   (0272) 264306
9-21 Princess Street, Manchester, M60 8AS   061-834 7201
80 Chichester Street, Belfast, BT1 4JY   (0232) 238451
71 Lothian Road, Edinburgh, EH3 9AZ   031-228 4181

HMSO's Accredited Agents
(see Yellow Pages)

and through good booksellers

FOREWORD

In preparing this new addition I have attempted to correct all the misprints and other minor errors in the last printing of the fifth edition, to revise entries in the light of recent advances where this seemed appropriate, and to include the new terms introduced since the last edition that a meteorologist might encounter in the scientific and technical meteorological and climatological literature, apart from those used only by a handful of expert specialists.

In the execution of this task I am glad to acknowledge the generous help of many of my former colleagues at Bracknell and the outstations who have provided critical comments on the entries in the old edition and suggested many new items for inclusion (often supplied with suitable wording). Any remaining defects are, of course, due only to me.

A great deal of Dr McIntosh's work remains unaltered, including the general structure and lay-out. However, I have decided to eliminate the climatological tables as being unsuitable for a *glossary*; such material, in much greater detail, is better sought in specific climatological publications. The Office has long ceased to have any active interest in either seismology or geomagnetism; entries relevant to those topics have therefore been either eliminated or drastically reduced. Similarly, certain entries dealing with technical, scientific or statistical matters that are not in themselves specifically meteorological have been abbreviated or eliminated.

Acronyms form a vexed issue; they proliferate, they are often ephemeral and related to temporary units of administration or policy steering committees, and are sometimes ambiguous and often ugly; they have been kept to a minimum.

SI units have been used almost exclusively in this edition except for those still used operationally in meteorology such as the knot and the millibar.

R.P.W. Lewis

PREFACE TO THE SIXTH EDITION

In 1916 the Meteorological Office published two pocket-sized companion volumes, the *Weather map* to explain how weather maps were prepared and used by forecasters, and the *Meteorological glossary* to explain the technical terms employed. That first edition of the *Meteorological glossary* contained 302 terms; the fifth edition, published in 1972, had something like seven times as many, and still the number of terms employed in meteorological research and day-to-day forecasting continues to increase so that it has been necessary to introduce well over 100 completely new entries in this sixth edition.

The first three editions of the *Meteorological glossary* (published in 1916, 1930 and 1939) were written collectively by the professional staff of the Office and this had led to some unevenness of treatment. When, in the mid 1950s, it was decided to have a new edition prepared incorporating information on the considerable advances that had taken place during and after the Second World War, it was thought that the unified approach to be expected from a single author was called for, and the Office was singularly fortunate in being able to secure the services of Dr D.H. McIntosh of the University of Edinburgh. Dr McIntosh is a physicist and meteorologist of wide experience who was himself a member of the Office for 15 years and was willing to consult and collaborate with the scientific staff of the Office to make the new edition (the fourth) as complete, informative and authoritative as possible.

In 1967 Dr McIntosh agreed to undertake the preparation of the fifth edition with co-operation from within the Meteorological Office, and it appeared in 1972 containing numerous revisions, new entries, and deletions of obsolete terms.

In 1986 Mr R.P.W. Lewis, a recently retired member of the Meteorological Office, was asked to prepare a further edition, incorporating the advances of the preceding 15 years. Mr Lewis was ideally qualified for this task, having been head of the Office's Library and Publications Branch for the last 13 years of a career that took him into a wide variety of areas of meteorology. His considerable experience and expertise is reflected in this latest edition which maintains the standards and updates the contents of its predecessors.

Meteorological Office, 1991

LIST OF SYMBOLS

1. Scalar quantities

a	radius of the earth
B_v,	radiance of black body
c	phase speed; specific heat
c_p	specific heat at constant pressure
c_v	specific heat at constant volume
e	vapour pressure
f	$2\Omega\sin\phi$ (Coriolis parameter)
g	gravitational acceleration
H	density scale height
h	pressure scale height
K	coefficient of eddy viscosity
k	wave number
l	mixing length
M	mean molecular weight of air
N	Brunt–Väisälä frequency
n	refractive index of a medium
p	pressure
p_s	surface pressure
q	quasi-geostrophic potential vorticity; specific humidity
R	specific gas constant; Bowen ratio
R^*	universal gas constant
Ra	Rayleigh number
Re	Reynolds number
Ri	Richardson number
Ro	Rossby number
r	humidity mixing ratio
S	entropy
T	temperature
t	time
U	relative humidity
u, v, w	velocity components in x, y, z directions
U_g, V_g	components of geostrophic wind
V	wind speed
x, y, z	rectangular coordinates: eastward, northward and vertical in a plane tangential to the earth's surface
Z	geopotential height
α	specific volume (ρ^{-1})
β	df/dy
Γ_d	dry adiabatic lapse rate
Γ_s	saturated adiabatic lapse rate
γ	ratio of specific heats (c_p/c_v); lapse rate
ζ	vertical component of relative vorticity
ζ_a	vertical component of absolute vorticity
η	viscosity
θ	potential temperature
θ_w	wet-bulb potential temperature
θ_{sw}	pseudo wet-bulb potential temperature
κ	$(\gamma-1)/\gamma = R/c_p$
λ	longitude
v	dynamic viscosity

Π	Exner function
ρ	density
σ	p/p_s; $-\alpha(\partial\theta/\partial p)$ (stability)
τ	Reynolds stress
Φ	geopotential
ϕ	latitude
ψ	stream function
Ω	angular velocity of the earth
ω	dp/dt (vertical component of velocity in pressure terms)

2. Vector quantities

i, j, k	unit vectors in direction of axes
V	wind velocity
V$_g$	geostrophic wind
V$_T$	thermal wind
ζ	vorticity $=$ curl**V**
$\boldsymbol\Omega$	angular velocity of the earth's rotation

3. Operators

$\dfrac{d}{dt}$ differentiation following a fluid element,

i.e. $\dfrac{\partial}{\partial t} + u\dfrac{\partial}{\partial x} + v\dfrac{\partial}{\partial y} + w\dfrac{\partial}{\partial z}$

$\left(\dfrac{d}{dt}\right)_G$ $\qquad \dfrac{\partial}{\partial t} + u_G\dfrac{\partial}{\partial x} + v_G\dfrac{\partial}{\partial y} + w_G\dfrac{\partial}{\partial z}$

∇	del-operator
∇^2	Laplacian operator

A

ablation. The disappearance of snow and ice by melting and evaporation. The chief meteorological factor which controls the rate of ablation is air temperature; subsidiary factors are humidity, wind speed, direct solar radiation and rainfall. The rate of ablation of a snow-field is also affected by such non-meteorological factors as size, slope and aspect of the snow-field, depth and age of snow, and nature of the underlying surface.

abroholos. A violent squall on the coast of Brazil, occurring mainly between May and August.

absolute extremes. See EXTREMES.

absolute humidity. An alternative for VAPOUR CONCENTRATION.

absolute instability. See STABILITY.

absolute instrument. An instrument with which measurements may be made in units of mass, length, and time (or in units of a known and direct relationship to these) and against which other non-absolute instruments may be calibrated. See also STANDARD.

absolute momentum. The vector $(\mathbf{M},\mathbf{N})$ where

$$\mathbf{M} = v + fx$$
$$\mathbf{N} = -u + fy$$

(u,v) being the horizontal velocity and f the CORIOLIS PARAMETER in the x,y directions.
   On the surface of the rotating earth, in the absence of forcing terms such as pressure gradients, $\mathbf{M}$ and $\mathbf{N}$ are the appropriate conserved quantities that replace the velocities v and u. The term 'pseudo-angular momentum' is also sometimes used.

absolute stability. See STABILITY.

absolute temperature. A temperature measured with respect to ABSOLUTE ZERO.

absolute vorticity. See VORTICITY.

absolute zero (of temperature). A temperature of $-273.15\ °C$, the zero on the kelvin (absolute) scale of temperature.

absorption. Removal of radiation from an incident beam, with conversion to another form of energy — electrical, chemical or heat.
   The absorption of radiation by gases is highly selective in terms of wavelengths and may depend also on pressure and temperature. Numerical expression is given by the law, variously known as Bouguer's or Lambert's law, applicable to monochromatic radiation:

$$I = I_0 e^{-\alpha m}$$

1

where I_0 is the intensity of incident radiation, I the intensity after passing through mass m of the absorbing substance and α the absorption coefficient. An alternative expression is

$$I = I_0 e^{-\alpha x}$$

where x is the path length through the absorbing substance. α has dimensions of M^{-1} for a mass absorption coefficient or L^{-1} for a length absorption coefficient.

The effectiveness of a gas as an absorber of solar or terrestrial radiation depends on the width and strength (absorption coefficient) of the absorption lines and bands, the concentration of the gas, and the wavelength positions of the bands relative to the maximum of the Planck curve (see RADIATION) at solar or terrestrial temperature, respectively. The relative energies involved in atmospheric absorption processes are represented in Figure 1, in which the Planck curves appropriate to solar and terrestrial radiation temperatures (6000 K and 245 K, respectively) are shown with equal areas to represent net balance of the two fluxes. Fine structure of the absorption bands is omitted. The main constituents of the atmosphere, N_2 and O_2, are almost completely transparent except in the far ultraviolet: minor constituents such as O_3, CO_2, H_2O and N_2O have intense absorption bands, mainly in the infra-red and longer wavelengths.

FIGURE 1.   (a) Black-body curves for 6000 K and 245 K, (b) atmospheric gaseous absorption spectrum for a solar beam reaching ground level, and (c) as (b) but for a solar beam reaching the temperate tropopause. The axes are chosen so that areas in (a) are proportional to radiant energy. Integrated over the earth's surface and over all solid angles, the solar and terrestrial fluxes are equal. Consequently, the two black-body curves are drawn with equal areas beneath them. An absorption continuum has been drawn beneath bands in (b). This is partly hypothetical because it is difficult to distinguish from the scattering continuum, particularly in the visible and near infra-red spectrum. Conditions are typical of mid-latitudes and for a solar elevation of 40° or diffuse terrestrial radiation.

2

The product of (a) and (c) in Figure 1 represents the energy absorbed by the STRATOSPHERE, and the product of (a) and {(b) − (c)} that absorbed by the TROPOSPHERE.

Since water has an appreciable absorption coefficient at wavelengths greater than about one micrometre, thick clouds are able to absorb over 20 per cent of incident solar radiation. Towards terrestrial radiation, clouds and fog behave as almost perfect black bodies. The surface of the earth is very variable in its absorption of solar radiation (see ALBEDO) but absorbs nearly all incident terrestrial radiation.

acceleration.   Rate of change of velocity with respect to time. It is a vector quantity having both magnitude and direction; thus uniform circular motion, involving change of direction albeit with constant speed, implies an acceleration directed towards the centre of the circle ('CENTRIPETAL ACCELERATION'). The dimensions are LT^{-2}.

In meteorology, the velocity and acceleration of the air are almost always measured relative to a set of axes fixed in the earth. Because the earth is rotating, such a set of axes, and the accelerations and velocities with respect to them, is not suitable for an immediate application of Newton's second law of motion. If the 'absolute' velocity and acceleration of the air (relative to the stars) are denoted by $\mathbf{V_a}$ and $d\mathbf{V_a}/dt$, while those relative to the earth are $\mathbf{V}$ and $d\mathbf{V}/dt$, then it may be shown that

$$\frac{d\mathbf{V_a}}{dt} = \frac{d\mathbf{V}}{dt} + 2\mathbf{\Omega} \times \mathbf{V} + \mathbf{c}$$

where $\mathbf{\Omega}$ is the ANGULAR VELOCITY of the earth and $\mathbf{c}$ is the CENTRIPETAL ACCELERATION of the air when at rest relative to the earth. The acceleration ($\mathbf{c}$) is directed towards the axis of rotation of the earth but is independent of $\mathbf{V}$ and may thus be conveniently combined vectorially with the acceleration towards the centre of the earth (due to gravity) into a single acceleration of $\mathbf{g}$ directed along the local vertical. It follows that the expression for the absolute value of the acceleration ($d\mathbf{V_a}/dt$) which needs to be equated to the vector sum of all the forces per unit mass acting on the air is

$$\frac{d\mathbf{V}}{dt} + 2\mathbf{\Omega} \times \mathbf{V} + \mathbf{g}.$$

acclimatization.   The process of physiological adjustment by animals and plants to stressful, but not necessarily extreme, environmental conditions.

accretion.   In meteorology, this usually refers to the growth of an ice particle by collision and COALESCENCE with water drops. The term is also used in the more general sense of growth of water drops, or ice particles, by collision and coalescence. See PRECIPITATION, ICE ACCRETION.

accumulated temperature.   The integrated excess or deficiency of temperature measured with reference to a fixed datum over an extended period of time. If on a given day the temperature is above the datum value for n hours and the mean temperature during that period exceeds the datum line by m degrees, the accumulated temperature for the day above the datum is nm degree-hours or $nm/24$ degree-days. By summing the daily entries arrived at in this way, the accumulated temperature above or below the datum value may be evaluated for periods such as a month, a season or a year.

In practice, daily values of accumulated temperature are derived not from hourly values but by a method involving the use of daily maximum (X) and minimum (N) temperatures; empirical formulae relating to X, N and the datum value (D) are used when D lies between X and N.

For the study of heating problems, engineers use various datum values according to the particular need; examples are 18.5 °C and 15.5 °C for heating and 5 °C for cooling. For datum values for agricultural applications see GROWING SEASON.

accuracy. In physical measurement, the closeness with which an observation of a quantity, or the mean of a series of observations, is considered to approach the unknown true value of the quantity. See also ERROR.

acid rain. A popular expression attached to the deposition of various airborne pollutants (see ATMOSPHERIC POLLUTION) which are perceived to have harmful effects on vegetation, soils, buildings and waters. Although originally applied only to rain and snow which was significantly acidic as a result of pollution, it came to be applied by the media to other deposited pollutants having ecological significance (e.g. to ozone and other photo-oxidants). Analysis of various deposits (for example, sediments in lakes) has shown that 'natural' rain which fell in pre-industrial times was slightly acid with a pH (see ATMOSPHERIC CHEMISTRY) of 5.0 or rather more; this acidity was probably due to dissolved atmospheric carbon dioxide with additional contributions from oxides of nitrogen produced in thunderstorms and to other acidic compounds arising from forest fires and volcanoes. By the late 1970s, however, the average pH of rainfall in densely populated industrial countries had fallen to about 4.0–4.5, with some exceptional occurrences where it was as low as 3.0 — about as acid as vinegar.

The mechanisms whereby rain becomes acidified and the biological effects of acid rain have been the subjects of intensive study and much has been learnt. However, even by 1987 many questions remain unanswered, especially on the 'effects' side, owing largely to the long time-records required and the slow appearance of some of the effects. It is clear, however, that the essential cause of the damage which has been observed in forests, soils and fish populations is directly relatable to airborne pollution although not necessarily in any simple way because of the complicating and sometimes synergistic influences of climatic stress, disease, soil type and other factors.

acoustic sounding. A technique for investigating the fine structure of the TROPOSPHERE by projecting high-intensity bursts of sound, each lasting a few milliseconds, in a well collimated beam from a loudspeaker (or array of loudspeakers) and recording the back-scattered echoes. Acoustic sounding provides information on temperature inversions, the eddy components of wind and temperature variations and also, through measurements of the Doppler shift (see DOPPLER EFFECT), on the more general wind structure. Acoustic sounders with transmitter and receiver either sharing the same antenna or placed so close to each other that their separation may be ignored are known as 'monostatic'; otherwise, they are 'bistatic'. Monostatic recorders provide information only on the temperature structure.

actinic rays. Radiation which effects chemical changes, as in photography. The term is also loosely used to signify ULTRAVIOLET RADIATION.

actinometer. An early name for an instrument which measures solar radiation, usually at normal incidence, as in the Linke-Fuessner and Michelson actinometers. The corresponding term for a recording instrument is 'actinograph'. The name 'actinometer' is also applied to an instrument which measures the intensity of ACTINIC RAYS. See also PYRANOMETER and PYRHELIOMETER.

actinon. Gas, of atomic mass 219 and atomic number 86, which is a radioactive isotope of RADON. It occurs in minute concentration in the atmosphere and plays a small part in the IONIZATION of the air at low levels.

adiabatic. An adiabatic process (thermodynamic) is one in which heat does not enter or leave the system.

Because the atmosphere is compressible and pressure varies with height, adiabatic processes play a fundamental role in meteorology. Thus, if a parcel of air rises, it expands against its lower environmental pressure; the work done by the parcel in so expanding is at the expense of its internal energy, and its temperature falls, despite the fact that no heat leaves the parcel. Conversely, the internal energy of a falling parcel is increased and its temperature raised, as a result of the work done on the air in compressing it.

Observation shows that such processes determine, to a large extent, the vertical temperature distribution within the TROPOSPHERE. It also supports the view that, to a first approximation, it is justifiable to treat the vertically moving, individual masses of air of indefinite size (termed 'parcels') as CLOSED SYSTEMS which move through the environment without unduly disturbing it or exchanging heat with it. Various non-adiabatic processes such as condensation, evaporation, radiation, and turbulent mixing also operate to produce temperature changes in the free atmosphere but their effects are generally negligible in comparison with those caused by appreciable vertical motion.

Such adiabatic or 'dynamical' temperature changes proceed at a definite rate. For dry (unsaturated) air the change in temperature per unit height change (i.e. the LAPSE rate) is given by the equivalent expressions:

$$\frac{\gamma-1}{\gamma} \frac{g}{R} \text{ or } \frac{g}{c_p}$$

when R is 287 J kg^{-1} K^{-1}, g 9.8 m s^{-2} and γ 1.4. The 'dry adiabatic lapse rate' (DALR) is about 0.98 °C per 100 m or, with sufficient accuracy, 1 °C per 100 m (5.4 °F per 1000 ft).

For a saturated rising parcel the fall of temperature is reduced by the latent heat liberated. The 'saturated adiabatic lapse rate' (SALR) is therefore less than that for unsaturated air by an amount which varies with temperature and pressure; at lower levels in temperate latitudes the SALR is about half that of the DALR. Since widespread vertical motion occurs in the TROPOSPHERE, the average lapse rate in this region lies between the DALR and SALR.

Two extreme types of process involving ascent of saturated air may be visualized:

(i) a reversible, adiabatic ascent at the SALR, in which all products of condensation — cloud, rain, hail or snow — are retained within the ascending air, partake of the temperature changes of the air, and are available for evaporation at the appropriate stages on subsequent descent of the air, which is also at the SALR, and

(ii) an irreversible and, strictly, non-adiabatic process, in which all products of condensation are removed during ascent and in which subsequent descent of the air is at the DALR.

The latter process corresponds much more closely to what happens in the atmosphere than does the former and is termed a pseudo-adiabatic process. Because of the loss of the heat content of the precipitated water, cooling on ascent in a pseudo-adiabatic process is at a rate slightly in excess of the SALR, but the difference between the rates is negligible; thus, saturated adiabatics and pseudo-adiabatics (lines on an AEROLOGICAL DIAGRAM representing the respective lapse rates) are, for practical purposes, identical. The important distinction between the pseudo-adiabatic and reversible processes lies in the different rates of temperature change undergone by the air on subsequent descent. See also ENTROPY, ISENTROPIC, ADIABATIC EQUATIONS.

adiabatic atmosphere. A hypothetical atmosphere characterized by the dry adiabatic lapse rate throughout. It is also termed 'neutral atmosphere' or 'convective atmosphere'. See also ADIABATIC.

adiabatic diagram. An alternative for AEROLOGICAL DIAGRAM, or THERMODYNAMIC DIAGRAM.

5

adiabatic equations.   The three (equivalent) relationships between the variables of state, pressure (p), specific volume (v), and temperature (T), of a PERFECT GAS in an ADIABATIC process.

They are:

(i) $pv^\gamma = $ constant
(ii) $Tv^{\gamma-1} = $ constant, and
(iii) $Tp^{-(\gamma-1)/\gamma} = $ constant

The third relationship is the one most often used in meteorology. For dry air at ordinary temperatures, $\gamma = 1.4$ and

$$\frac{\gamma-1}{\gamma} = \frac{c_p-c_v}{c_p} = \frac{R}{c_p} = 0.286.$$

The same value may be used as an approximation for moist air. See SPECIFIC HEAT.

adiabatic region.   That region of the atmosphere (the TROPOSPHERE) where the temperature LAPSE rate is determined mainly by ADIABATIC motion of the air; it is also termed the 'convective region'.

adsorption.   The penetration of a substance, e.g. gas or thin film of liquid, into the surface layer of a solid with which it is in contact.

advection.   The process of transfer (of an air-mass property) by virtue of motion. In particular cases, attention may be confined to either the horizontal or vertical components of the motion. The term is, however, often used to signify horizontal transfer only.

advection fog.   Fog formed by the passage of relatively warm, moist and stable air over a cool surface. It is associated mainly with cool sea areas, particularly in spring and summer, and may affect adjacent coasts. It may occur also over land in winter, particularly when the surface is frozen or snow-covered — sometimes in conjunction with RADIATION FOG. The term is also used to describe pre-existing fog transferred from a distant source which may not necessarily be cooler, e.g. the inland spread of sea fog due to a developing sea-breeze circulation.

advective change of temperature.   That contribution to local temperature change which is caused by either (or both) horizontal or vertical ADVECTION of air. The horizontal component of change, which is generally the more effective in the TROPOSPHERE, is proportional to the horizontal temperature gradient at the level concerned and to the wind speed in the direction of this gradient; the vertical component of change is proportional to the vertical wind velocity and to the static STABILITY of the air and depends also on whether or not the air is saturated.

aerobiology.   The study of airborne micro-organisms, pollens, spores, etc. especially as sources of infection.

aerodrome meteorological minima.   Limiting meteorological conditions prescribed for the purpose of determining the usability of an aerodrome either for take-off or for landing.

aerodynamic roughness, smoothness.   A physical boundary is 'aerodynamically rough' when fluid flow is turbulent down to the boundary itself. Over such a boundary the velocity profile and surface drag are independent of the fluid viscosity (η) but depend on a ROUGHNESS LENGTH (z_0) which is related to the height and spacing of the roughness elements of the surface. A surface is 'aerodynamically smooth' if there exists a layer adjacent to it in which the flow is laminar and in which the velocity profile and surface drag are related to the fluid viscosity. A surface which is aerodynamically smooth at low speed of flow may become aerodynamically rough at a higher speed. A surface may thus be described as rough or smooth only in terms of the associated flow; alternatively, the flow itself may be described as aerodynamically rough or smooth.

In meteorology, nearly all surfaces are aerodynamically rough for any measurable wind speed.

aerodynamics. The study of the forces acting on bodies in motion through the air.

aerogram. An AEROLOGICAL DIAGRAM, due to A. Refsdal, in which the abscissa is log T and the ordinate $T \log p$.

aerological diagram. A graphical representation of the observations of pressure, temperature and humidity, made in a vertical sounding of the atmosphere.

The reference lines which facilitate the plotting of a sounding and its assessment after plotting are isobars, isotherms, dry adiabatics, saturated (pseudo-) adiabatics and lines of constant saturation humidity mixing ratio. Each of the diagrams in common use — the tephigram, emagram and Stüve diagram — has its own particular advantage. Each has, exactly or very nearly, the property of being a true thermodynamic ('equivalent') diagram in that equal area represents equal energy at any point of the diagram; possession of this property simplifies energy and height (geopotential) calculations.

The TEPHIGRAM (T–ϕ gram) — see Figure 2 — has rectangular Cartesian coordinates in which the abscissa is temperature (T) and the ordinate ENTROPY (ϕ) (entropy is now normally designated S, not ϕ), i.e. log θ, where θ is the dry-bulb potential temperature; the dry adiabatics ($\theta =$ constant) are therefore straight lines perpendicular to the isotherms. The basic coordinates in the EMAGRAM are T and $\log p$ (where p is pressure); in a 'rectangular' model the T and $\log p$ axes are perpendicular to each other; in the 'oblique' emagram these axes meet at an angle of 45°, thus making the dry adiabatics (slightly curved) and isotherms meet at an angle of about 90°. This latter is a decided advantage in assessing the static STABILITY of an ascent curve since the normal range of lapse rate of temperature in the troposphere is between the isothermal and dry adiabatic rates. In the STÜVE DIAGRAM rectangular coordinates of T and p^κ ($\kappa = 0.286$) are used; the dry adiabatics are then straight lines (as are the isobars and isotherms) but the property of strict equivalence of energy and area is sacrificed.

aerology. A word denoting the study of the atmosphere, but generally used in the sense of a study limited to the atmosphere above the surface layers.

aeronomy. A term sometimes used to denote that branch of atmospheric physics which is concerned with those regions, upwards of about 50 km, where DISSOCIATION and IONIZATION are fundamental processes.

aerosol. In meteorology, an aggregation of minute particles (solid or liquid) suspended in the atmosphere. See NUCLEUS.

afterglow. See ALPINE GLOW.

ageostrophic wind. The VECTOR difference between the actual wind and the GEOSTROPHIC WIND (see Figure 3); it is also called the 'geostrophic departure' and 'geostrophic deviation'.

FIGURE 2. Aerological diagram (tephigram) illustrating pressure-temperature plots (with dew-points shown as dashed lines) of two ascents made at Crawley (51° 05′N, 0° 13′W). (a) Ascent made in Arctic air at 1200 GMT on 6 December 1980, and (b) ascent made in maritime tropical air at 1200 GMT on 17 November 1980.

Actual wind = geostrophic wind + ageostrophic wind

The ageostrophic wind is of fundamental importance in that it is necessarily associated with CONVERGENCE or DIVERGENCE and vertical motion in the atmosphere. The ISALLOBARIC WIND is a particular example of a component of the ageostrophic wind; an ageostrophic component is also present in the SURFACE WIND and GRADIENT WIND.

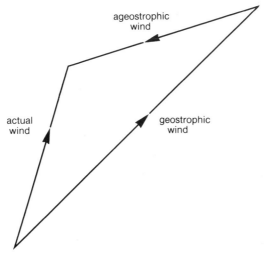

FIGURE 3.   Ageostrophic wind.

aggregation.   The process of growth of snowflakes or ice crystals by collision and adherence.

agroclimatology.   The study of those aspects of climate which are relevant to the problems of agriculture. Such study involves types of data, e.g. EARTH TEMPERATURE and ACCUMULATED TEMPERATURE, which are often not considered in more general CLIMATOLOGY.

agrometeorology.   The scientific study of weather and climate in relation to agriculture. The application of such knowledge.

agro-met station.   A STATION at which measurements are made that are especially relevant to AGROMETEOROLOGY. (An agro-met station is not necessarily identical with an 'agricultural meteorology station' as defined by WMO.)

air.   The mixture of gases which form the earth's ATMOSPHERE. In the absence of dust and water vapour, the composition of the air up to a height of about 20 km is taken to be as shown in Table I; the percentage composition by volume and weight is given, i.e. VOLUME FRACTION and MIXING RATIO.

Of the gases shown in Table I only carbon dioxide and ozone have appreciable local variations of concentration. There are also, mainly in the low atmosphere, minute variable quantities of such gases as radon, actinon, thoron, sulphur dioxide, hydrogen chloride, methane and nitrous oxide.

Ordinary (moist) air may be regarded as a mixture of dry air and water vapour. The concentration of water vapour in surface air varies from a small fraction of 1 per cent to over 3 per cent; in general the concentration decreases with increasing altitude.

TABLE I.   *Composition of dry air*

	Molecular weight	Proportional composition	
		By volume *per cent*	By weight *per cent*
Dry air	28.966	100.0	100.0
Nitrogen	28.013	78.09	75.54
Oxygen	31.999	20.95	23.14
Argon	39.948	0.93	1.27
Carbon dioxide	44.010	0.03	0.05
Neon	20.183	1.8×10^{-3}	1.2×10^{-3}
Helium	4.003	5.2×10^{-4}	7.2×10^{-5}
Krypton	83.800	1.0×10^{-4}	3.0×10^{-4}
Hydrogen	2.016	5.0×10^{-5}	4.0×10^{-6}
Xenon	131.300	8.0×10^{-6}	3.6×10^{-5}
Ozone	47.998	1.0×10^{-6}	1.7×10^{-6}

aircraft to satellite data relay (ASDAR).   A system whereby observations of temperature, wind and position made from an aircraft are transmitted to a communications satellite for relay to receiving stations on the ground. ASDAR is particularly useful in providing observations from intercontinental aircraft routes over the oceans where conventional sources of data are sparse or non-existent, and from the immediate vicinity of airports where the availability of up-to-the-minute information is important for aircraft safety.

airfield colour code.   A code which summarizes actual or forecast landing conditions at an airfield. Different combinations of cloud base and visibility are each given the name of a different colour.

airglow.   General term for the radiation which is emitted continuously by the upper atmosphere. The day, twilight and night emissions are termed DAYGLOW, TWILIGHT GLOW and NIGHTGLOW, respectively, the last being the most extensively studied.

airlight.   The increase in apparent brightness of a distant object viewed in daylight, owing to scattering of light towards the observer by particles held in suspension in the atmosphere, and by air molecules, between the observer and object. Airlight, and therefore object brightness, increases with object distance owing to increase in the number of scattering agents. A critical point is reached, limiting daylight VISIBILITY, at which the brightness of a suitable object is just indistinguishable from its background. See also KOSCHMIEDER'S LAW, CONTRAST THRESHOLD OF THE EYE.

air mass.   A body of air in which the horizontal gradients of temperature and humidity are relatively slight and which is separated from an adjacent body of air by a more or less sharply defined transition zone (FRONT) in which these gradients are relatively large.

The horizontal dimensions of air masses are normally hundreds or even thousands of kilometres. The term is, however, also used in relation to phenomena of much smaller scale, e.g. the sea-breeze.

Homogeneity in a body of air is produced by prolonged contact, in a 'source region', with an underlying surface of uniform temperature and humidity. The main source regions are those in which occur the permanent or semi-permanent anticyclones, with rather indeterminate boundaries, which are a prominent feature of the GENERAL CIRCULATION — the subtropical, polar, and winter continental anticyclones. Slow transformation of the air-mass properties acquired at the source region is effected on subsequent movement of the air from the region, mainly through its contact with a different surface, but to an appreciable extent also by radiation and large-scale vertical motion. The synoptic meteorology of middle latitudes, in particular, is dominated by considerations of the properties originally acquired by the air mass and the manner of their recent modification — whether becoming warmer or cooler, more or less moist, or more or less stable.

Air masses are classified into groups designated as 'polar' (P) or 'tropical' (T), 'maritime' (m) or 'continental' (c), defining the basic temperature and humidity characteristics, respectively; more generally, a twofold classification in terms of both elements, e.g. Tm or Pc, is used. Further divisions are sometimes made into Arctic or Antarctic (A) air and into classes of more local significance, e.g. Mediterranean air.

air-mass analysis. Synoptic identification of AIR MASSES and location of boundaries (FRONTS) between adjacent air masses.

The identification of an air mass implies the determination of its source region and would appear to entail the retrospective tracking of the air back to such a region. In practice, such a procedure is very seldom required because of the continuity provided by a series of synoptic charts. An air mass moving from its source region usually has, in its early stages, a well defined front at its junction with the adjacent air mass. Further identification of the limits of the air masses is largely made in terms of the movement of this front which has a high degree of continuity with time. Cloud type, visibility, and lapse rate of temperature are elements in which there is normally a discontinuity at an air-mass boundary. Recourse is sometimes made, in identifying air masses, to such derived parameters as POTENTIAL TEMPERATURE and WET-BULB POTENTIAL TEMPERATURE which are conservative, or quasi-conservative, for certain specific processes to which the air may be subjected.

air-mass climatology. Description of climate in terms of the frequencies and properties of the different types of AIR MASS which affect a specified region in a specified period.

air-mass thunderstorm. A THUNDERSTORM which is formed by convection within an AIR MASS, usually by heating of the lower layers. By implication, it is a storm in whose formation neither a front nor large-scale dynamical lifting of the air mass plays an important part.

air meter. An instrument for measuring the flow of air. It consists of a light 'windmill' in which inclined vanes are carried on the spokes of a wheel arranged to rotate about a horizontal axis. A system of counters is provided to show the number of rotations of the wheel. Calibration is effected in terms of a speed unit so that the instrument acts as a convenient portable ANEMOMETER. Both 'sensitive' (low-speed) and 'high-speed' air meters are used.

air pocket. An obsolescent term for a region of descending air in which an aircraft experiences a proportionate decrease of lift.

Air pockets are usually experienced in association with convective-type storms ('downdraughts') and, in strong and squally winds, on the leeward side of hills, buildings and other obstructions.

Aitken nucleus. See NUCLEUS.

albedo. A measure of the reflecting power of a surface, being that fraction of the incident RADIATION (total or monochromatic) which is reflected by the surface.

Typical values of total albedo (per cent) of various surfaces are: forest 5 to 10, wet earth 10, rock 10 to 15, dry earth 10 to 25, sand 20 to 30, grass 25, old snow 55 and fresh snow 80. The albedo of a water surface varies from about 5 per cent when the incident and reflected beams are near the normal, to 70 per cent at near-grazing incidence. The albedo of clouds is difficult to measure but is known to depend on cloud type and thickness; estimates of an average value in time and space vary from 50 to 65 per cent.

The albedo of the earth–atmosphere system as a whole ('planetary albedo') is estimated to be about 40 per cent. This signifies that about four-tenths of the incident solar radiation is returned to space, without change of wavelength, by reflection from clouds and the earth's surface and by back-scattering from air molecules and dust. A similar value, with real variations up to about 5 per cent, is inferred from photometric comparisons of the earth-lit and sunlit segments of the moon.

albedometer. An instrument for measuring the ALBEDO of a surface.

Aleutian low. A depression, centred near the Aleutian Islands in the North Pacific Ocean, which is a conspicuous feature of the northern hemisphere mean-sea-level pressure chart in winter. The depression has an average central pressure below 1000 mb in January, and represents the aggregate of the many depressions which affect this region in winter.

aliasing. A type of error that arises when measurements of a variable are made at finite intervals (spatial or temporal) instead of continuously. Values deriving from a high-frequency harmonic component can be interpreted as due to a low-frequency component, and vice versa, with consequential errors in the analysis of the data. See

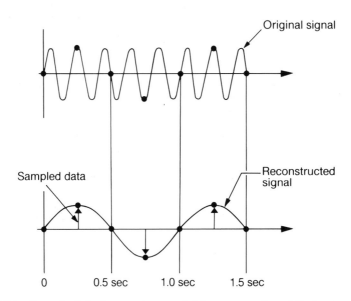

FIGURE 4. Example of aliasing error in a sampled-data system using a 5 Hz signal sampled at four samples per second.

alidade. An instrument for measuring the angular elevation of an object, e.g. a cloud feature or a searchlight spot. The object is sighted by the observer along a rod whose angular position (0–90°) with respect to the horizontal is obtained by reading from a scale of degrees. Both fixed and portable (or 'hand') alidades are used.

12

alpha (or **α**) **particle.** A particle emitted spontaneously from the nuclei of certain radioactive elements. It is identical with a HELIUM nucleus, comprising two neutrons and two protons, and therefore carries a positive charge of two units.

α particles are of such low penetrative power (only a few centimetres in air) that the particles emitted by radioactive materials in the earth's crust are insignificant in forming IONS. α particles are, however, also emitted from the radioactive gases, mainly, RADON and THORON in the lower layers of the atmosphere, and are responsible for a significant part of the ionization of the air at these levels, over land. See also BETA PARTICLE, GAMMA RADIATION.

alpine glow. A series of phenomena seen in mountainous regions around sunrise and sunset.

Two principal phases are generally recognized:

(i)   The true alpine glow. At sunset this phase begins when the sun is 2° above the horizon; snow-covered mountains in the east are seen to assume a series of tints from yellow to pink, and finally purple. As this phase is due mostly to direct illumination by the sun it terminates when the mountain tops pass into the SHADOW OF THE EARTH. The alpine glow is most striking when there are clouds in the western sky and the illumination of the mountains is intermittent.

(ii) The afterglow. This begins when the sun is well below the horizon, i.e. by 3° or 4°. The lighting is faint and diffuse with no sharp boundary and occurs only when the PURPLE LIGHT is manifest in the sky.

Alter shield. A form of RAIN-GAUGE shield consisting of a funnel-shaped construction of spaced slats surrounding the gauge. The purpose of the shield is to improve the representativeness of the gauge catch, especially of snow, in windy conditions. See also EXPOSURE.

alti-electrograph. A balloon-borne device for obtaining records of the vertical component of the electric field inside thunderstorms. The direction and magnitude of the current flow between two vertically separated points is recorded on a disc of pole-finding paper. A chemical reaction at the electrodes pressing on the paper makes a trace, the width of which varies with current.

altimeter. An instrument for determining the altitude (generally of an aircraft) with respect to a datum level. The two main types are (i) a radio altimeter and (ii) a pressure altimeter.

The radio altimeter converts the time taken by a radio pulse to be reflected from the underlying surface into a geometric height; it is in general sufficiently accurate only at low levels, and false readings are occasionally produced by large raindrops.

The pressure altimeter is an ANEROID BAROMETER which is calibrated directly in height units on the basis of the 'altimeter equation':

$$z = \frac{R\overline{T_v}}{g} \log_e \frac{p_0}{p_1} \approx 67.4 \ \overline{T_v} \ \log_{10} \frac{p_0}{p_1} \quad \text{(metres)}$$

$$\approx 221.1 \ \overline{T_v} \ \log_{10} \frac{p_0}{p_1} \quad \text{(feet)}$$

where z is the height, R the specific gas constant for dry air, $\overline{T_v}$ the mean virtual air temperature in the air column at the bottom and top of which the pressures are p_0 and p_1, respectively and g the gravitational acceleration.

The assumptions made in the instrument graduation are that p_0 has the value appropriate to mean-sea-level pressure of the ICAO STANDARD ATMOSPHERE, and that the variation of mean virtual temperature with height corresponds to that in the standard atmosphere.

Corrections may be required to take account of the fact that actual conditions differ from those of the standard atmosphere. The first concerns p_0 and involves an ALTIMETER SETTING; the required correction is the same at all altitudes. Thus an ICAO altimeter, for example, shows the height interval in the ICAO standard atmosphere between the altimeter-setting pressure and the ambient pressure. The second correction involves a positive (negative) correction when actual $\overline{T}_v$ is higher (lower) than that of the standard atmosphere; this correction is negligible at low altitudes. See also D-VALUE.

altimeter setting. The altimeter setting, designated QNH in the aircraft Q-CODE, is defined as that value of pressure, for a particular aerodrome and time, which, when set on the sub-scale of a standard ALTIMETER (based, for example, on the ICAO STANDARD ATMOSPHERE), will cause the altimeter to read the height of the aerodrome when the aircraft is at rest on the aerodrome, assuming that adjustment has been made for the height of the cockpit above aerodrome level. For procedure for obtaining QNH see *Observer's handbook* [1].

altitude. The angular height of an object above the horizon: synonymous with angle of ELEVATION.

In meteorology, altitude generally signifies height above mean sea level (geometric metres or feet); in dynamical meteorology, however, height is usually expressed in GEOPOTENTIAL metres or feet. In aviation, altitude signifies geometric height of an aircraft above mean sea level.

altocumulus (Ac). One of the CLOUD GENERA. (Latin, *altum* height, and *cumulus* heap.)

'White or grey, or both white and grey, patch, sheet or layer of cloud, generally with shading, composed of laminae, rounded masses, rolls, etc., which are sometimes partly fibrous or diffuse and which may or may not be merged; most of the regularly arranged small elements usually have an apparent width of between one and five degrees' [2, p. 31]. See Figure 5. See also CLOUD CLASSIFICATION.

J.A. Walton

FIGURE 5. Altocumulus translucidus with cumulus humilis beneath.

14

altostratus (As).　One of the CLOUD GENERA. (Latin, *altum* height, and *stratus* spread out.)

'Greyish or bluish cloud sheet or layer of striated, fibrous or uniform appearance, totally or partly covering the sky, and having parts thin enough to reveal the sun at least vaguely, as through ground glass. Altostratus does not show halo phenomena' [2, p. 35]. See also CLOUD CLASSIFICATION.

ambient pressure, temperature, etc.　The pressure or temperature, etc. in that part of the atmosphere which immediately surrounds a specified physical entity such as cloud or a thermal.

amorphous clouds.　A term used in respect of a more or less continuous layer of low clouds without regular features and generally associated with rain, as for example NIMBOSTRATUS.

amplitude.　For a true sinusoidal wave variation, the magnitude of the maximum departure of the quantity concerned from its mean value. For a more complex type of oscillation, the amplitude is usually taken as being half the mean difference between maxima and minima. See also WAVE MOTION.

anabatic wind.　A local wind which blows up a slope which is heated by sunshine. It is a feature which is much less common than its converse, the KATABATIC WIND.

anafront.　A FRONT (warm or cold) along which the warm air is ascending relative to the cold air. Frontal activity is generally well marked with such fronts.

analogue.　In synoptic meteorology, a past synoptic situation which resembles the current situation over an appreciable area.

Analogues are usually selected from the same time of year as the current situation. The sequence of weather which followed an analogue was in the past sometimes used as the basis of a weather forecast, in both short-range and long-range forecasting. In the former case the analogues referred to the surface (and upper-air) synoptic charts which were drawn as a routine. In the latter case they might have referred to charts which showed, for example, the distribution of the values of monthly temperature anomalies (departures from average) at stations distributed over a large area.

analysis.　In synoptic meteorology, the co-ordination by means of ISOPLETHS or representative symbols of the elements plotted on a surface or upper-air chart ('surface analysis' and 'upper-air analysis', respectively), generally for the purpose of making a weather forecast.

Surface analysis comprises both ISOBARIC ANALYSIS (the drawing of isobars) and AIR-MASS ANALAYSIS (the identification of air masses and drawing of fronts); other types of isopleth, notably ISALLOBARS, may be drawn on the surface chart as aids in analysis. In upper-air (isobaric) analysis, CONTOURS are drawn, sometimes also STREAMLINES, isopleths of temperature and isopleths of wind direction and speed.

anaprop.　See ANOMALOUS RADIO PROPAGATION.

anelastic approximation.　The EQUATION OF CONTINUITY may be written

$$\text{div}(\rho\mathbf{V}) = -\frac{\partial\rho}{\partial t}.$$

The anelastic approximation is the replacement of the right-hand side of the equation by zero, yielding

$$\text{div}(\rho\mathbf{V}) = 0.$$

15

This approximation eliminates SOUND WAVES, including LAMB WAVES, from the solutions of the equations of motion; its use gives a better representation of internal gravity waves than does that of the HYDROSTATIC APPROXIMATION.

It has been used in numerical models of cumulus development. It is a valid approximation if the depth of the circulations considered are much less than the density SCALE HEIGHT (H_z) where

$$H_z^{-1} \equiv -\frac{1}{\rho_z} \frac{d\rho_z}{dz}.$$

anemogram. The record made by an ANEMOGRAPH.

anemograph. An instrument for recording in visual form the speed, and sometimes also the direction, of the wind. Each of the basic types of ANEMOMETER may be used as the wind-speed sensor of the anemograph, and when combined with a suitable WIND VANE used as the wind-speed and direction sensors.

For normal purposes the cup-contact and cup-generator anemographs are most commonly used. The cup-contact anemometer can record the mean wind speed over a given time interval by displaying its contacts as marks on a chart moving with time, the marks being readily counted by means of the scale markings on the chart. The cup-generator anemometer can record instantaneous wind speed by feeding its output into a moving coil pen-recording meter.

The pressure-tube anemograph, now obsolescent, makes use of the difference of pressure set up between two pipes, one of which is kept facing the wind by the action of a wind vane, while the other is connected to a system of suction holes on a vertical tube. The difference of pressure so produced is arranged to raise a float carrying a pen, the height of which above the zero position is made proportional to wind speed by suitable design of the float. The instrument takes a certain time to respond to changes of wind speed, but gusts and lulls with periods down to a few seconds are indicated accurately, though they may not be distinguishable on the record. Either 'direct recording' at a point underneath the instrument head, or 'remote recording' at a considerable distance from the instrument, may be arranged.

See also EXPOSURE.

anemometer. An instrument for determining the speed of the wind. One or other of three properties of the wind is used in such instruments:

(i) its kinetic energy, which causes rotation as in cup anemometers and anemometers of the windmill type (AIR METER),
(ii) its pressure, as in the pressure-tube ANEMOGRAPH and pressure-plate anemometer and
(iii) its cooling power, as in the hot-wire anemometer.

Cup anemometers consist of three or four cups, conical or hemispherical in shape, mounted symmetrically about a vertical axis. In the cup-contact anemometer, the closing of electrical contacts produces an intermittently audible note, the rate of recurrence of which is proportional to the wind speed. In the cup-generator anemometer, the rotating cups are made to generate a voltage which registers on a dial calibrated in knots, metres per second or miles per hour. In the cup-counter or run-of-wind anemometer, the integrated distance of travel of the air is registered on a counter.

In the pressure-plate anemometer, the deflection of a flat plate placed in the wind is measured; its use is confined mainly to atmospheric turbulence measurement. In the hot-wire anemometer, the current required to maintain constant the electrical resistance (and so the temperature) of a fine platinum wire which is exposed to the wind may be used as a measure of the wind speed; alternatively, a large resistance may be placed in series with the wire, the current kept constant and the varying potential drop across the wire used as a measure of the wind. This instrument is used when rapidity of response to wind fluctuations is important.

See also EXPOSURE.

aneroid barometer. The aneroid ('without liquid') BAROMETER was invented by Lucien Vidie in about 1843. In its simplest form it consists of a shallow capsule of thin corrugated metal which is exhausted of air. The faces are kept apart by the stiffness of the metal or by a separate spring. Compensation for temperature is provided by a bimetallic link or (over a limited range of pressure) by a small quantity of residual air in the capsule. In some instruments several capsules are employed. The relative movements of the faces due to changes of atmospheric pressure are conveyed and magnified through a train of levers to a chain which actuates a pointer on a dial.

An aneroid barometer is light, portable and convenient. Since, however, it may be subjected to errors introduced by imperfect elasticity of the metal etc., and to changes of zero (both generally small in modern instruments), it requires occasional checking against a mercury barometer and in the past has normally been used only when the use of a mercury barometer was impracticable. Recently, however, aneroid barometers which are as accurate and reliable as a Kew-pattern barometer have been developed. See PRECISION ANEROID.

angels, radar. Radar reflections obtained on certain occasions of apparently clear air. One type, comprising clouds of point-target echoes, has been definitely associated with birds. It is probable that other types are associated with abrupt changes of refractive index of the air.

angle of elevation. See ALTITUDE.

ångström unit. A measure of wavelength, equal to 10^{-10} m, denoted by Å.

angular momentum. The angular momentum (or moment of momentum) per unit mass of a body rotating about a fixed axis is the product of the linear velocity of the body and the perpendicular distance of the body from the axis of rotation. The dimensions are $L^2 T^{-1}$.

A point P of the earth at latitude ϕ has a west–east (zonal) velocity of $\Omega\, a \cos \phi$ (see Figure 6). Thus, at this latitude, the absolute angular momentum of air of relative zonal velocity u (and not too far removed from the surface of the earth)

$$= (u + \Omega\, a \cos \phi)\, a \cos \phi$$

$$= u\, a\, \cos \phi + \Omega\, a^2 \cos^2 \phi.$$

The first of these terms is the 'relative angular momentum' of the air, reckoned positive for positive u. The second term is the angular momentum of the coinciding point of the earth.

The transfer of angular momentum effected by the atmosphere and the manner in which the entire earth–atmosphere system conserves absolute angular momentum over a long period of time are of fundamental significance in the explanation of the GENERAL CIRCULATION of the atmosphere.

angular velocity. The angular velocity of a moving line is the time rate of change of the angle between the line and a fixed line in a plane containing two successive positions of the moving line. Angular velocity is represented by a vector normal to this

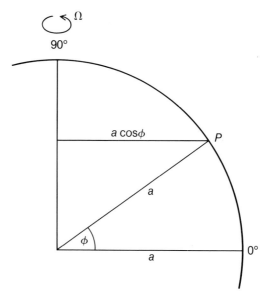

FIGURE 6.   Angular momentum.

plane. A suitable convention is adopted as to which direction of rotation is considered positive. The angular velocity of a moving point about a fixed point is the angular velocity of the line joining the two points. The angular velocity of a moving point about a fixed axis is the rate of change of the angle between a plane drawn through the axis and the moving point, and a fixed plane passing through the axis. The angular velocity of a solid body about an axis is the angular velocity of any point of the solid body about that axis. The dimension is T^{-1}.

Angular velocity is a vector quantity which is normally measured either in revolutions per unit time or in radians per unit time. Since there are 2π radians per revolution, ω (radians per second) is related to N (revolutions per minute) by the expression $\omega = \pi N/30$.

angular velocity of the earth.   The ANGULAR VELOCITY of the earth (Ω) may be represented by a vector parallel to the axis of the earth and directed northwards. This vector may be resolved into components $\Omega \cos \phi$ about the line directed towards local north (rotation in a vertical plane) and $\Omega \sin \phi$ about the local vertical (rotation in a horizontal plane) — see Figure 7. The total vector is of magnitude 2π radians per sidereal day $= 7.292 \times 10^{-5}$ rad s^{-1}.

$\Omega \sin \phi$ is the only component which is significant in large-scale motion, which is almost entirely horizontal. The rotation is anticlockwise in the northern hemisphere and clockwise in the southern hemisphere; it is thus cyclonic in both cases.

anomalous audibility.   Audibility of sound waves over a region which is separated from the source of the waves by a 'zone of silence'. See AUDIBILITY.

anomalous radio propagation.   Under normal atmospheric conditions, radio waves up to a few metres in length, transmitted in nearly horizontal beams (for example, radar), attain ranges on the earth's surface similar to the distance of the optical horizon, being limited by the curvature of the earth. In abnormal atmospheric conditions, in which there is an abnormal decrease of REFRACTIVE INDEX with height through the lower atmosphere, the waves will be subject to 'anomalous propagation' in

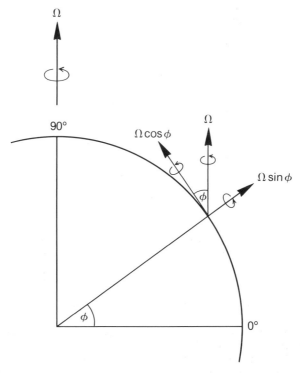

FIGURE 7.   Angular velocity of the earth.

which they suffer a marked downward refraction sufficient to extend their normal range. This effect is also known as 'super refraction'.

The extreme form of this effect occurs in conditions of a large inversion of temperature, or a large decrease of humidity with height, or both, extending through a shallow atmospheric layer near the earth's surface. This is called a 'radio duct', and in these circumstances radio detection ranges are greatly enhanced. For radar, where measurements of the range are very important, the effect is to increase interference and ground clutter by the generation of false echoes. Such conditions may arise in anticyclonic SUBSIDENCE inversions or when air is advected from a warm, dry, land surface over a cooler sea. The maximum wavelength that can be propagated by a radio duct is a function of the thickness of the duct and the change of REFRACTIVE INDEX through the duct. The frequency of occurrence of the phenomenon is greater for shorter wavelengths. It is often referred to as anaprop.

anomalous sound propagation.   The propagation of sound waves along a path, from source to receiver, other than close to the earth's surface, as the result of which ANOMALOUS AUDIBILITY occurs. See AUDIBILITY.

anomaly.   The departure of an element from its long-period average value for the place concerned. The space distribution of such anomalies at a specified time is known as an 'anomaly pattern'.

The term is also used in other senses; for example, a place that is relatively warm for its latitude (as western Norway is in winter) is sometimes said to have a positive temperature anomaly.

Antarctic air. An AIR MASS, originating over the Antarctic continent, which is cold and dry in all seasons. It is sometimes designated polar continental (*Pc*) air.

Antarctic Circle. The parallel of latitude 66° 33′S, south of which lies the 'Antarctic zone', or 'southern polar zone'.

Antarctic front. A FRONT which develops and persists around the Antarctic continent in about latitudes 60–65° S, and divides ANTARCTIC AIR from the maritime POLAR AIR to the north. The front is often found to exist round a large part of the hemisphere.

anthelion. A colourless mock sun (PARHELION) appearing at the point of the sky opposite to and at the same altitude as the sun. The phenomenon is rare. Rather more frequently oblique arcs crossing at that point are reported. The phenomenon is no doubt caused by the reflection of light from ice crystals, but the exact explanation is in doubt. Also termed 'counter sun'.

anticorona. An alternative for GLORY.

anti-crepuscular rays. See CREPUSCULAR RAYS.

anticyclogenesis. The initiation or strengthening of anticyclonic circulation round an existing ANTICYCLONE.

anticyclolysis. The disappearance or weakening of anticyclonic circulation round an existing ANTICYCLONE.

anticyclone. That atmospheric pressure distribution in which there is a high central pressure relative to the surroundings. It is characterized on a synoptic chart by a system of closed isobars, generally approximately circular or oval in form, enclosing the central high pressure (see Figure 8). The term 'anticyclone' was selected by Sir Francis Galton in 1861 as implying the possession of characteristics opposite to those found in the cyclone or depression. Thus, the circulation about the centre of an anticyclone is clockwise in the northern hemisphere (anticlockwise in the southern), and the weather is generally quiet and settled.

Two contrasting types of anticyclone, 'warm' and 'cold', are recognized. The former has a warm troposphere (though sometimes cold in a shallow layer near the earth's surface), high tropopause and cold stratosphere; the anticyclonic circulation is deep and the feature slow moving. The preferred region is the subtropical belt at 30–40° latitude, for example the Azores and Bermuda regions, but with frequent extensions across western Europe; typical warm anticyclones also occur in higher latitudes in association with BLOCKING. In contrast, the cold anticyclone has a relatively cold baroclinic troposphere, low tropopause, warm stratosphere and shallow circulation. It often forms in the cold air behind a depression and moves fairly rapidly in a direction between south and east (northern hemisphere), sometimes then slowly transforming to a warm type. Persistent radiational cooling in winter over high-latitude continents produces semi-permanent cold anticyclones, or 'continental anticyclones', as in Siberia and North America.

The dynamical structure of the anticyclone is one of horizontal convergence at high levels, horizontal divergence at low levels and slowly subsiding air throughout a large part of the troposphere. The SUBSIDENCE proceeds at a maximum rate in early stages of formation of the anticyclone and is greatest in mid-troposphere. The subsidence results, through dynamical warming of the air, in a decrease in relative humidity and an increase of static STABILITY of the air, often with the formation of an 'anticyclonic inversion' of temperature. These processes often result in fine, cloudless weather.

FIGURE 8.   Anticyclone centred off the north-east coast of the United Kingdom at 1200 GMT on 4 November 1987.

There are, however, important exceptions: in summer, sea and coastal fog and, in winter, widespread stratocumulus ('ANTICYCLONIC GLOOM') forming in moist air at the base of the inversion, or radiation fog, forming in conditions of little cloud, are commonly associated with an anticyclone.

anticyclonic gloom. Conditions of poor illumination occurring with an overcast sky of stratus or stratocumulus beneath an inversion in an ANTICYCLONE. It is particularly associated with the accumulation, in and below the cloud, of the pollution products of industrial areas.

antisolar point. That point, below or above the horizon, towards which the extension of the line from the sun to an observer's eye is directed.

anti-trades. The anti-trades, or counter-trades, are upper winds which in some low-latitude areas prevail above the TRADE WINDS and which, more or less opposite in direction to the trade winds, are responsible for poleward transport of air aloft. The transition to anti-trades, marked by a temperature inversion, occurs at some 1–1.5 km in high trade-wind latitudes in eastern oceanic regions and at increasingly high levels equatorwards and to the west.

antitriptic wind. A class of winds in which the effect of FRICTION or eddy VISCOSITY predominates over the CORIOLIS ACCELERATION and the CENTRIPETAL ACCELERATION effects. The air movement is from high to low pressure, a quasi-steady state being established by the effect of friction or eddy viscosity. Such winds occur mainly in low latitudes where the Coriolis acceleration is very small, or in small-scale systems (e.g. sea-breeze) in higher latitudes.

antitwilight (or **antitwilight arch**). Alternatives for COUNTERGLOW.

anvil cloud. Cloud having at its top a projecting point or wedge like an anvil. The form is usually assumed by the tops of fully developed CUMULONIMBUS clouds. The anvil is considered to indicate in nearly all cases the presence of ice crystals or snowflakes; it quickly assumes a fibrous or nebulous appearance and is often stretched out from the main cloud columns by changes of wind with height. The term 'anvil cirrus' is applied to an anvil which becomes separated from the main cloud.

aphelion. That point of the orbit of a planet or comet which is farthest from the sun. Aphelion for the earth occurs on about 1 July; the sun–earth distance is then 1.5 per cent greater than the yearly mean distance.

apogee. That point of the orbit of a satellite, natural or artificial, which is farthest from the earth.

apparent form of the sky. The somewhat 'flattened' appearance of the sky presented to most observers. Among the effects of this appearance is the tendency to overestimate the elevation angles of objects in the sky and the angles they subtend at the eye of the observer, e.g. that of the moon near the horizon.

Appleton layer. A layer of the IONOSPHERE at some 300 km height, now usually termed the F_2-layer. See also F-LAYER.

applied meteorology. The application of meteorological knowledge in a wide variety of activities, for example, industry, transport, hydrology and agriculture, for the purpose of using meteorological conditions to the best advantage. Forecasting, or climatology, or the application of the results of meteorological research in specific problems may be involved. For certain operations, notably in agriculture, the application of meteorological knowledge may include the altering of meteorological conditions on a small scale, as in the construction of SHELTER-BELTS.

APT.   Abbreviation for AUTOMATIC PICTURE TRANSMISSION.

aqueous vapour.   An alternative for WATER VAPOUR.

Arago's point.   A 'neutral' point of the sky at which the normally observed polarization of the light from a clear sky disappears; that discovered by Arago in the early 19th century is some 20° above the ANTISOLAR POINT. Other neutral points were subsequently discovered by Babinet and Brewster about 20° above and below the sun, respectively. Day-to-day and systematic diurnal and seasonal variations of the position of these points, probably associated with haze variations, have been discovered. See also POLARIZATION.

arcs of contact.   Upper and lower 'arcs of contact' to the 22° HALO occasionally form, the lower arc being very rare. At high solar elevations the arcs may appear concave towards the sun. At low solar elevations the higher arc is convex towards the sun. The points of contact may have the appearance of 'mock suns' (PARHELIA) and may display brilliant colour.

Arctic air.   An AIR MASS originating in the snow- and ice-covered Arctic. If such an air mass travels almost directly south to affect the British Isles in winter or spring it is associated with a northerly wind, snow showers (especially on north-facing coasts and hills), low temperatures, exceptional visibility and steep lapse rate.

Arctic Circle.   The parallel of latitude 66° 33′N, north of which lies the 'Arctic zone', or 'northern polar zone'.

Arctic front.   A FRONT which separates ARCTIC AIR to the north from polar maritime air or polar continental air to the south. A section of the Arctic front is often found in the area from south Greenland to north of Norway in winter and spring.

Arctic sea smoke.   If, when cold air moves over warm water, the vapour pressure at the water surface exceeds the saturation vapour pressure at the air temperature, then evaporation from the water surface proceeds at a higher rate than can be accommodated by the air. The excess water vapour over that required to saturate the air condenses and, in the unstable conditions present in the layer near the surface, the condensed water is carried continuously upwards to evaporate into the drier air above. 'Steam' or 'smoke' thus appears to rise from the water surface. If an inversion exists near the water surface, fog may be confined below the inversion and become dense.

The phenomenon occurs, for example, over inlets of the sea in high latitudes; over newly formed openings in pack ice; over lakes and streams on calm, clear nights; and over damp ground heated by bright sunshine in cool conditions. Alternative names are 'frost smoke', 'sea smoke', 'steam fog', 'warm-water fog', 'water smoke' and 'the barber'.

arcus (arc).   A supplementary cloud feature (Latin for arch).

'A dense, horizontal roll with more or less tattered edges, situated on the lower front part of certain clouds and having, when extensive, the appearance of a dark menacing arch.

This supplementary feature occurs with CUMULONIMBUS and, less often, with CUMULUS' [2, p. 23]. See also CLOUD CLASSIFICATION.

argon. The most abundant of the INERT GASES, with VOLUME FRACTION and MASS FRACTION, relative to dry air, of 0.93×10^{-2} and 1.27×10^{-2}, respectively. Its molecular weight is 39.948. Its inertness and relatively high density render it a suitable tracer (by means, for example, of the argon:nitrogen ratio) of the degree of the GRAVITATIONAL SEPARATION of the atmospheric constituents.

arid climate. A climate in which the rainfall is insufficient to support vegetation is termed arid. Köppen and Geiger in their *Klimakarte der Erde* [3] use the following formulae for the limits of rainfall for arid and semi-arid climates:

Rainfall mainly in cold season	$R = 2t$
Rainfall evenly distributed throughout year	$R = 2t + 14$
Rainfall mainly in hot season	$R = 2t + 28$

where t is the mean annual temperature in degrees Celsius. If the annual rainfall (in centimetres) is less than R and greater than $R/2$ the climate is steppe or semi-arid; if it is less than $R/2$ the climate is desert or arid. See also DESERT, STEPPE.

arithmetic mean. See also MEAN.

artificial ice nucleus. A freezing or sublimation NUCLEUS generated for the purpose of CLOUD SEEDING.

artificial precipitation. See CLOUD SEEDING.

ascendant. See GRADIENT.

A-scope indicator. See RADAR METEOROLOGY.

ASDAR. Abbreviation for AIRCRAFT TO SATELLITE DATA RELAY.

ash. The incombustible solid material released when a substance is burned; a small proportion escapes to the atmosphere to contribute to ATMOSPHERIC POLLUTION.
     The proportion of total ash that emerges into the atmosphere and also the average size of emerging particles depend on the velocity of the flue gases and are, in the absence of a grit arrester, much greater for industrial than for domestic chimneys. Emerging particle sizes range from 2 mm downwards, all except the smallest particles being deposited near the source. Measurement of such ash and other deposited material is made by the DEPOSIT GAUGE.

aspect. The aspect of sloping ground is the geographic direction in which the line of greatest downslope points. The 'aspect angle' is, in the northern hemisphere, the angle between this line and geographic south, usually reckoned positive eastwards, negative westwards.

aspirated psychrometer. A PSYCHROMETER in which a high rate of ventilation is provided artificially as in the ASSMANN PSYCHROMETER or WHIRLING PSYCHROMETER.

Assmann psychrometer. A PSYCHROMETER in which a definite rate of ventilation is secured by drawing the air over the thermometer bulbs by means of a fan driven by a motor and in which the thermometers are mounted in a polished metal frame as a protection against solar radiation.

atmosphere. The gaseous envelope which is held to the earth by gravitational attraction and which, in large measure, rotates with it. The internal motions of the atmosphere caused by solar radiational heating constitute, together with their physical effects, the main concern of meteorology. The term is also used of the gaseous envelopes of planets and stars.

The composition of the earth's atmosphere is discussed under AIR. Of special importance in meteorology are the time and space variations of water vapour. A distinction is drawn, in terms of atmospheric composition, between the HOMOSPHERE and the HETEROSPHERE.

The atmosphere's outermost fringe has been termed the 'exosphere', at the lower limit of which (about 700 km) the escape to space of neutral particles was calculated to begin. S. Chapman has, however, suggested that, since the particles at such levels are mainly ionized and so are controlled not by diffusion but by the earth's magnetic field, the atmosphere may be considered to extend to a height of several or even many earth radii, where the density falls to that of interplanetary gas.

Division of the atmosphere into the regions TROPOSPHERE, STRATOSPHERE, MESOSPHERE, and THERMOSPHERE, according to characteristic temperature lapse rates, is illustrated in Figure 9; the approximate pressure and density variations with height are also shown. The terms OZONOSPHERE, CHEMOSPHERE and IONOSPHERE are also used of certain regions in which specified processes occur. See also STANDARD ATMOSPHERE.

atmospheric boil. An alternative for SHIMMER.

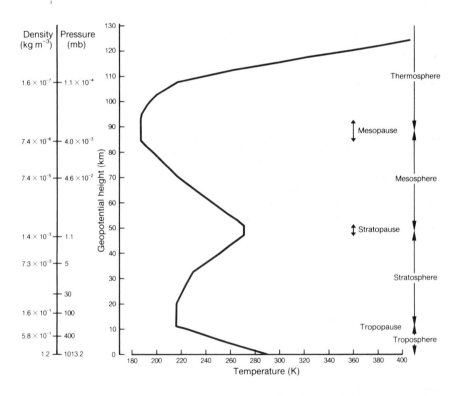

FIGURE 9. Average temperature structure of the atmosphere from 0 to 130 km. Data are taken from the *U.S. Standard Atmosphere, 1976* [4], which depicts idealized middle-latitude year-round mean conditions. From 0 to 32 km this corresponds with the ICAO standard atmosphere [5] as adopted by the World Meteorological Organization.

atmospheric chemistry. A term which is generally taken to comprise the study of the chemical composition of the air, of aerosols, and of rainfall. It includes the processes of PHOTOCHEMISTRY which are important at levels above about 40 km.

Monthly samples of air and rainfall, collected at a network of stations which was augmented and made world-wide during the International Geophysical Year, are chemically analysed for the presence and concentration of inorganic substances including Na, K, Mg, Ca, NH_3, Cl, S and the ions NO_3 and CO_3. The pH value of the sample (the logarithm, to base 10, of the inverse of the concentration of hydrogen ions in the solution) has also been determined.

Widely different results, difficult to interpret, have been obtained with the air sampling measurements, which have been concerned mainly with the contained particulate matter. More consistent results, though with unexplained space and time variations, have been obtained with the rainfall sampling measurements. In particular, a deficiency of Cl relative to Na and other ions which occur in sea water has been found.

In addition to the above, surface air has been sampled for CO_2 and O_3. Also, various measurements have been made of the nature and amount of the radioactive material from nuclear explosions which is contained in the troposphere and stratosphere, or is deposited as dust on the earth's surface, or is washed out of the atmosphere by precipitation.

atmospheric electricity. The various electrical phenomena which occur naturally in the lower atmosphere; the IONOSPHERE is conventionally excluded, except in so far as it reacts on the electrical properties of the lower atmosphere. While the THUNDERSTORM is the most familiar manifestation of atmospheric electricity, substantial electrical effects exist also in fine-weather regions and have been continuously measured at various places since the latter part of the 19th century. It is now generally held that the two classes of phenomena are closely linked.

The fine-weather electric field recorded in surface POTENTIAL GRADIENT measurements is directed downwards to earth and has a mean value of about 100 volts per metre. Such a field implies a negative charge on the earth's surface of about 1 coulomb per 1000 km^2 (3 electrostatic units per m^2) in fine-weather regions.

Balloon measurements show that in such regions the potential gradient decreases with height, rapidly at first, then more slowly; a normal positive SPACE CHARGE is implied in the low atmosphere, with a surplus of positive over negative ions. Integration of the measured electric field with height shows that, relative to the earth's surface, a positive potential of about 4×10^5 volts exists at a height of about 15 km, with little further addition to this value with further height increase owing to the high conductivity of the air at this and higher levels.

Various ionizing radiations render the air electrically conducting, though only feebly so near the earth's surface — see ION. A so-called 'conduction current' flows in fine-weather regions under the action of the field which prevails there (positive ions down to negatively charged earth, negative ions up); this is partially offset by a 'convection current' which transfers positive space charge bodily upwards from the low atmosphere. The measured air–earth current averages about 2×10^{-12} A m^{-2}. Its space and time variations are considerably smaller than those of the potential gradient; in particular, decrease of the electric field with increasing height is compensated by increasing conductivity, so maintaining an almost constant current, in accordance with Ohm's law.

Calculation shows that the air–earth current would, within a few minutes, destroy the electric field which gives rise to it and so would itself die out but for the action of a continuous compensating mechanism. The origin of the required 'supply current', for long a matter of doubt and controversy, is now thought to lie in rain storms and thunder-clouds, as was originally suggested by C.T.R. Wilson. Various mechanisms operate in the 'disturbed weather' regions:

(i)   Large electrical charges accumulate at the base of thunder-clouds; when such clouds are low, the electric field between cloud and ground often exceeds the value required for brush discharge, especially from elevated or pointed conductors. Since thunder-cloud bases carry a predominant negative charge, positive ions are mainly discharged from earth, leaving there a net negative charge.

(ii)   Air-to-ground lightning flashes convey a predominantly negative charge to ground.

(iii)   Precipitation carries a net positive charge to ground.

(iv)   Aircraft measurements above thunder-clouds indicate that negative ions flow downwards from the high atmosphere to the positively charged thunder-cloud tops, leaving a net positive charge on the high atmosphere.

The sum of effects (i) and (ii) exceeds effect (iii), resulting in the conveying of a net negative charge to ground in disturbed regions; this charge and the positive charge conveyed upwards above thunderstorms quickly spread over the conducting earth and high atmosphere, respectively, and give rise to the downward-directed field 'leakage current' observed in fine-weather regions. Net (space and time) balance exists between the supply and leakage currents. A major piece of supporting evidence for this mechanism is the existence of a systematic diurnal variation of the fine-weather field in phase with integrated thunderstorm activity over the world (maximum about 1800 GMT, minimum about 0400 GMT), observed in those regions (e.g. oceanic) where complications of diurnal variation of low-level conductivity are absent.

atmospheric optics.   The optical phenomena of interest to meteorologists include, among many other examples, the BLUE OF THE SKY, SUNRISE AND SUNSET COLOURS, the production of RAINBOWS, CORONAE, HALOES, MIRAGES, PARHELIA, the fading of daylight, and the TWINKLING of stars. See, for example, Humphreys[6], and Minnaert [7].

atmospheric pollution.   Airborne gases, particulates and aerosols at concentrations that are significantly higher than normal or higher than in pre-industrial times. Such pollution is in a sense 'foreign' to the living world and often harmful or unsightly. Occasionally, it can be beneficial, for example in providing sulphur or nitrogen necessary for plant growth to soils deficient in these species. The major constituents of atmospheric pollution are various oxides of sulphur and nitrogen, carbon dioxide and monoxide, CHLOROFLUOROCARBONS (CFCs), ozone, ammonia, hydrochloric acid, various gaseous hydrocarbons, sulphates, nitrates, smoke, heavy metals, and radioactive material. Carbon dioxide is included despite being an important constituent of the 'natural' atmosphere, because its addition in large amounts arising from the burning of fossil fuels has a strong influence on the GREENHOUSE EFFECT. Natural sources of a number of these constituents include volcanoes, forest fires, and lightning (which produces nitrogen compounds). Major sources of pollution are the burning of fossil fuels (coal and oil) in electricity generating stations, the waste products of oil refineries and the manufacturing industry in general, domestic heating by coal fires (much diminished in advanced societies), and the internal combustion engine; other sources include animal waste from farming (which produces ammonia), the use of fertilizers and pesticides, aerosol sprays and refrigerants (CFCs), accidental releases from nuclear reactors, rubbish tips and sewage works, and slash-and-burn methods as used in some tropical regions.

Transformations of the primary pollutants to secondary pollutants by chemical and photochemical reactions, or by radioactive decay, after their initial emission are often important. Thus, hydrocarbons and other compounds emitted by the internal combustion engine react together under the influence of strong sunlight to yield toxic gases including ozone; these processes are particularly important in areas of light winds and limited vertical mixing when 'photochemical smog' is the result. Similarly, CFCs in the high atmosphere take part in photochemical reactions that can lead to a reduction of ozone and an increase of ultraviolet radiation at the earth's surface.

Pollution emitted in a plume may be brought down to ground level by atmospheric turbulence where some constituents can undergo gradual deposition by sedimentation, impaction, or chemical adsorption — 'dry deposition'. Polluted cloud or fog droplets when driven by the wind may be deposited on to vegetation with high efficiency — 'occult deposition'. Pollution that has dispersed into an air mass that later produces precipitating clouds is likely to be washed down to earth in rain or snow — 'wet deposition'. The wet and occult deposition (and to a lesser extent dry deposition) is markedly episodic because of the complicated relationships in space and time between the original emissions and the three-dimensional patterns of atmospheric flow that lead to ultimate deposition on the surface of the earth. In Europe as a whole, roughly two-thirds of the total sulphur deposit is due to dry deposition, the remainder being due to wet and occult; the last two are particularly important in areas of high rainfall, especially in hilly country to the east or north-east of large sources.

Local dry deposition can be much reduced by the building of very high chimneys, and various cleansing techniques can remove much of the airborne total acid content at the expense of disposing into the environment the resulting waste products (e.g. dumping sludge into mines and pits).

There is little doubt that atmospheric pollution can cause damage to plants and animals as well as to inanimate objects such as the surface of buildings. However, the relationship between such damage and deposition is not always clear; neither is the relationship between the magnitude of the amount of pollutant emitted at a source and the resulting deposition of primary and secondary pollutants when considered on a single-event basis.

Atmospheric pollution may be measured by drawing air through liquid absorbents or solid filters followed by quantitative chemical analysis. Recent techniques involve the use of controlled chemical reactions in a continuous stream of polluted air with measurement of associated chemiluminescence, and can produce a virtually continuous record of the concentration of a gaseous pollutant. All advanced countries maintain a network of stations monitoring atmospheric pollution and results are co-ordinated by various national and international organizations.

atmospheric pressure. See PRESSURE.

atmospherics. Natural electrical impulses, mainly originating in LIGHTNING discharges, which cause crashing or grinding noises in a radio receiver; the phenomenon is also termed 'static'. The impulses which originate within a radius of some thousands of kilometres reach a particular receiver on a multitude of paths suffering reflections from the lower ionosphere and the earth's surface, and constitute a varying background noise level in the receiver. Since the impulses decrease in strength with increasing distance from their place or origin, the noise level is greatest in low latitudes where thunderstorms are most frequent.

While the phenomenon constitutes a serious difficulty in long-distance radio transmission, it is used to advantage in geophysics in various ways: first, by suitable arrangement of apparatus, in the locating of thunderstorms; second, by examination of the wave-forms, in the study of the nature of lightning discharges; third, in the determination of diurnal and seasonal variations of thunderstorm activity; fourth, in the detection of sudden enhancements of atmospherics and thus of associated solar flares (see SEA).

The term 'sferics' is the accepted contraction of the word 'atmospherics' for meteorological purposes. See also SFERICS FIX.

atmospheric tides. Effects directly analogous to the simple gravitational tides familiar in the oceans are produced in the atmosphere by the action of the moon, the elements chiefly involved being pressure and wind. The tides are most conveniently identified at the earth's surface from the pressure variations, which are positive or negative with respect to the mean pressure level according to whether the air is heaped up over, or drawn away from, a locality. Two pressure maxima (one at the longitude corresponding to that 'under' the tide-producing body and the other at the antipodal point) and two pressure minima (at longitudes 90° from the pressure maxima) occur at any given epoch. At any given place the effect is such as to produce a semi-diurnal oscillation. When the pressure data are arranged in lunar time, there emerges a systematic effect of 'correct' phase and very small amplitude (0.09 mb near the equator, decreasing towards either pole).

Although the gravitational tidal action of the sun is smaller than that of the moon by the factor 2.4, the amplitude of the atmospheric oscillation which is governed by solar time is much the greater; the related effect on surface pressure is clearly visible on low-latitude barograms — see DIURNAL VARIATION. The explanation is that the solar-controlled oscillations are caused by a combination of thermal and gravitational action, the former being the more important; conventionally both agencies are considered in discussions of atmospheric tides.

The main solar tidal components are those of 24-hour and 12-hour periods. They are termed the S_1 and S_2 components respectively, the former being much the more variable in phase and amplitude. The mean amplitude of S_1 is about half that of S_2.

In middle latitudes the amplitude of the solar tidal wind is a small fraction of 1 m s^{-1} at the surface and it changes little throughout the troposphere. Between the low and high stratosphere it increases from less than 1 m s^{-1} to about 7 m s^{-1} and there is a further large increase to some $15-30 \text{ m s}^{-1}$ at $80-90$ km; at these and higher levels the tidal component is a very significant, and sometimes dominant, part of the total wind.

The RESONANCE hypothesis, that the large magnitude of S_2 requires for its explanation the existence of an atmospheric FREE PERIOD very close to 12 hours, is not supported by more recent theory according to which, sufficient explanation of the magnitude of S_2 lies in the semi-diurnal nature of the temperature wave which affects the whole of the atmosphere. The smaller value of S_1 is explained by the structure of the atmosphere, which tends to suppress this oscillation.

atmospheric window. A term applied to that region of the ABSORPTION spectrum of water vapour which extends from about 8.5 to 11 μm. Ground radiation in this range of wavelengths is, in contrast to ground radiation of other wavelengths, little absorbed by water vapour and, in the absence of cloud, escapes to space. See LONG-WAVE RADIATION, GREENHOUSE EFFECT.

attached thermometer. A thermometer attached to a mercury BAROMETER, the thermometer bulb being within the metal tube surrounding it. A reading of this thermometer is required in order that an appropriate 'temperature correction' may be applied to the barometer reading if the thermometer has a reading other than the STANDARD TEMPERATURE of the barometer.

In some marine barometers the attached thermometer is incorporated in the GOLD SLIDE.

attachment. In meteorological literature, often used with particular reference to the disappearance of free ELECTRONS by their attachment to neutral oxygen atoms or molecules, thus forming negative IONS. The rate of the process is expressed by an 'attachment coefficient' with dimensions $L^3 T^{-1}$.

attenuation. In geophysics, the depletion of electromagnetic energy (e.g. solar radiation, radio waves, radar waves) which is effected by the earth's atmosphere and its constituents. The rate of attenuation is represented by the 'attenuation coefficient' (σ) which includes the effects both of ABSORPTION and of SCATTERING of the radiation, and is defined for monochromatic radiation by the equation, analogous to Beer's law, which applies to absorption only:

$$I = I_0 e^{-\sigma x}$$

where I_0 is the intensity of radiation emitted at the source (or incident on the top of the atmosphere in the case of solar radiation) and I is the intensity after path length x through the absorbing and scattering medium.

The extent to which the above law may be applied over the whole spectrum, or a substantial part of the spectrum, depends on the degree of variation of attenuation coefficient with wavelength.

See also ABSORPTION, EXTINCTION COEFFICIENT, TURBIDITY.

audibility. The audibility of a sound in the atmosphere may be measured by the distance from its source at which it remains just audible.

The extent to which the loudness of sound decreases with distance from source at a rate other than the 'theoretical' inverse square of distance law is determined by the structure of the atmosphere. The factors mainly involved are the vertical distribution of temperature and wind and probably also the degree of atmospheric turbulence. It is, in turn, possible to infer much about the height variations of temperature and wind in the atmosphere from observations of the space distribution of audibility and time of arrival of sound waves on occasions of natural or artificial explosions.

The REFRACTION of sound waves in a calm atmosphere is governed by the lapse rate of temperature. The waves are bent upwards in conditions of strong lapse, and audibility is correspondingly low; conversely, it is high during a surface inversion because of downward bending of the waves sufficient for them to return to ground. Audibility is greater downwind than upwind, especially when there is a rapid increase of wind with height. A turbulent atmosphere is considered to limit audibility because of the associated mass exchange in the vertical and the accompanying dissipation of the sound energy. Audibility is generally greater, sometimes by a factor of 10 or 20, by night than by day because the conditions required for great audibility are much more characteristic of night than of day. For the same reason great audibility is a prominent feature of polar latitudes.

The 'anomalous audibility' associated with the so-called ANOMALOUS SOUND PROPAGATION is produced not by low-level meteorological conditions but by those at high levels. This feature, observed with large explosions, is characterized by a 'zone of silence' beginning at a distance from the source greater than that reached by the surface-propagated wave, with an outer zone of audibility at still greater distances. Typical distances from the source are 100 and 200 km, respectively; the zone boundaries may be approximate circles with the source as centre or may depart apppreciably from this. The cause of the outer zone is the refraction of sound waves in the high atmosphere to an extent which is sufficient to bend them back to earth. Calculations made from observations of the distribution of audibility and the time of arrival of sound waves on occasions of large explosions show that the major cause of the refraction is a region of high atmospheric temperature at a height of some 50 km, and that the effect of the refraction caused by high-level winds is mainly to make the shapes of the zone boundaries less circular.

aureole. A luminous bluish-white area, of small angular radius, immediately surrounding the sun or moon and bounded by a brownish-red ring. See CORONA.

The term 'aureole' is also used by some authors for the bright area, with no definite boundary, seen round the sun when the latter is not veiled by thin cloud.

aurora. This term (Latin for 'dawn') is applied to the phenomenon in which visible light is emitted by the high atmosphere.

In the 'auroral zones' of maximum frequency, some 20° from the geomagnetic poles, aurora is visible on almost every clear night; the northern auroral zone lies just north of Norway, south of Iceland and Greenland, over northern Canada and north of Siberia. The estimated mean annual frequency of nights of visible aurora, but for the intervention of cloud, twilight and moonlight, is about 150 in northern Scotland and about 10 in southern England; the corresponding frequencies of overhead displays are about 10 and 1, respectively. The terms 'aurora borealis' or 'northern lights' and 'aurora australis' or 'southern lights' apply, respectively, to the northern and southern hemispheres.

In the British Isles, aurora is usually seen near the northern horizon as a 'glow', or as a quiet 'arc', a grey-white feature with relatively sharp lower border. In a great display, during which the aurora may extend so far equatorwards as to be visible in the tropics, other auroral forms appear, with much movement and colour, of which the most characteristic are yellow-green and red. Greenish 'rays' may cover most of the sky polewards of the magnetic zenith, ending in an arc which is usually folded and sometimes with red lower border, the display then resembling a moving 'drapery'. If the display passes overhead, the parallel rays moving along the lines of force appear, by perspective, to converge at the magnetic zenith, thus producing a 'corona'. The later stages of a great display are usually marked by 'flaming', in which light surges upwards from the horizon. The display ends with a polewards recession of the auroral forms, the rays often then degenerating into diffuse surfaces of white light. Numerous reports of audible aurora and of aurora reaching almost to the ground in great displays are generally discredited.

The distribution of aurora in time and space has been determined by visual observation and by simultaneous photography from a number of stations. More recently, 'all-sky' automatic cameras and radio-echo equipment have also been used. The observations have shown that aurora is most frequent (in places equatorwards of the auroral zone) towards midnight and near the equinoxes, that auroral processes occur also during the day, and that auroral light originates at heights varying between 70 and 1000 km, with a marked peak frequency at about 100 km.

In spectral measurement of auroral light, molecular and atomic nitrogen and oxygen (atmospheric) and atomic hydrogen lines are the chief of those identified; the hydrogen line is subject to Doppler shift (see DOPPLER EFFECT) towards shorter wavelengths, implying the entry of high-speed protons into the atmosphere. The characteristic yellow-green and red colours are atomic-oxygen emission lines at 5577 Å and 6300 Å, respectively. The latter is emitted at higher levels; when red coloration appears at low levels, the emission is in a band of molecular nitrogen.

Aurora, like geomagnetic disturbance with which it is very closely associated, is caused by the entry into the high atmosphere of a stream of charged solar particles which are deflected by the earth's permanent magnetic field and precipitate over limited regions of the atmosphere. The various auroral forms arise from the primary and secondary collision processes of the solar particles with atmospheric gases and free electrons, and probably also in part from electric discharge processes which result from the generation of powerful electric fields in the high atmosphere.

See also GEOMAGNETISM.

austausch. A German term signifying the mixing and exchange implicit in atmospheric TURBULENCE. The term 'austausch coefficient' is an alternative for EXCHANGE COEFFICIENT.

autobarotropy. The (idealized) atmospheric state in which surfaces of constant pressure and density (or specific volume) remain always in coincidence. See BAROTROPIC.

autoconvective lapse rate. That temperature LAPSE rate which defines a state of constant atmospheric density with height; for lapse rates in excess of this the density increases with height. The lapse rate is given by the expression g/R, and has the value 3.42 °C per 100 metres, i.e. about 3½ times that of the dry adiabatic lapse rate. Lapse rates well in excess of the autoconvective lapse rate have been observed near the earth's surface (approximately the lowest 15 metres) during intense INSOLATION.

The implication contained in the term that this lapse rate represents critical conditions for the 'automatic' establishment of convection is erroneous. Because ascending air cools on expansion, the critical condition is not one of increasing density with height, as in an incompressible fluid, but a lapse rate in excess of the adiabatic lapse rate. See ADIABATIC.

autocorrelation. Autocorrelation, also termed 'serial correlation' and 'lag correlation', signifies CORRELATION within a series of observations often spaced at equal intervals of space or time. The measure of internal correlation — the 'autocorrelation coefficient' or 'serial correlation coefficient' — is often employed as a measure of the degree of PERSISTENCE within geophysical TIME SERIES; such coefficients, evaluated for various interval lags, may also be combined in the CORRELOGRAM as a means of testing for PERIODICITY.

The autocorrelation coefficient for a lag of L intervals of time is given by:

$$ r_L = \frac{\sum\limits_{i=1}^{N-L} (x_i, x_{i+L})}{(N-L)\sigma_x^2} $$

where x_i, x_{i+L} are departures from the mean of the series of N observations and σ_x is the STANDARD DEVIATION of the series.

The autocorrelation coefficient for successive daily values of pressure or temperature at places in the British Isles is about 0.8.

automatic data-processing. A systematic sequence of operations on data, performed largely automatically, the object of which is to extract or revise information, usually by electronic digital computer systems.

A computer system may be regarded as comprising four units — programmer, operator, software and hardware. The programmer devises the instructions for the computer's operation, the programmed instructions being collectively known as software. The operator is responsible for the operation of the programs; the mechanical and electronic components which he/she operates are termed the hardware.

The hardware of a large modern computer generally consists of a central processor (arithmetic unit and control unit), main and auxiliary storage and various peripheral devices for input and output. The main storage is a fast-access memory and often consists of integrated circuit chips; the auxiliary storage is an additional slow-access store, often consisting of magnetic drums or disks, which may be needed for more complex operations. The control unit directs the sequence and timing of operations, interpreting the coded instructions from the main memory and stimulating the appropriate circuits to execute the instructions.

The programmer, when presented with a problem, first breaks it down in some form of 'flow diagram'. This shows diagrammatically the sequence of operations to be followed by the computer in solving the problem. The sequence of instructions is then coded in precise detail; accuracy and completeness of the instructions are tested in trial runs on the computer and any necessary amendments are made.

Three categories of program may be distinguished. First, a 'control' type of program, supplied with the computer, is read into the machine. Its main functions are to control the progress of all other programs which may be operating simultaneously and to allocate storage space and the input and output devices required. The second category of program covers an operation common to many problems handled by a particular type of computer; a procedure of this kind, made permanently available for insertion in larger programs as required, is often called a 'library routine'. Finally there is the program devised for the problem in hand, perhaps incorporating one or more library sub-routines, usually known as an 'application program'.

A program may be written in a 'low-level language', i.e. one appropriate to a particular type of computer, or in a 'high-level language'. A low-level language program, though detailed and rather difficult to devise, is economical in computer time; it is therefore appropriate to a repetitive operation such as occurs in numerical weather forecasting. A high-level language program is more easily written and may be appropriate to more than one type of computer but may be less efficient in its use of computer time. Widely used high-level languages include FORTRAN and PASCAL in science, and COBOL and 'C' in business.

The operator provides the computer with the data and programmed instructions. The input material may be on magnetic tape, paper tape, or magnetic disk; other devices such as character-recognition units may be used. Output from a computer may be on tape or cards, or in the form of figures and letters from a printer, or as lines or plotted characters on a chart, or as a display on a visual display unit.

automatic picture transmission (APT). As used in meteorology, this term denotes the automatic transmission of images of the earth and clouds by weather SATELLITES. Conventionally the term is restricted to analogue transmissions from sun-synchronous (i.e. polar-orbiting) satellites.

A scanned image is transmitted from the satellite to a ground station where the image is rebuilt on a facsimile machine. A ground station usually obtains three pictures, covering about 6 million km^2 of the earth's surface, on each transit of the satellite.

automatic weather station. A STATION, often situated in an isolated location, at which measurements of meteorological elements are made by automatic methods which do not require local human supervision and control. Various sensors may be used, producing output in either digital or analogue form. Output from individual sensors may be accumulated locally, transmitted continuously or at fixed times to a central controlling station, or made available for interrogation on demand.

autumn. See SEASONS.

available potential energy. As defined and used by E.N. Lorenz, this term denotes that (small) part of the total POTENTIAL ENERGY of the atmosphere available for conversion to KINETIC ENERGY under adiabatic flow. It is the difference between the total potential energy of the atmosphere at a given time and the hypothetical minimum total potential energy that would result after adiabatic redistribution of the mass of the atmosphere such that the density stratification was everywhere horizontal. The available potential energy is some ten times greater than the total kinetic energy but is only about 1 per cent of the total potential energy.

The concept is strictly applicable only to the atmosphere as a whole and is significant in studies of the GENERAL CIRCULATION.

avalanche wind. An avalanche often causes, in advance and at the sides of the descending mass of snow or ice, a very high wind capable of causing destruction at some distance from the avalanche itself. It is known as the 'avalanche wind' or 'avalanche blast'.

average. An alternative for ARITHMETIC MEAN.

aviation forecast. A meteorological forecast issued for a defined period for the purpose of aviation, generally covering conditions over a specific route, or for a specific area and at a terminal or terminals.

The main elements included are the weather, clouds (type, base, thickness), visibility, upper winds, surface winds at terminals, height of the 0 °C isotherm, and icing and turbulence risks. See also SIGNIFICANT-WEATHER CHART.

Avogadro's law. At normal temperature and pressure the weight of any gas in grams which is numerically equal to its MOLECULAR WEIGHT occupies 22.4 litres. The law implies that equal volumes of gases, under the same conditions of temperature and pressure, contain the same number of molecules.

Avogadro's number. The number (N) of molecules per MOLE of a gas, equal to $6.02252 (\pm 0.00028) \times 10^{23}$. See AVOGADRO'S LAW.

azimuth. The azimuth of an object is the horizontal angle between the observer's MERIDIAN and the line joining observer and object. It is normally measured in degrees clockwise from true (geographic) north.

Azores anticyclone. Part of the subtropical high-pressure belt of the northern hemisphere. On mean surface-pressure charts the anticyclone in summer has a central pressure of about 1024 mb at about 35° N, with the axis of a ridge extending across northern France and northern Germany; in winter the central pressure is almost 1024 mb at about 30° N, with the axis of the ridge then lying across southern Spain. Fluctuations of several millibars in strength, and small fluctuations in position and shape, have been observed on decadal and longer time-scales, especially in winter.

B

Babinet's point.   See ARAGO'S POINT.

Babinet's principle.   The DIFFRACTION produced by a drop, for example a drop in a water cloud, is the same as that produced by a small aperture in a screen.

back-bent occlusion.   A FRONT which is formed in the rear quadrant of a DEPRESSION and which is attributed to the air-mass contrast originally developed at the OCCLUSION. As the low-pressure centre moves along the occlusion (usually eastwards) the line of air-mass contrast moves south-eastwards behind the centre as a back-bent occlusion. Secondary cold fronts also appear in the rear quadrant of depressions and do not necessarily arise from back-bent occlusions.

backing.   The changing of the wind direction in an anticlockwise rotation in either hemisphere. The opposite to veering.

baguio.   A local name by which the TROPICAL CYCLONES experienced in the Philippines are known. A number of the cyclones or typhoons of the western Pacific cross the Philippines; in addition, there is a class of cyclone which is especially associated with these islands, occurring from July to November.

balance equation.   An equation expressing the balance between the horizontal non-divergent motion and the corresponding pressure field. The equation may be written

$$f\nabla^2\psi + \nabla\psi.\nabla f + 2\left\{\frac{\partial^2\psi}{\partial x^2}\ \frac{\partial^2\psi}{\partial y^2} - \left(\frac{\partial^2\psi}{\partial x\partial y}\right)^2\right\} = g\nabla^2 z$$

where z is the height of the pressure surface.

Since the flow in the middle troposphere is usually close to a state of non-divergence, the wind at such levels may be approximately derived from the STREAM FUNCTION ψ found by solving the balance equation numerically on a computer.

The equation is obtained by taking the DIVERGENCE of the EQUATIONS OF MOTION and omitting the lower-order terms involving divergence. It has in the past proved very important when applied to NUMERICAL WEATHER PREDICTION since it makes it possible to use the PRIMITIVE EQUATIONS without introducing spurious high-frequency disturbances of large amplitude; however, it has now been largely superseded by modern INITIALIZATION techniques.

ballistics.   The science governing the propulsion and flight of artillery projectiles. Ballistics falls naturally into two parts — interior (the chemistry, physics and mechanics within the gun) and exterior (the aerodynamic and meteorological effects upon the projectile in flight). The flight path of an artillery projectile is affected by the density and air motion along the trajectory. Gunners are provided with a BALLISTIC TEMPERATURE and a BALLISTIC WIND which enable them to offset the elevation and bearing of the gun so that the projectile will hit the desired target.

ballistic temperature.   A measure of the difference between the idealized standard temperature profile and the observed temperature distribution encountered by the ballistic projectile. A ballistic density is similarly defined. The ballistic temperature, density and wind are used when laying the gun to introduce corrections which cancel out the effects of the actual atmosphere on the trajectory.

ballistic wind. An average wind computed from the actual winds as they vary with height along the ballistic trajectory. Combined with the ballistic temperature and density, the ballistic wind enables the artillerman to adjust the bearing and elevation of the gun so that the projectile will hit the target. Also known as 'equivalent constant wind'.

ball lightning. A rare form of LIGHTNING in which a persistent and moving luminous white or coloured sphere is seen; the explanation of this form of lightning is controversial as was formerly its existence.

Reports of the sphere dimensions vary from a few centimetres to about a metre but are most commonly from 10 to 20 cm. Duration varies from a few seconds to several minutes. Many reported cases follow a brilliant lightning flash and may be physiological in nature (after-image); other reported cases have, however, occurred without a preceding flash. Sometimes more than one sphere is seen by an observer, or a sphere is reported in the same locality by various observers. The speed of travel is generally about walking pace. Spheres have been reported to vanish harmlessly, to bounce from the ground or an obstruction, or to pass into or out of rooms leaving, in some cases, sign of their passage as, for example, a hole in a window pane.

balloon sounding. Exploration of the earth's atmosphere by means of a balloon inflated with a gas which is lighter than air (hydrogen or helium). Such a balloon may be followed by theodolite (PILOT BALLOON) or, if a reflector is attached, by radar (RADAR WIND) to obtain upper-air winds. Automatic instruments may be attached to obtain readings of temperature and relative humidity at various pressure levels (RADIOSONDE). The maximum altitude attained by a balloon depends on its FREE LIFT. Radiosonde balloons generally reach from 18 to 30 km.

Very large balloons carrying considerable equipment loads are used on occasion for geophysical research, e.g. into cosmic rays. Others carry a special device so that they may remain for a considerable period at a predetermined level (CONSTANT-LEVEL BALLOON). Others are used for launching small rockets — see ROCKOON SOUNDING. Early research balloons were manned, with corresponding limitation in height attained. Among the early ascents made for meteorological purposes were those by J. Welsh in 1852 and by J. Glaisher between 1862 and 1866. Captive balloons are sometimes used to obtain meteorological data in the lower atmosphere.

banner cloud. A stationary cloud attached to, and extending downwind from, an isolated mountain peak, the cloud appearance being one of an extended flag. The requirements are an isolated sharp peak, strong wind and relatively moist air. Among well known examples of the cloud are those associated with Mt. Everest and the Matterhorn.

The physical explanation normally advanced for the cloud formation is the lifting of air in a lee eddy from a lower level than on the windward side. An aerodynamic pressure reduction in the lee of the peak may contribute to the cloud formation in a manner analogous to the formation of aircraft wing-tip CONDENSATION TRAILS.

bar. A unit of atmospheric pressure equal to the pressure of 29.530 inches or 750.062 millimetres of mercury under the standard conditions of temperature 0 °C (density of mercury 13 595.1 kg m^{-3}) and gravitational acceleration 9.80665 m s^{-2}.

$$1 \text{ bar} = 10^3 \text{ MILLIBARS} = 10^6 \text{ DYNES cm}^{-2} = 10^5 \text{ NEWTONS m}^{-2}.$$

See STANDARD DENSITY, STANDARD GRAVITY, SI UNITS.

barat. A strong, north-westerly squall on the north coast of the island of Sulawesi, most frequent from December to February.

baroclinic. A baroclinic atmosphere is one in which surfaces of pressure and density (or specific volume) intersect at some level or levels. The atmosphere is always, to some extent, baroclinic. Strong baroclinicity implies the presence of large horizontal temperature gradients and thus of strong THERMAL WINDS.

baroclinic instability. A type of dynamic instability, associated with a strongly BAROCLINIC region of the atmosphere, which is considered to be responsible for at least part of the development of wave disturbances within the strong westerly wind flow which frequently occurs in the upper air in middle and high latitudes. Growth of the disturbances is characterized by ascent of the warmer, and descent of the colder, air masses, representing a decrease of potential energy and an associated release of kinetic energy. Theory indicates that the degree of instability of the disturbances depends, among other factors, on their wavelength.

baroclinic leaf. An elongated leaf-shaped cloud pattern, as seen in a satellite picture, which forms within the jet-stream zone. It often develops into a COMMA CLOUD.

baroclinic wave. A wave depression which forms in a strongly BAROCLINIC region of the atmosphere. BAROCLINIC INSTABILITY is important in the intensification of such waves.

barogram. The record made by a BAROGRAPH.

barograph. A recording BAROMETER. That in common use is essentially an ANEROID BAROMETER which is arranged to give a continuous recording. Those used by the Meteorological Office are known as the 'open scale' and the 'small pattern', which differ in the type of aneroid element and lever mechanism used. In ship barographs, use is made of an increase of lag coefficient and anti-vibration mounting.
  Two non-portable types of barograph, used at some observatories, are the 'float barograph', in which the movements of the mercury in a barometer are communicated to a recording pen by means of a float, and the 'photobarograph', in which the position of the mercury meniscus is recorded photographically.

barometer. An instrument for measuring atmospheric pressure. The mercury barometer, for example the FORTIN BAROMETER or the KEW-PATTERN BAROMETER, is satisfactory for normal use but is difficult to transport. In these instruments the height of the mercury column in a glass tube, one end closed and uppermost, above the level of mercury in an open vessel (cistern) in which the open end of the tube is immersed, is used as a measure of atmospheric pressure. Barometer 'corrections' must be applied to take account of departures of mercury and scale temperature, and of local GRAVITY, from the 'standard' values assumed in the calibration of the instrument.
  A non-mercury type of barometer in common use is the ANEROID BAROMETER. See also STANDARD DENSITY, STANDARD GRAVITY, STANDARD TEMPERATURE.

barometric characteristic. In synoptic meteorology, an observation of the shape of the BAROGRAPH trace during the three hours before the observation. Such observations are plotted on synoptic charts in internationally agreed symbols which simulate the general form of the barograph trace.

barometric tendency. In synoptic meteorology, an observation of the barometric change during a specified period (usually three hours) prior to the observation. Such observations, plotted on synoptic charts, are the basis on which ISALLOBARS are drawn.

barothermograph. An instrument which records temperature and pressure simultaneously, two pens being used to record separate traces of temperature and pressure.

The term has also been loosely used in connection with the obsolete Dines METEOROGRAPH in which temperature was recorded as a function of pressure; in this instrument the recording pen was made to move in one direction by a change of pressure, and in a direction almost at right angles by a change of temperature.

barotropic. A barotropic atmosphere is that hypothetical atmosphere in which surfaces of pressure and density (or specific volume) coincide at all levels. The concept of barotropy, though idealized, gives a useful first approximation in some types of atmospheric problem. The contrasting atmospheric state is the BAROCLINIC.

barotropic wave. A wave-like disturbance in a BAROTROPIC atmosphere. The real atmospheric flow is well approximated (on the synoptic scale) by barotropic waves at about 600 mb where the horizontal divergence is least. Forecasting of the movement of such waves is based on the conservation of absolute VORTICITY by individual air particles.

bathythermograph. An instrument which provides a continuous trace of sea-temperature variations with depth.

A liquid-in-glass thermometer moves a metal point in one direction across a smoked glass slide while the slide itself is made to move, by metal bellows which are sensitive to pressure (depth) variations, in a direction at right angles to this. The instrument is lowered from a ship to a selected level and is then recovered. The instrument is often used from ships which are under way but then seldom to depths below about 300 m.

beaded lightning. An alternative for PEARL-NECKLACE LIGHTNING.

bearing. The true (geographic) bearing of an object is synonymous with its AZIMUTH. Magnetic bearing is the corresponding angle measured clockwise from magnetic north. Approximate bearings may be named as the compass points, N, NE, etc.

Beaufort notation. A code of letters indicating the state of the weather, past or present. The code was originally introduced by Admiral Beaufort for use at sea but is equally convenient for use on land. Additions have been made to the original schedule. A phenomenon of moderate intensity is indicated by the appropriate lower case letter: if of slight intensity the suffix $_o$ is added, if of greater intensity a capital letter is used. Continuity is indicated by repetition of the letter, intermittency by prefixing the letter i, and showers by prefixing the letter p. Thus, for example, ir_o denotes intermittent slight rain, and pS heavy snow shower. See *Observer's handbook* [1, pp. 72–79].

Beaufort scale. Wind force is estimated on a numerical scale ranging from 0, calm, to 12, hurricane, first adopted by Admiral Beaufort. The specification of the steps of the scale originally given had reference to a man-of-war of the period 1800–50 and therefore now possesses little more than historic interest.

Details of the Beaufort scale are contained in Table II. The velocity equivalents were originally based on the empirical relationship between estimated number and measured velocity, $V = 1.87 \sqrt{B^3}$, where V is in miles per hour, and B is the corresponding Beaufort number. The pressure equivalents were derived from the formula $p = 0.003 V^2$, where p is in pounds per square foot and V is in miles per hour.

Beaumont period. A period of 48 consecutive hours during which the dry-bulb temperature in the screen has been 10 °C or above and the relative humidity 75 per cent or above on at least 46 of the 48 hourly observations; the aim of defining such a period is to provide warning of an imminent outbreak of potato blight.

Bénard cell.   When a fluid is carefully heated from below or cooled from above in the laboratory, a cellular pattern of CONVECTION may be established in which the motion in the centres of the cells (upward or downward) is opposed to that on the periphery. Such cells are known as Bénard cells. There is some evidence that such cells develop at times in layered cloud after dusk by radiative cooling at the cloud top.

Bergeron (or Bergeron–Findeisen) theory.   That theory which attributes the initiation of precipitation from a cloud to the growth by sublimation of a small concentration of ice crystals among predominantly supercooled water droplets. See PRECIPITATION.

berg wind.   A local name for the offshore FÖHN-type wind in South Africa.

Bermuda anticyclone.   That cell of the semi-permanent subtropical high-pressure belt which is frequently centred near Bermuda in the western part of the North Atlantic Ocean.

Bernoulli's theorem.   In an inviscid fluid in steady motion the sum per unit mass of the kinetic energy ($v^2/2$), the potential energy possessed by virtue of being in a pressure field (p/ρ), and the gravitational potential energy (gz) is constant. That is:

$$v^2/2 + p/\rho + gz = \text{constant.}$$

Here v is fluid velocity and z is height above a selected reference level.

beta (or β) particle.   A swiftly moving ELECTRON emitted spontaneously by certain radioactive elements. Beta particles have moderate penetrative power amounting to some metres in air near ground level; their emission by natural radioactive materials in the ground contributes significantly (about one-fifth) to the IONIZATION of air at low levels over land. See also ALPHA PARTICLE, GAMMA RADIATION.

beta- (or β-) plane approximation.   The assumption that the earth may be treated as flat for the purpose of numerical calculation with the Coriolis parameter (f) being treated as a constant equal to f_0 except where it appears differentiated in advection terms, in which case df/dy is conventionally denoted by β.

bi-directional vane (or bivane).   An instrument comprising two sensitive vanes of similar characteristics, free to turn about vertical and horizontal axes, respectively. The instrument is used, with direct pen or with electrical recording, to indicate simultaneous variations of the horizontal and vertical components of the wind in low-level turbulence measurements. See *Geophysical Memoir* No. 65 [8].

billow clouds.   Parallel rolls of cloud, separated by relatively narrow, clear spaces. The phenomenon has been explained in various ways: as waves formed at a surface of discontinuity; as convection cells initiated by shearing motion at an almost horizontal boundary between two air streams; or by cellular motion initiated by the development of static instability in a shallow layer of air. See Figure 17.
   The term is often loosely applied to clouds of the variety UNDULATUS.

bimetallic thermograph.   A THERMOGRAPH in which the sensitive element is a curved strip formed by welding together two metals differing in their coefficients of expansion. The changes of curvature undergone by the strip with changes of temperature are used to actuate a recording pen through a lever mechanism.

TABLE II — *Beaufort scale:*

Force	Description	Specifications for use on land	Specifications for use at sea
0	Calm	Calm, smoke rises vertically.	Sea like a mirror.
1	Light air	Direction of wind shown by smoke drift, but not by wind vanes.	Ripples with the appearance of scales are formed, but without foam crests.
2	Light breeze	Wind felt on face; leaves rustle; ordinary vane moved by wind.	Small wavelets, still short but more pronounced. Crests have a glassy appearance and do not break.
3	Gentle breeze	Leaves and small twigs in constant motion; wind extends light flag.	Large wavelets. Crests begin to break. Foam of glassy appearance. Perhaps scattered white horses.
4	Moderate breeze	Raises dust and loose paper; small branches moved.	Small waves, becoming longer; fairly frequent white horses.
5	Fresh breeze	Small trees in leaf begin to sway; crested wavelets form on inland waters.	Moderate waves, taking a more pronounced long form; many white horses are formed. Chance of some spray.
6	Strong breeze	Large branches in motion; whistling heard in telegraph wires; umbrellas used with difficulty.	Large waves begin to form; the white foam crests are more extensive everywhere. Probably some spray.
7	Near gale	Whole trees in motion; inconvenience felt when walking against wind.	Sea heaps up and white foam from breaking waves begins to be blown in streaks along the direction of the wind.
8	Gale	Breaks twigs off trees; generally impedes progress.	Moderately high waves of greater length; edges of crests begin to break into spindrift. The foam is blown in well-marked streaks along the direction of the wind.
9	Strong gale	Slight structural damage occurs (chimney pots and slates removed).	High waves. Dense streaks of foam along the direction of the wind. Crests of waves begin to topple, tumble and roll over. Spray may affect visibility.
10	Storm	Seldom experienced inland; trees uprooted; considerable structural damage occurs.	Very high waves with long overhanging crests. The resulting foam, in great patches, is blown in dense white streaks along the direction of the wind. On the whole the surface of the sea takes a white appearance. The 'tumbling' of the sea becomes heavy and shock-like. Visibility affected.
11	Violent storm	Very rarely experienced; accompanied by widespread damage.	Exceptionally high waves (small and medium-sized ships might be for a time lost to view behind the waves). The sea is completely covered with long white patches of foam lying along the direction of the wind. Everywhere the edges of the wave crests are blown into froth. Visibility affected.
12	Hurricane	—	The air is filled with foam and spray. Sea completely white with driving Spray; visibility very seriously affected

Specifications and equivalent speeds

Force	Mean pressure (at standard density)* mb	Equivalent speed at 10 m above ground					
		Knots		Miles per hour		Metres per second	
		Mean	Limits	Mean	Limits	Mean	Limits
0	0	0	<1	0	<1	0.0	0.0–0.2
1	0.01	2	1–3	2	1–3	0.8	0.3–1.5
2	0.04	5	4–6	5	4–7	2.4	1.6–3.3
3	0.13	9	7–10	10	8–12	4.3	3.4–5.4
4	0.32	13	11–16	15	13–18	6.7	5.5–7.9
5	0.62	19	17–21	21	19–24	9.3	8.0–10.7
6	1.1	24	22–27	28	25–31	12.3	10.8–13.8
7	1.7	30	28–33	35	32–38	15.5	13.9–17.1
8	2.6	37	34–40	42	39–46	18.9	17.2–20.7
9	3.7	44	41–47	50	47–54	22.6	20.8–24.4
10	5.0	52	48–55	59	55–63	26.4	24.5–28.4
11	6.7	60	56–63	68	64–72	30.5	28.5–32.6
12	≥7.7	—	≥64	—	≥73	—	≥32.7

* The pressure due to the wind on any object exposed to it arises from the impact of the air on the windward side and suction on the leeward side; the mean pressure depends on the shape and size of the object. The values given apply to a disc of approximate area 0.1 to 10 m^2. See STANDARD DENSITY.

binomial distribution. When the PROBABILITY of occurrence of an event in a single trial is p, then the number of occurrences in samples each consisting of n independent trials accords with the binomial distribution of MEAN DEVIATION np and STANDARD DEVIATION $\sqrt{\{np(1-p)\}}$.

Thus the probability $p(x)$ that an event of probability p occurs x times in n independent trials is

$$p(x) = \frac{n!}{x!(n-x)!} \cdot p^x(1-p)^{n-x}.$$

The binomial distribution finds a useful application in cases of two contrasting alternatives, e.g. rain or no rain or testing of two forecasting systems, provided that the result of each trial is independent of the results of earlier trials. See FREQUENCY DISTRIBUTION, NORMAL (FREQUENCY) DISTRIBUTION, POISSON DISTRIBUTION.

bioclimatology. The study of climate in relation to life and health.

biosphere. That part of the earth's envelope, comprising the seas, lower atmosphere and surface layer of the earth's crust, in which living organisms exist in their natural state.

bise. A cold and relatively dry north-east wind which blows often in France and Switzerland.

Bishop's ring. A dull reddish-brown ring which is seen round the sun in a clear sky. In the middle of the day the inner radius of the ring is about $10°$, the outer $20°$; when the sun is low the ring is larger. Bishop's ring was first seen after the great eruption of Krakatoa in 1883 and remained visible till the spring of 1886. It was also seen after the eruptions of Soufrière in St. Vincent and Mt. Pelée in Martinique in 1902, after the north Siberian meteorite in 1908, and at the time of nearest approach to the earth of Halley's comet on 18 and 19 May 1910. The phenomenon is attributed to DIFFRACTION associated with fine dust in the high atmosphere.

Bjerknes–Jeffreys theorem. That a state of hydrostatic equilibrium of the earth's atmosphere (i.e. with winds everywhere equal to zero) is impossible unless there are no variations of density on surfaces of constant geopotential. This result follows simply from the VORTICITY EQUATION.

black-body radiation. See RADIATION.

black-bulb thermometer. A mercurial MAXIMUM THERMOMETER with blackened bulb, mounted in an evacuated outer glass sheath and exposed horizontally to the sun's rays for the purpose of ascertaining the maximum temperature 'in the sun'. On account of the difficulty of obtaining comparable results with different instruments and of interpreting the indications of an individual instrument, black-bulb thermometers are not now recommended as a means of measuring solar radiation. Such measurements are carried out with a PYRANOMETER or PYRHELIOMETER.

black frost. A condition in which the temperature of the ground cools to a sub-freezing temperature but does not reach a value so low as the FROST-POINT of the adjacent air. There is then no deposit of HOAR FROST on the ground or on terrestrial objects which thus remain 'black' in appearance. The phenomenon is associated with relatively dry air.

black ice. A popular alternative for GLAZE, often used with reference to its occurrence on road surfaces. A thin sheet of ice, relatively dark in appearance, may be formed when light rain or drizzle falls on a road surface which is at a temperature below 0 °C. It may also be formed when supercooled fog droplets are intercepted by bridges, trees, etc.

Blaton's equation.   If ψ is the angle between the direction of the horizontal wind and a reference direction (e.g. north), K_H the horizontal curvature of the trajectory, K_S the horizontal curvature of the streamline, and V the wind speed, then

$$K_H = K_S + \frac{1}{V}\frac{\partial \psi}{\partial t}$$

where $\partial\psi/\partial t$ is the local rate of turning of the horizontal wind direction. In a steady state $(\partial\psi/\partial t = 0)$, $K_H = K_S$; and streamlines and trajectories coincide. Blaton's equation is purely kinematic.

blizzard.   A term originating in North America meaning a very cold, strong to gale force wind that is laden with snow at least some of which has been raised from snow-covered ground. In general parlance in the United Kingdom it is loosely used to mean any remarkably heavy fall of snow. In forecasts issued by the Meteorological Office a 'blizzard' is defined as the simultaneous occurrence of moderate or heavy snowfall with winds of at least force 7 (approximately 15.5 m s^{-1}), causing drifting snow and reduction of visibility to 200 m or less, while a 'severe blizzard' implies winds of at least force 9 (approximately 22.6 m s^{-1}) and reduction of visibility to near zero. In the USA, an 'ordinary blizzard' implies a temperature of less than −7 °C, a wind of greater than 15 m s^{-1} and visibility of less than 150 m, while a 'severe blizzard' implies a temperature of less than −12 °C, a wind speed greater than 20 m s^{-1} and zero visibility.

blocking.   The term applied in middle-latitude synoptic meteorology to the situation in which there is interruption of the normal eastward movement of depressions, troughs, anticyclones and ridges for at least a few days.

A blocking situation is dominated by an anticyclone whose circulation extends to the high troposphere. The normal ZONAL FLOW from the west is transformed into MERIDIONAL CIRCULATIONS branching polewards and equatorwards. For Europe and the North Atlantic the longitude most favoured for blocking is about 10° W. Over this region the percentage frequency of days of blocking situation has a maximum in April and a minimum in August.

blowing dust or sand.   'Dust or sand, raised by the wind to moderate heights above the ground. The horizontal visibility at eye level is sensibly reduced' [2, p. 120].

blue moon, sun.   A rare phenomenon in which more intense particle SCATTERING of red light than of blue light makes the directly viewed luminary appear blue (or green).

A conspicuous event of this kind occurred in the British Isles on 26 September 1950 and was due to scattering by smoke particles which originated in earlier forest fires in Alberta, Canada. The smoke was shown by aircraft reconnaissance to be at heights between about 10 and 13 km. Application of scattering theory to the differential extinction of light measured in different parts of the visible spectrum indicated that the predominant radius of the scattering particles was about 0.5 μm.

blue of the sky.   The blue of the sky is caused by the SCATTERING of sunlight by the individual molecules of the air — so-called RAYLEIGH SCATTERING. For such small particles, Rayleigh showed that the proportion of scattered light is greater for shorter than for longer wavelengths (inversely proportional to the fourth power of the wavelength). Thus the light which reaches the observer after scattering is rich in the blue and violet short waves (to the latter of which the eye is less sensitive); the light which reaches the observer directly is deficient in the short waves — hence the yellow or red disc of the sun.

The light scattered by dust particles suspended in the air normally has no wavelength dependence — but see BLUE MOON, SUN. Such particles therefore introduce a white tinge to the blue sky to a degree which increases with the dust concentration. Thus, for example, the sky is of a deeper blue in polar than in anticyclonic air, and also as seen from a mountain top than from a low level.

bogus observation. An observation invented by a forecaster and added to the original set of synoptic and other observations in order to force a computer-controlled OBJECTIVE ANALYSIS scheme to produce an analysis much closer than it would otherwise have done to the true situation as seen by the forecaster. Bogus observations are usually placed in regions of sparse or faulty data.

boiling-point. That temperature at which, under conditions of equilibrium between a plane surface of a liquid and its overlying vapour, the existing (saturation) VAPOUR PRESSURE is equal to the external pressure. The variation of the boiling point of water with external pressure is given under HYPSOMETER.

bologram. The record of a BOLOMETER.

bolometer. An instrument for the determination of the intensity of RADIATION, employing a blackened conductor whose change of resistance with temperature gives a measure of the quantity required. The instrument is often employed in the investigation of the distribution of energy in the spectrum, especially in the infra-red region; it is then called a 'spectrobolometer'.

Boltzmann's constant. The universal constant (k) given by the ratio R^*/N where R^* is the universal GAS CONSTANT and N is AVOGADRO'S NUMBER.

k has the value 1.3805×10^{-23} J K^{-1} and appears in equations involving the expression of energy in terms of temperature.

bora. A cold, often very dry, north-easterly wind which blows, sometimes in violent gusts, down from the mountains on the eastern coast of the Adriatic. It is strongest and most frequent in winter and on the northern part of the coast, and occurs when pressure is high over central Europe and the Balkans and low over the Mediterranean. If associated with a depression over the Adriatic it is accompanied by heavy cloud and rain or snow. The term is also applied to cold, squally, downslope winds in other parts of the world.

boreal climate. In the KÖPPEN CLASSIFICATION, a CLIMATE characterized by a snowy winter and warm summer, with a large annual range of temperature, such as occurs over the European, Asian and American continents between about latitudes 40 and 60° N.

Bouguer's halo. A rare type of HALO, white in colour and centred on the ANTISOLAR POINT with an inner radius of about 35° (also called ULLOA'S CIRCLE).

Bouguer's law. See ABSORPTION.

boundary layer. That layer of a fluid adjacent to a physical boundary in which the fluid motion is much affected by the boundary and has a mean velocity less than the free-stream value. The depth of the boundary layer in the earth's atmosphere varies markedly with static stability, from a few hundred metres in stable conditions to 1 or 2 km in convective conditions. See also FRICTION LAYER.

Bourdon tube. A curved tube of elliptical cross-section, in which changes of volume cause changes of curvature; these are used to actuate a pointer or recording pen. It may be used as a barometer or pressure gauge, being evacuated and responding to variations of external pressure; or, for example, it may be connected to a steel thermometer bulb, when it responds to variations in the volume of the mercury which fills the bulb, the connecting capillary tubing and the Bourdon tube.

Boussinesq approximation. The neglect of departures of density from a long-term average state except when they are coupled to the acceleration of gravity in the buoyancy force.

For a more precise discussion, consider a compressible fluid (e.g. the atmosphere) in which all the motions derive from small perturbations of a basic state which is a function of height (z) only; this basic state is indicated by suffix zero. Values of pressure, temperature, density and potential temperature are then given by, for example,

$$p(x,y,z,t) = p'(x,y,z,t) + p_0(z).$$

If f is any one of the variables of state (p, ρ, etc.) then we may define a set of SCALE HEIGHTS (H_f) by

$$H_f^{-1} = \left| \frac{1}{f_m} \frac{df_0}{dz} \right|$$

where f_m is the mean value of f_0. The Boussinesq approximation is equivalent to the assumption that the motion is confined to a layer whose thickness is very much less than the smallest H_f. The momentum and thermodynamical equations for adiabatic flow may then be written

$$\frac{d\mathbf{V}'}{dt} + \frac{1}{\rho_0} \nabla p' - \frac{g\theta'}{\theta_0} \mathbf{k} = 0$$

$$\operatorname{div} \mathbf{V}' = 0$$

$$\frac{d\theta'}{dt} + w' \frac{d\theta_0}{dz} = 0.$$

A derivation is given by Holton [9, pp. 160–162] and further detailed discussion by Spiegel and Veronis [10].

Bowen ratio. The ratio (R) of the amount of sensible heat to that of latent heat (see HEAT) lost by a surface to the atmosphere by the processes of conduction and turbulence.

Even over water surfaces, very variable values of R are found (including negative values, which signify sensible heat transfer from atmosphere to surface). An average value of about $+0.1$ is considered to hold for oceans, implying that 90 per cent of the heat energy received by oceans is used in evaporation.

Boyden index. An INSTABILITY INDEX (I) given by

$$I = Z - T_{700} - 200$$

where Z is the 1000–700 mb thickness (in decametres) and T_{700} is the 700 mb temperature (in degrees Celsius).

45

Thunderstorms become increasingly likely the further I increases above a threshold value of 94. Allowance for the movement (for example by the 700 mb wind) of the isopleths of I over an area enables some account to be taken of advective changes.

Boyle's law. At constant temperature, the volume of a given mass of gas (or the SPECIFIC VOLUME) is inversely proportional to the pressure on the gas.

brave west winds. A nautical expression denoting the prevailing westerly winds of temperate latitudes. The region of strong westerly winds of the southern hemisphere (latitudes 40–50° S) is termed the ROARING FORTIES.

breakaway depression. A term applied to a WARM-FRONT WAVE or WARM-OCCLUSION DEPRESSION because of the tendency of such secondary depressions to move eastwards away from the parent depression after formation.

breeze. A wind of moderate strength. (See BEAUFORT SCALE.) The word is generally applied to winds, caused by convection, which occur regularly during the day or night; they include LAND- AND SEA-BREEZES, MOUNTAIN BREEZE, VALLEY WIND.

Brewster's point. See ARAGO'S POINT.

briefing, meteorological. Oral explanation by a meteorologist, to a specialized audience, of existing and expected meteorological conditions.

bright band. A conspicuous quasi-horizontal region of enhanced reflection obtained in radar scanning of a precipitation belt in the vertical plane which is due to the melting of snowflakes (see MELTING BAND). Other bands that are bright, which occasionally appear at higher levels, are referred to as 'upper bands'; they are thought to mark regions of appreciable changes of size or shape of solid precipitation elements and have been ascribed, for example, to 'precipitation streaks' or VIRGA whose shapes are changed by vertical wind shear as they are moved through the radar beam by the wind. See also RADAR METEOROLOGY.

brightness. The radiated power per unit solid angle per unit area received by a radiometer from an emitting source. The units are watts per steradian per square metre.

brightness temperature. The temperature of the black body that has the same BRIGHTNESS as the radiating surface that is under discussion. When photographic images of the earth and its clouds are produced from observations of infra-red radiation made by meteorological satellites, different brightness temperatures are indicated by different shades of grey such that warm surfaces (e.g. tropical oceans) are nearly black and very cold surfaces (e.g. high clouds) are nearly white.

British Rainfall Organization. This term was first used after the death of G.J. Symons in 1900 to describe the voluntary organization of rainfall observers which he had built up since about 1860. In 1919 the work of the Organization was transferred to the Meteorological Office. The term continued in use on the title page of *British Rainfall* and elsewhere for several decades but is now (since about 1958) no longer used.

British Summer Time (BST). The standard of time in common use in the British Isles during a period in summer which was defined by the Summer Time Act, 1925. British Summer Time is one hour in advance of Greenwich Mean Time; 0900 GMT is thus the same as 1000 BST.

Brocken spectre. When an observer stands on a hill which is partially enveloped in mist and in such a position that his/her shadow is thrown on to the mist he/she may get the illusion that the shadow is a person seen dimly through the mist. The illusion is that this person or 'spectre' is gigantic and at a considerable distance. The Brocken is a mountain in Germany. See also GLORY.

Brückner cycle. The name usually given to a feature described by E. Brückner in 1890 who by drastic smoothing was able to discern a series of 25 'cycles' covering the years 1020 to 1890 in which comparatively cool, rainy periods alternated with warmer, drier periods. The individual 'cycles', which were of very small amplitude, varied in duration from 20 to 50 years with an average length of 34.8 years and have not survived subsequent investigation.

Brunt–Väisälä frequency. The frequency N defined by the equation

$$N^2 = \frac{g}{\theta} \cdot \frac{\partial \theta}{\partial z}$$

where $\partial \theta / \partial z$ is the vertical gradient of potential temperature.

With certain simplifying assumptions, N is the angular frequency of small vertical oscillations of a parcel of air about its equilibrium position in a stable atmosphere.

brush discharge. Discharge of electricity from sharp points on a conductor. See ST. ELMO'S FIRE.

Buchan spells. A total of nine periods during the year which were advanced by Alexander Buchan in 1867, on the basis of some 50 years' observations, as constituting fairly reliable periods of unseasonal cold (six cases) or warmth (three cases) in south-east Scotland. The periods were: cold, 7–14 February, 11–14 April, 9–14 May, 29 June–4 July, 6–11 August and 6–13 November; warm, 12–15 July, 12–15 August, 3–14 December.

While evidence has since been advanced in support of similar spells in other parts of Europe, the reality of the spells as statistically significant features has also since been disputed. The reliability of the spells is certainly too low to permit their direct application to long-range weather forecasting.

bumpiness. A type of disturbance experienced by an aircraft in flight characterized by rapid variations, including generally alterations of sign, of the vertical component of velocity.

Bumpiness is associated generally with either convection currents in an unstable atmosphere, or a flow of air across surface irregularities, or both. It is much more common and intense over the land than over the sea. It is generally most marked in the lowest kilometre of the atmosphere (see FRICTION LAYER) but may extend to much higher levels above hilly country. Different types of aircraft may experience different types of bumpiness when flying through identical atmospheric conditions. See also AIR POCKET, CLEAR-AIR TURBULENCE.

buoyancy. The buoyancy of a balloon or airship is the total load, including the envelope and fittings, that can just be supported. This buoyancy arises from the difference between the density of the light gas inside the envelope and that of the heavier air outside. The vessel will just rest in equilibrium when the total weight is the same as that of the air displaced; the buoyancy of the vessel is thus the difference between the weight of the gas in the envelope and that of the volume of air displaced (principle of Archimedes).

Analogous buoyancy forces act on parcels or 'bubbles' of air which are a. a different temperature from that of the surrounding air and are fundamental in the process of free CONVECTION. Where T' and T are the temperatures (kelvin) of the air parcel and the environment, respectively, the buoyancy force acting per unit mass is given by $g\{(T'/T)-1\}$. The force is reckoned positive upwards $(T' > T)$ and negative downwards $(T' < T)$.

buran. A strong north-easterly wind which occurs in the USSR and central Asia. It is most frequent in winter, when it is very cold, and often raises a drift of snow, but strong north-easterly winds in summer are also termed buran. The winter snow-bearing wind is also termed 'purga'.

Burger number. The ratio of the ROSSBY RADIUS OF DEFORMATION (λ) to the characteristic length scale of the motion (L), i.e. λ/L. When the Burger number is small, the flow is predominantly horizontal and geostrophic. The reciprocal L/λ is sometimes referred to as a 'rotational Froude number'.

Buys Ballot's law. A rule in synoptic meteorology, enunciated in 1857 by Buys Ballot, of Utrecht, which states that if, in the northern hemisphere, one stands with one's back to the wind, pressure is lower on one's left hand than on one's right, whilst in the southern hemisphere the converse is true. This law implies that, in the northern hemisphere, the winds blow anticlockwise round a depression, and clockwise round an anticyclone; the converse is true in the southern hemisphere. This is a statement of the direction of the GEOSTROPHIC WIND.

C

calendar. For meteorological purposes it is usual to adhere to the civil calendar, and to publish summaries of climatological data for ordinary civil months or weeks. Previously, there were certain advantages in publishing data summaries for epochs other than the ordinary year starting on 1 January: the 'grower's year' (beginning on 6 November) and the 'water year' (beginning on 1 October) were examples. However, the advent of AUTOMATIC DATA-PROCESSING has enabled summaries of data to be produced for any desired period immediately on demand so that regular publication of summaries for a small selection of periods has become obsolete.

See also SEASONS.

calibration. Originally the name given to the process of finding the calibre (area of cross-section) of a tube. When a tube is not of uniform cross-section, marks so spaced that the volume of the tube between consecutive marks is everywhere the same are calibration marks.

The use of the word has now been extended to include the determination of absolute values appropriate to selected fixed points of an instrument, by comparison with primary or secondary STANDARD instruments.

calm. Absence of appreciable wind: on the BEAUFORT SCALE of wind force calm is accorded the figure 0 and has a wind speed equivalence of less than one knot.

calorie (or **gram-calorie**). A unit of heat, being the heat required to raise the temperature of 1 gram of water by 1 degree Celsius. This quantity of heat depends, however, on the initial temperature of the water. The 15 °C calorie (cal_{15}), defined as the heat required to raise the temperature of 1 gram of water from 14.5 to 15.5 °C, and the International Table calorie (cal_{IT}) are the calories most commonly used. It was decided at the Ninth General Conference of Weights and Measures (1948) that the JOULE should replace the calorie as the unit of heat, or the joule equivalent be given. The dimensions are ML^2T^{-2}. See also KILOCALORIE.

$$1\ cal_{15} = 4.1855\ J$$
$$1\ cal_{IT} = 4.1868\ J.$$

calvus (cal). A CLOUD SPECIES. (Latin for bald or stripped.)

'CUMULONIMBUS in which at least some protuberances of the upper part are beginning to lose their cumuliform outlines but in which no cirriform parts can be distinguished. Protuberances and sproutings tend to form a whitish mass, with more or less vertical striations' [2, p. 20]. See also CLOUD CLASSIFICATION.

canalization. See FUNNELLING.

candela. Luminous intensity unit in the SI SYSTEM. The magnitude of the candela is such that the LUMINANCE of the black-body radiator at the temperature of solidification of platinum is 60 candela per square centimetre.

canopy. The collective name given to the three-dimensional structure formed by the stems and foliage of a group of individual plants or trees.

cap cloud. A name frequently given to the transient patches of cloud which sometimes form on or just above the tops of growing cumulus clouds, and are soon absorbed into them. It is also used for clouds on hilltops. It is technically known as PILEUS.

CAPE. An abbreviation for convective available potential energy. See STABILITY.

capillary potential. A concept used in SOIL MOISTURE studies, being the force of attraction exerted by soils on contained water, or the equivalent force required to extract the water from the soil against the capillary forces (SURFACE TENSION) acting in the soil pores. It is generally expressed in the pressure unit of atmospheres or the equivalent height of a specified liquid column (mercury or water).

Since the capillary potential increases very rapidly with increasing dryness of soil, a logarithmic measure, pF, is often used, defined as the logarithm to base 10 of the capillary potential expressed in centimetres of water. Thus a capillary potential of 1 atmosphere corresponds to approximately pF 3, since a pressure of 1 atmosphere supports a column of water about 1000 cm in length. See also FIELD CAPACITY, WILTING POINT.

capillatus (cap). A CLOUD SPECIES. (Latin for having hair.)

'CUMULONIMBUS characterized by the presence, mostly in its upper portion, of distinct cirriform parts of clearly fibrous or striated structure, frequently having the form of an anvil, a plume or a vast, more or less disorderly mass of hair. Cumulonimbus capillatus is usually accompanied by a shower or by a thunderstorm, often with squalls and sometimes with hail; it frequently produces very well-defined VIRGA' [2, p. 20]. See also CLOUD CLASSIFICATION.

carbon dating. A technique of estimating the age of carbon-containing fossil materials, based on the measurement of their RADIOACTIVITY per unit mass and comparison with that of materials of known age. The technique is relevant, for example, to problems of past climatic changes. It is based on the fact that assimilation of RADIOACTIVE CARBON (^{14}C) ceases at the time of death of living material, the radioactivity of the material then decreasing at a rate which is determined by the radioactive HALF-LIFE of ^{14}C. Thus, for example, a period of 5500 years would be assumed to have elapsed since the death of a fossil sample whose specific radioactivity is one-half that of a living sample. Basic assumptions, inherent in the method, are

(i)   that a steady state has long existed between the rate of production of ^{14}C and its rate of disappearance, and
(ii)   that the strong latitudinal variation of production of ^{14}C by cosmic rays is eliminated in its lifetime by world-wide mixing in the atmosphere and oceans.

A modification of the basic technique has been used in order to estimate a possible secular change (see SECULAR TREND) of CARBON DIOXIDE (CO_2) concentration in the atmosphere. The method adopted in this case is to measure the specific radioactivity of a recent or living sample and also that of wood, for example, of known greater age since death (say, 100 years). From these measurements the relative proportions of ^{14}C and CO_2 in the atmosphere at the two epochs are inferred.

Available data suggest that the basic assumption of a steady state between ^{14}C production and disappearance has recently been invalidated, to an extent which is not negligible, by the amount of ^{14}C produced in the atmosphere by nuclear-weapons testing. For this reason, the use of plant or animal materials which have been alive during the period since about 1950 is best avoided in the application of the techniques.

carbon dioxide. A gas, of chemical formula CO_2, with VOLUME FRACTION and MASS FRACTION, relative to dry air, of 3×10^{-4} and 5×10^{-4}, respectively. It is created by animal life and by the oxidation of carbon compounds (as in fuel burning), and is used by plants; it is also destroyed by photochemical processes in the high atmosphere. Owing to a large but temperature-dependent solubility of CO_2 in water, the oceans act as a great reservoir of the gas. The measured amount in the atmosphere has increased significantly during this century. Because of the important part played by CO_2 in the radiative equilibrium of the atmosphere, such secular changes of CO_2 amount are advanced as a possible contributory factor in effecting climatic changes. The CO_2 absorption band of chief importance lies between 12.5 and 17.5 μm, with peak absorption at about 15 μm.

carbon monoxide. A poisonous and colourless gas, of chemical formula CO, which is formed by the incomplete combustion of carbon-containing material. Its presence in minute concentration in the atmosphere has been observed spectroscopically.

Cartesian coordinates. A system of coordinates in which the x, y, z axes are mutually at right angles (rectangular system). In meteorology, a 'right-handed' system is usually employed in which the xy plane is horizontal, positive x and y to east and north, respectively, and positive z vertically upwards.

cascade impactor. A device used for sampling the aerosol content of gases. The gas is forced through a series of nozzles of decreasing size, each one being directed on to a slide which captures part of the aerosol content of the gas. The distribution of size on each slide is related to the nozzle diameter and the free-stream velocity; the smaller nozzles lead to more efficient capture of smaller particles by virtue of the greater impaction speed.

castellanus (cas). A CLOUD SPECIES, previously termed 'castellatus'. (Latin, *castellum* castle.)
  'Clouds which present, in at least some portion of their upper part, cumuliform protuberances in the form of turrets which generally give the clouds a crenelated appearance. The turrets, some of which are taller than they are wide, are connected by a common base and seem to be arranged in lines. The castellanus character is especially evident when the clouds are seen from the side.
  This term applies to CIRRUS, CIRROCUMULUS, ALTOCUMULUS and STRATOCUMULUS' [2, p. 18]. See also CLOUD CLASSIFICATION.

CAT. Abbreviation for CLEAR-AIR TURBULENCE.

catchment area. Defined for administrative purposes as the area within the jurisdiction of a Catchment Board under the Land Drainage Act of 1930. The term is also commonly used with the same meaning as DRAINAGE AREA.

CAVT. Abbreviation for CONSTANT ABSOLUTE VORTICITY TRAJECTORY.

ceiling. The 'ceiling' of a specified mass (e.g. balloon, aircraft, thermal) is the maximum height in the atmosphere which the mass can attain.
  The term is also used in the USA, and fairly generally in aviation circles, to specify the lowest height above the ground at which all cloud layers at and below that level cover more than half of the sky.

ceilometer. An alternative name for CLOUD-BASE RECORDER.

celestial sphere. The imaginary sphere on the inner surface of which the heavenly bodies appear to lie, the observer being situated at the centre of the sphere. See also EQUATOR, POLE.

Celsius scale. A scale of temperature based on one introduced in 1742 by Celsius, a Swedish astronomer and physicist, who divided the interval between the ice- and boiling-points of water into 100 parts, the lower fixed point being marked 100. The present system, whereby the ice-point is marked 0 and the boiling-point 100, was introduced later. Alternative names are the centigrade scale and, less commonly, the centesimal scale. See TEMPERATURE SCALES.

centigrade. An alternative, though now less favoured, name for the Celsius scale of temperature.

centre of action. A term, introduced by Teisserenc de Bort in 1881, which generally signifies an area covered by a large-scale low- or high-pressure system, which dominates the circulation, and so has a big influence on weather conditions over a large area for a considerable period of time.

The term has, however, also been used with other meanings. G.T. Walker, for example, defined a centre of action as an area in which conditions of pressure, temperature, rainfall or ice were strongly correlated with similar conditions in other parts of the world; an 'active' centre was one in which the conditions were highly correlated with conditions which occurred at other, 'passive' centres at a later time.

centrifugal force. A body rotating in a circle round a central point is subject to a 'centrifugal force' acting outwards from the centre, of magnitude $\omega^2 R$ or $V^2 R$ per unit mass where ω = angular velocity, V = linear velocity and R = radius of curvature of path.

This force arises in meteorology in two main ways:

(i)   The observed force of GRAVITY is the vector sum of the force of gravitation directed towards the earth's centre and the centrifugal force acting on the earth and atmosphere due to rotation round the earth's axis.
(ii) Air moving on a curved path with respect to the earth's surface is subject to a centrifugal force, and corresponding CENTRIPETAL ACCELERATION, giving rise to the cyclostrophic component of the GRADIENT WIND.

centripetal acceleration. That ACCELERATION of a body moving on a curved path, equal and opposite to the CENTRIFUGAL FORCE per unit mass, which is directed to the instantaneous centre of curvature of path of the body.

ceraunometer. An instrument designed to count and to give warning of the occurrence of LIGHTNING flashes within a specified area.

c.g.s. system. A system of units based on the centimetre, the gram and the second. This is the system in which the various derived units have usually been expressed in the sciences, including meteorology.

The c.g.s. system is being replaced by SI UNITS.

chaff. Very fine needles or ribbon (of copper, metallized nylon or glass fibre) used for wind measurement. If large numbers of these needles are ejected, for example from a METEOROLOGICAL ROCKET, they form a cloud which acts as a passive wind sensor and which can be tracked by radar. The finest chaff, only 0.025 mm in diameter, is suitable for wind measurement up to 85 km.

chain lightning.   An alternative for PEARL-NECKLACE LIGHTNING.

chance expectation.   The probability of a certain result, given that the processes producing the result have no known, recognizable or definable connection with any of the possible values of this result. See PROBABILITY.

Chapman layer.   An atmospheric layer conforming in properties to a model proposed by S. Chapman, who investigated the distribution in height of the product (electrons, molecules, free radicals, etc.) of a process depending on absorption of solar RADIATION.
   The formation of a particular product in the high atmosphere by incident solar radiation is subject to two counteracting influences in that, with increasing depth of penetration, the amount of gas capable of absorbing the radiation increases because of increasing gas density downwards, while the amount of radiation decreases because of absorption at higher levels. There is thus a level, for a particular absorbing gas, at which the rate of formation of the product is a maximum and at which is centred a layer of characteristic 'shape'.

characteristic.   See BAROMETRIC CHARACTERISTIC.

Charles's law.   At constant pressure, the volume of a given mass of gas (or the SPECIFIC VOLUME) is directly proportional to the absolute temperature.

chemosphere.   Term sometimes applied to that region of the ATMOSPHERE, extending mainly over the height range 40–80 km, in which PHOTOCHEMISTRY is important.

chinook.   A warm and dry west wind, of the FÖHN type, which occurs on the eastern side of the Rocky Mountains. Its arrival is usually sudden, with a consequent large temperature rise and rapid melting of snow.

chi-square distribution.   If x_1, ..., x_n are items chosen at random from a NORMAL (FREQUENCY) DISTRIBUTION of mean zero and standard deviation unity, then

$$\sum_{j=1}^{n} x_j^2$$

falls into the chi-square distribution with n-1 DEGREES OF FREEDOM. This distribution has been studied extensively, and tables of the important PERCENTILES have been published to facilitate the application of the CHI-SQUARE TEST and other tests of SIGNIFICANCE.

chi-square statistic.

$$\chi^2 = \Sigma\{(O-E)^2/E\}.$$

For description see CHI-SQUARE TEST.

chi-square test.   A statistical test of the agreement between an observed distribution of frequencies and the distribution expected according to some hypothesis. If O is the observed and E the expected frequency in each of n classes, then the CHI-SQUARE STATISTIC, often denoted by χ^2, is

$$\Sigma\{(O-E)^2/E\}.$$

On certain assumptions, of which the most important is that the lowest expected frequency should not be less than 5 (so that $(O-E)$ is normally distributed), the value of χ^2 can be shown to fall into the CHI-SQUARE DISTRIBUTION, and the published tables can be used to estimate the SIGNIFICANCE of the value found. The appropriate number of DEGREES OF FREEDOM is found by subtracting from the number of classes the number of constants found from the data that are used in estimating the expected frequencies.

The chi-square test has two special advantages, namely:

(i)   Apart from some adjustments in a 2×2 table, it is valid whatever the statistical distribution of the elements in question, provided only that the lowest expected frequency exceeds 5.
(ii) The contributions of the classes to χ^2 tend to be very unequal in size; thus when a significant value of χ^2 is found, it is easy to identify the cells making the important contributions.

A significantly high value of χ^2 casts doubt on the hypothesis used for finding the expected frequencies, and examination of the contribution to χ^2 made by the various classes will often point the way to a more satisfactory hypothesis. A significantly low value of χ^2, i.e. one likely to be exceeded by chance in over 95 per cent of cases, is a much rarer occurrence but is just as effective in casting doubt on the original hypothesis; it usually signifies that the frequencies of certain classes are not independent, but are linked in some way.

If the application of the chi-square test is complicated by the occurrence of classes with expected frequencies of less than 5, the difficulty can often be overcome by pooling the observed and expected frequencies of such classes with others.

chlorofluorocarbons (CFCs).   Chemical compounds containing carbon, hydrogen, chlorine and fluorine, the proportions of each element varying from one compound to another. Chlorofluorocarbons are often known by their trade names of which Freon is one of the commonest. They have been widely used as refrigerants and spray-can propellants. Although extremely stable under normal conditions of use at ground level, when they diffuse upwards into the high atmosphere they take part in complex photochemical reactions involving oxygen and ozone which tend to lead to a reduction of the total amount of ozone in the OZONE LAYER. Because of the environmental hazards posed by such a reduction, steps are being taken to check the indiscriminate use of CFCs and to seek alternatives. See also GREENHOUSE EFFECT.

chromosphere.   That part of the atmosphere of the SUN above the REVERSING LAYER. Some 10 000 km thick, it consists of faintly luminous gases. Though visible directly only during a solar eclipse it is studied at other times by means of a spectroheliograph and is found to be a region of various types of disturbance including the SOLAR FLARE.

circle of inertia.   In horizontal motion of a body on the rotating earth, subject to no force except gravity, the path described by the body, except near the equator, is approximately a circle which is known as the 'circle of inertia'. Inward acceleration (V^2/r) is balanced by the deviating force (fV) due to the earth's rotation and the path is thus of radius r, given by $r = V/f$, where V is the velocity of the body and f the CORIOLIS PARAMETER. Circular motion ($r = $ constant) is not strictly adhered to because of the variation of f with latitude.

circular frequency distribution.   A distribution in a series of vector quantities (e.g. winds) such that, when the individual vectors are drawn on a polar diagram, the lines of equal frequency of the vector end-points are circles centred on the end-point of the vector mean wind of the series. If the frequency of wind components about the mean is distributed in accordance with the NORMAL (FREQUENCY) DISTRIBUTION, the distribution is termed 'normal circular'. Frequency distributions of wind data often depart drastically from the circular form, which should be used only after its appropriateness has been confirmed.

circulation. The circulation (C) round a closed curve is defined as the line integral round the curve of the velocity vector component along the curve, i.e.

$$C = \oint \mathbf{V}.\, \mathrm{d}\mathbf{s} = \oint V_T \mathrm{d}s$$

where $\mathbf{V}$ is the fluid velocity, d$\mathbf{s}$ the vector line element of the curve and V_T the component of $\mathbf{V}$ tangential to the curve. Cyclonic circulation is reckoned positive, anticyclonic negative. The dimensions are L^2T^{-1}.

circulation, general atmospheric. See GENERAL CIRCULATION.

circulation index. A numerical measure of the strength of the atmospheric circulation as, most commonly, in the ZONAL INDEX.

circulation theorem of Bjerknes. The (relative) circulation (C) along a closed curve is defined by the line integral $\oint \mathbf{V}$.ds where $\mathbf{V}$ is the wind velocity (relative to the rotating earth) and ds a (vector) line element of the curve. The distribution in space of mass may be represented by a set of surfaces of constant specific volume α ($= \rho^{-1}$) such that α increases by one unit from one surface to the next. (These surfaces are known as isosteric or isopycnic surfaces.) Similarly, the distribution of pressure may be represented by a set of isobaric surfaces. In general, the two sets of surfaces intersect and define a continuous family of isosteric-isobaric unit tubes, or solenoids. The number (N) of these isosteric-isobaric solenoids enclosed by the closed curve may be shown to be given by

$$N = \oint \alpha \, \mathrm{d}p.$$

Bjerknes's circulation theorem states that the rate of change of C along a closed curve always consisting of the same fluid particles is given by

$$\frac{\mathrm{d}C}{\mathrm{d}t} = N - 2\,\Omega\,\frac{\mathrm{d}A_E}{\mathrm{d}t}$$

where Ω is the angular velocity of the earth and A_E the area of the projection of the curve on the equatorial plane. (The absolute circulation (C_a) is given by $C + 2\,\Omega\,A_E$ so that the theorem may also be written dC_a/dt, whence for a barotropic system the absolute circulation remains constant.)

Bjerknes's circulation theorem may be used to provide intuitive understanding of a wide range of atmospheric motions including LAND- AND SEA-BREEZES.

circumzenithal arc. One of the HALO PHENOMENA in the form of a short arc centred on the zenith and convex to the sun, at or near the highest point of the 46° halo (if present). The arc is formed by REFRACTION in suitably orientated ice crystals and may show vivid rainbow colouring.

cirrocumulus (Cc). One of the CLOUD GENERA.
'Thin, white patch, sheet or layer of cloud without shading, composed of very small elements in the form of grains, ripples, etc., merged or separate, and more or less regularly arranged; most of the elements have an apparent width of less than one degree' [2, p. 27]. See also CLOUD CLASSIFICATION.

cirrostratus (Cs).   One of the CLOUD GENERA.

'Transparent, whitish cloud veil of fibrous (hair-like) or smooth appearance, totally or partly covering the sky, and generally producing halo phenomena' [2, p. 29]. See also CLOUD CLASSIFICATION.

cirrus (Ci).   One of the CLOUD GENERA. (Latin for lock or tuft of hair.)

'Detached clouds in the form of white, delicate filaments or white or mostly white patches or narrow bands. These clouds have a fibrous (hair-like) appearance, or a silky sheen, or both' [2, p. 25]. See Figure 10. See also CLOUD CLASSIFICATION.

J.A. Walton

FIGURE 10.   Cirrus.

CISK.   Abbreviation for CONDITIONAL INSTABILITY OF THE SECOND KIND.

Clausius–Clapeyron equation.   The equation which expresses the change of pressure with temperature in a state of equilibrium between two phases of the same substance.

The form of the equation familiar in meteorology is

$$\frac{1}{e'} \cdot \frac{\mathrm{d}e'}{\mathrm{d}T} = \frac{L}{R_v \mathrm{T}^2}$$

where e' is the saturation water vapour pressure, T the temperature, L the latent heat and R_v the specific gas constant for water vapour. This equation is integrated to give the saturation vapour pressure as a function of temperature. See VAPOUR PRESSURE.

clear-air turbulence (CAT).   Air TURBULENCE of a type other than that associated with airflow close to rough ground or that encountered in or near convective clouds. It does not always occur in clear air and may be associated with thin cirrus.

Clear-air turbulence has been observed mainly in the high troposphere and low stratosphere, especially in the vicinity of JET STREAMS. Its chief practical significance lies in the acceleration, varying in intensity up to several times g, which may be imparted to high-speed aircraft. Investigation of the horizontal and vertical anomalies in the flow pattern that constitute the turbulence is made difficult by their small scale.

Evidence suggests, however, that favourable conditions for the development of such anomalies include high static stability, and large horizontal or vertical wind shear; orographic effects may also be an important contributory factor, and much CAT is believed to be associated with HELMHOLTZ INSTABILITY.

clear ice. An alternative for GLAZE. See also ICE FORMATION ON AIRCRAFT.

clear sky, day of. In a resolution adopted at the International Meteorological Meeting at Utrecht in 1874, a 'day of clear sky' was defined as one on which the average CLOUDINESS at the hours of observation is less than two-tenths of the sky.

On 1 January 1949 OKTAS were first adopted, and since then a 'day of clear sky' has been defined in the United Kingdom as one on which the average cloudiness at the hours of observation is less than two oktas.

climagram. A climatic diagram comprising a plot of monthly values of two selected meteorological elements (ordinate and abscissa), the plotted points being joined by a line which represents the annual variation of the relationship of the elements.

climate. The climate of a locality is the synthesis of the day-to-day values of the meteorological elements that affect the locality. Synthesis here implies more than simple averaging. Various methods are used to represent climate, e.g. both average and extreme values, frequencies of values within stated ranges, frequencies of weather types with associated values of elements. The main climatic elements are precipitation, temperature, humidity, sunshine, wind velocity, and such phenomena as fog, frost, thunder, gale; cloudiness, grass minimum temperature, and soil temperature at various depths may also be included. Climatic data are usually expressed in terms of an individual calendar month or season and are determined over a period (usually about 30 years) long enough to ensure that representative values for the month or season are obtained.

The climate of a locality is mainly governed by the factors of:

(i)   latitude,
(ii) position relative to continents and oceans,
(iii) position relative to large-scale atmospheric circulation patterns,
(iv) altitude, and
(v)   local geographical features.

A broad classification is made into CONTINENTAL CLIMATE and MARITIME CLIMATE.

Among well known classifications of greater refinement (though inevitably imperfect and so subject to criticism) are those of W. Köppen (in 1923) (see KÖPPEN CLASSIFICATION) and C.W. Thornthwaite (in 1931 and 1948).

climatic change and variability. Geological and botanical evidence of past climates is to the effect that, on the geological time-scale, periods when there was little or no ice anywhere in the world have alternated with much shorter periods when glaciation was widespread — see ICE AGE. The difference between mean world temperatures in these extreme conditions is estimated at about 8 °C, the range being greatest in high latitudes.

Since the last GLACIAL PHASE of the Quaternary Ice Age, varying with locality from about 8000 to 40 000 years ago, there have been large fluctuations in climate in what is, technically, still an 'interglacial' ice-age period. Conspicuous among the fluctuations of European climate, paralleled to a variable extent elsewhere, are the CLIMATIC OPTIMUM of about 5000 to 2000 BC and the 'Little Ice Age' of about AD 1550 to 1850.

Other probable fluctuations are a warm, dry period in the 9th to 12th centuries AD and a stormy period in the 12th to 14th centuries AD. The evidence in the historical period includes: in Europe, fluctuations of Alpine glaciers and in the traffic across Alpine passes; in Asia, variations in the level of the Caspian Sea and other salt lakes; in North America, variations in the rate of growth of the sequoia trees of California, some of which are over 3000 years old.

It is now generally agreed that in most parts of the world instrumental observations of the past 100 to 200 years contain SECULAR TREND(S) of statistical significance, implying real climatic change(s) during this period. The most conspicuous feature has been a warming tendency from the late 19th century to the mid 20th century over both land and sea on a scale which is almost world-wide and averages about 0.5 °C. It is supported by evidence such as the recession of glaciers and of Arctic sea ice; in the British Isles the warming was particularly marked in the winter months but was reversed after 1940 by the occurrence of a number of very cold winters, outstanding among them that of 1962/63, the coldest in central England since 1739/40. Regional climate is often found to vary on interdecadal time-scales.

Among the suggested causes of climatic variation are changes of solar RADIATION, astronomical (earth orbit) changes, CONTINENTAL DRIFT, POLAR WANDERING, mountain building, volcanic eruptions, changes of CARBON DIOXIDE content of the atmosphere, and changes of heat storage by the oceans. It is as yet quite uncertain which of these or other factors is most important in effecting climatic changes.

climatic optimum. That period, lasting from about 5000 to 2000 BC, when average temperatures are considered to have reached their highest level, probably on a world-wide scale, since the last ICE AGE. During this period European temperatures are thought to have averaged up to about 2 or 3 °C higher than at present. Periods equally warm are, however, believed to have occurred in each of the several interglacial phases of the last million years.

climatic zones. The word CLIMATE is derived from a Greek word meaning 'to incline' and the original zones of climate were zones in which the inclination of the sun's rays at noon was the same, that is, zones of latitude. The accumulation of meteorological data has shown that winds and rainfall, as well as temperature, have a zonal arrangement, but that the true climatic zones do not run strictly parallel to lines of latitude. Eight principal zones are distinguished: near the equator a zone of tropical rain climate, then two subtropical zones of STEPPE and DESERT climate, then two zones of temperate rain climate, then, in the northern hemisphere only, an incomplete zone of BOREAL CLIMATE with a great annual range of temperature and finally, two polar caps of snow climate. The equatorial zone is divided into the equatorial rain-forest zone, which extends over the Atlantic and Pacific Oceans as the DOLDRUMS, with rain in all seasons, and a belt of SAVANNA climate on either side with a well marked alternation of dry and rainy seasons, the latter occurring in the 'summer' months. The subtropical zones include most of the world's great deserts — the Sahara and Kalahari, and the deserts of Arabia, Arizona, South America and Australia; over the oceans they include the TRADE WIND belts and the HORSE LATITUDES. The temperate zones are divided into the Mediterranean climates with mild, rainy winters and hot, dry summers, and the temperate rain belts with rain in all seasons. On the eastern margins of the continents, especially in Asia, the subtropical desert zone and the Mediterranean climate are replaced by areas with a MONSOON climate.

climatological station. A STATION from which climatological data are obtained. Such stations are classified by the World Meteorological Organization as 'principal' (readings taken hourly or observations made at least three times daily in addition to hourly tabulations from autographic records), or 'ordinary' (observations made at least once daily, including daily readings of extreme temperature and of amount of precipitation). Stations may also be classified as 'precipitation stations' (in normal British terminology, 'rainfall stations') where this is the only element observed; as 'climatological stations for specific purposes' where only a specific element or elements are observed; or as 'agricultural meteorological stations' where meteorological and biological data, or data otherwise contributing to the establishment of the relationship between the weather and the life of plants and animals, are observed.

climatology. The study of CLIMATE.

climatotherapy. The treatment of disease by suitable climatic environment, i.e. applied BIOCLIMATOLOGY.

closed system. A closed (thermodynamic) system is one in which there is no exchange of matter between the system and its environment though there is, in general, exchange of energy. The atmosphere as a whole may, to a high degree of approximation, be considered a closed system.

cloud. An aggregate of very small water droplets, ice crystals, or a mixture of both, with its base above the earth's surface. The limiting liquid-particle diameter is about 200 μm, larger drops than this comprising DRIZZLE or RAIN.

With the exception of certain rare types (NACREOUS and NOCTILUCENT) and the occasional occurrence of CIRRUS in the lower stratosphere, clouds are confined to the troposphere. They are formed mainly as the result of vertical motion of air, as in convection, or in forced ascent over high ground, or in the large-scale vertical motion associated with depressions and fronts. Cloud may result, in suitable lapse rate and moisture conditions, from low-level turbulence or from mixing or from other minor causes.

At temperatures below 0 °C cloud particles frequently consist entirely of supercooled water droplets down to about −10 °C in the case of layer clouds and to about −25 °C in the case of convective clouds. At temperatures below these very approximate limits and above about −40 °C (temperature of HOMOGENEOUS NUCLEATION), many clouds are 'mixed' but with ice crystals predominating in the lower part of the temperature range.

cloud amount. The amount of sky estimated to be covered by a specified cloud type (partial cloud amount), or by all cloud types (total cloud amount). In either case the estimate is made to the nearest OKTA (eighth) and is reported on a scale which is essentially one of the 'nearest eighth', except that figures 0 and 8 on the scale signify a completely clear and completely cloudy sky, respectively, with consequent adjustment to other figures near either end of the scale. See also CLOUDINESS.

cloud base. That lowest zone in which the type of obscuration perceptibly changes from that corresponding to clear air or haze to that corresponding to water droplets or ice crystals. (Conference of Directors of Members of the International Meteorological Organization in 1947.)

Height of cloud base is reported as height above ground level. In forecasts it is often expressed in PRESSURE ALTITUDE.

cloud-base recorder. An instrument for measuring and automatically recording the height of the cloud base. Modern forms use a vertically pointing pulsed beam of light produced by a laser. The consequential reflection from the cloud is picked up by a detector, co-located with the transmitter, which measures the time-lag between the emitted and received pulses. Older forms of recorder measure the angle of elevation of a modulated infra-red light beam, scanning in the vertical plane, at the instant when light scattered by the base of the cloud is detected by a photoelectric cell looking vertically upwards, at a known distance from the light source. In both types of instrument, the signals from the detector are converted by suitable electronic circuitry into measurements of cloud base which can be output and converted into the movement of a pen across a chart.

cloudburst. A popular term for a very sudden and very heavy shower, often accompanied by thunder and hail. It is associated with strong upward and downward currents.

cloud ceiling. The height above aerodrome level of the lowest layer of cloud of more than 4 oktas.

cloud chamber. The 'Wilson cloud chamber' is an apparatus in which supersaturation and condensation are produced, for example in moist air, by sudden ADIABATIC cooling of the air contained in a chamber.

cloud classification. Various methods of CLOUD classification are made, as follows:

(i) In the publication *International cloud atlas* [2, p. 11], division is made into CLOUD GENERA (ten basic characteristic forms), with subdivision as required into (a) CLOUD SPECIES (cloud shape and structure); (b) CLOUD VARIETIES (cloud arrangement and transparency); (c) Supplementary features and accessory clouds — see INCUS, MAMMA, VIRGA, PRAECIPITATIO, ARCUS, TUBA, PILEUS, VELUM and PANNUS; (d) Growth of a new cloud genus from a 'mother-cloud', indicated by the addition of 'genitus' to the new cloud and mother-cloud genera, in that order, if a minor part of the mother-cloud is affected, and of 'mutatus' if much or all of the mother-cloud is affected, e.g. stratocumulus cumulogenitus, or stratus stratocumulomutatus.

(ii) A classification is made in terms of the level (*étage*) — high, middle, or low — at which the various cloud genera are usually encountered. In temperate regions the approximate limits are: high, 5–13 km (16 500–45 000 ft); middle, 2–7 km (6500–23 000 ft); low, 0–2 km (0–6500 ft). The high clouds are Ci, Cc and Cs (see CLOUD GENERA for significance of abbreviations); the middle clouds are Ac, As (the latter often extending higher) and Ns (usually extending both higher and lower); the low clouds are Sc, St, Cu, and Cb (the last two often also reaching middle and high levels).

For synoptic purposes, a ninefold cloud classification is made in each of these three latter divisions of cloud genera, the corresponding codes being designated C_H, C_M and C_L respectively. The purpose is to report characteristic states of the sky rather than individual cloud types.

(iii) Less formal classifications are made (a) in terms of the physical processes of cloud formation, notably into HEAP CLOUDS and LAYER CLOUDS (or 'sheet clouds') and (b) in terms of cloud composition, namely ICE-CRYSTAL CLOUDS, WATER(-DROPLET) CLOUDS and MIXED CLOUDS.

cloud discharge. A lightning flash confined within a thunder-cloud. See LIGHTNING.

cloud genera. The ten characteristic cloud types, comprising CIRRUS (Ci), CIRRO-CUMULUS (Cc), CIRROSTRATUS (Cs), ALTOCUMULUS (Ac), ALTOSTRATUS (As), NIMBOSTRATUS (Ns), STRATOCUMULUS (Sc), STRATUS (St), CUMULUS (Cu), and CUMULONIMBUS (Cb). See CLOUD CLASSIFICATION.

cloudiness. Amount of sky covered by cloud, irrespective of type, i.e. total CLOUD AMOUNT.

Charts showing the distribution of mean cloudiness over the earth in each month and the year are given in the *Manual of meteorology* [11, pp. 146–171].

cloud physics. 'The study of the physical processes which govern the formation, nature, size distribution and number (per unit volume) of the individual particles which together constitute CLOUD, FOG or PRECIPITATION' [12]. It is often taken to include the study of cloud particle electrification processes.

cloud searchlight. Light projected vertically, by remote control, from a searchlight at distance L (about 300 metres) from an observer causes a spot of light to be thrown on to the lowest cloud overhead. The observer measures the angle of elevation (E) of the light spot by means of an ALIDADE and deduces the cloud base (h) from the relationship $h = L \tan E$. Fixed and portable versions of the instrument are used in the Meteorological Office.

cloud seeding. The attempted modification of the physical processes occurring within natural clouds by injecting the clouds with a seeding agent such as:

(i)   'dry ice' (solid carbon dioxide) at very low temperature to cause local cooling commonly to below $-40\,°C$ and spontaneous freezing,
(ii) silver iodide to act as ICE NUCLEI, and
(iii) hygroscopic salt nuclei or a spray of fine water drops to stimulate COALESCENCE.

Injection of the seeding agent into the cloud has been carried out from aircraft, balloons, rockets and shells, and from the ground. In general, the purpose of the cloud seeding has been to stimulate precipitation. Attempts have also been made to clear stratus clouds and to inhibit hail and thunderstorms. Some success has been claimed in at least the first two aims, but the existence of a method of cloud seeding of commercial value has not yet been established.

clouds, particle distribution in. Measurements have shown water droplets in clouds to have a 'median volume diameter' of about 15 μm, with a range from about 1 to 100 μm; median volume diameter is defined as the drop diameter such that half the total water present is contained in drops of larger diameter. Rather larger median values have been found in convective clouds (15–20 μm) than in layer clouds (10–15 μm). Still larger systematic differences with locality have been reported (larger drops in air free from pollution), probably because of associated differences in concentration of effective condensation nuclei.

Measured water-droplet concentrations are generally in the range 1×10^8 to $8 \times 10^8\,m^{-3}$, but values in altocumulus clouds are smaller.

ICE CRYSTALS occur in clouds in various forms, determined by such conditions as temperature and degree of supersaturation with respect to ice. Clouds composed of ice crystals are very tenuous in comparison with water clouds; a typical range for the concentration of individual particles or clusters of particles is from 1×10^5 to $5 \times 10^5\,m^{-3}$. See also RAINDROPS.

cloud species. A subdivision of CLOUD GENERA in terms of cloud shape and structure. The 14 species comprise FIBRATUS, UNCINUS, SPISSATUS, CASTELLANUS, FLOCCUS, STRATIFORMIS, NEBULOSUS, LENTICULARIS, FRACTUS, HUMILIS, MEDIOCRIS, CONGESTUS, CALVUS and CAPILLATUS.

None, one, or more than one of the cloud species may be allotted to any specific example of cloud genus. See also CLOUD CLASSIFICATION.

cloud street.   An extended line of cumulus cloud parallel to the wind direction, often in an otherwise lightly clouded sky. Various sources of thermals spaced across wind may give rise to parallel cloud streets. It appears that such streets may also be produced in an air mass in which the convection layer has a well marked top and in which the wind direction in the layer is almost constant.

clouds, water content of.   The amount of water, in the liquid or solid state, which is contained in unit volume of cloud. It is usually expressed in grams per cubic metre.

   The water content of convective cloud may be computed on the basis of various assumptions. It is normally assumed that the water which is condensed on ADIABATIC expansion is retained within the rising air. The computed values increase with the temperature of the CONDENSATION LEVEL and decrease with the amount of ENTRAINMENT of ambient air which is assumed. In each theoretical model the computed maximum values occur towards the top of the cloud; such a value for mid-latitude summer conditions is, on the assumption that there is no mixing of cloud and environment, about 5 g m^{-3}. Measurements have supported the theoretical distribution of water content with height in a convective cloud. While measured values have approached the maximum theoretical values in some instances, most measured values have been much lower. Because of the large dependence of water content on cloud-base temperature and degree of vertical development of cloud, average water-content values of convective clouds are of little significance. There is considerable spatial variability of water content at any level in convective clouds.

   Median values of water content of low-level layer clouds in middle latitudes are about 0.2 g m^{-3}, and of medium-level clouds about 0.1 g m^{-3}. Variability is, however, rather great since values up to about five times the median values have been measured.

cloud varieties.   A subdivision of CLOUD GENERA in terms of the arrangement of the cloud elements and the cloud transparency. The nine varieties comprise INTORTUS, VERTEBRATUS, UNDULATUS, RADIATUS, LACUNOSUS, DUPLICATUS, TRANSLUCIDUS, PERLUCIDUS and OPACUS.

   None, one, or more than one of the cloud varieties may be allotted to any specific example of cloud genus.

   See also CLOUD CLASSIFICATION.

cloudy day.   Defined as a day on which the average CLOUDINESS at the hours of observation is more than six OKTAS. (Before 1 January 1949 the criterion was based on an average cloudiness greater than eight-tenths of the sky.) Such days have sometimes been described in the United Kingdom as 'overcast days'.

coalescence.   In meteorology, usually used to denote the growth of water drops by collision. The term is also often used for the growth of an ice particle by collision with water drops (ACCRETION). See PRECIPITATION.

coalescence efficiency.   That fraction of the number of collisions between water drops which results in the union of the drops and the formation of larger drops. See also COLLISION EFFICIENCY, COLLECTION EFFICIENCY.

coherence.   A term used for the tendency, almost universal in meteorology, for measurements which are close together in space and time to be more alike than other similar but more widely separated measurements.

   If the strictly periodic components of variation, derived from the diurnal or annual changes, are removed, the meteorological data are yet found to exhibit coherence on a wide range of scales in space and time. This fact must be allowed for if standard statistical methods are to give reliable results when applied to coherent data. See also PERSISTENCE.

col. That atmospheric pressure distribution, saddle-backed in shape, which occurs between two ANTICYCLONES and two DEPRESSIONS, arranged alternately (see Figure 11). A col which is markedly elongated along the high-pressure axis is an 'anticyclonic col', along the low-pressure axis a 'cyclonic col'; where there is no marked elongation, the col is 'neutral'.

Light, variable winds are a feature of all types of col. The weather of the neutral col is dominated by the characteristics of the particular air mass but is often thundery in summer and dull or foggy in winter. The general characteristics of the anticyclonic and cyclonic cols are those of the anticyclone and depression, respectively.

Since small pressure changes suffice to cause appreciable movement of a col, it is generally not an abiding feature of the synoptic pressure distribution. The cyclonic col occupies a region through which an approaching depression may readily pass.

colatitude. The colatitude at any point is the complementary angle of the LATITUDE (ϕ), i.e. $90° - \phi$; it is also termed 'polar distance'.

cold anticyclone. See ANTICYCLONE.

cold conveyor belt. Cold air that travels westward (relative to an advancing DEPRESSION) just ahead of the surface WARM FRONT beneath the WARM CONVEYOR BELT and ascends near the depression centre reaching the middle TROPOSPHERE near the apex of the WARM SECTOR.

cold dome. A term applied, in upper-air analysis, to a closed centre of low pressure on a FRONTAL CONTOUR CHART. Such a region on the chart indicates the isolation of cold air at that level from the main body of cold air which is seen on the chart, usually at higher latitudes.

The term is also applied to a mass of cold air resting on the earth's surface in the general shape of an inverted bowl.

cold front. A FRONT whose movement is such that the colder air mass is replacing the warmer.

The passage of a cold front is normally marked at the earth's surface by a rise of pressure, a fall of temperature and dew-point, and a veer of wind (in the northern hemisphere). Rain occurs in association with most cold fronts and may extend some 100 to 200 km ahead of or behind the front. Some cold fronts give only a shower at the front, while still others give no precipitation. Thunder, and occasionally a LINE SQUALL, may occur at a cold front.

The average slope of a cold frontal surface is about 1 in 50. A cold front moves, on average, at about the speed of the GEOSTROPHIC WIND component normal to the front and measured at it. See also ANAFRONT, KATAFRONT.

cold-front wave. A secondary wave DEPRESSION which forms on an extended COLD FRONT along which there is an appreciable THERMAL WIND more or less parallel to the front. Such wave depressions are common and may be members of a depression 'family', each member of which may go through a typical life cycle of development and decay. The formation of a wave is aided by the distortion of the front, as for example by a range of hills or by movement of an upper-air disturbance over the front. Initial movement of the wave depression is parallel to the WARM-SECTOR isobars.

cold-occlusion depression. A SECONDARY DEPRESSION which forms at the point where a COLD FRONT and a WARM FRONT unite to form a cold OCCLUSION. The formation of this type of secondary has been linked with a THICKNESS pattern in which the main features are a weak gradient near the centre of the primary depression and

FIGURE 11. Col over the British Isles at 1200 GMT on 10 May 1988.

marked DIFFLUENCE at the point of occlusion. There is no well defined pattern of behaviour of the secondary after its formation.

cold pole. That region of the earth's surface, one in each hemisphere, where the lowest air temperature has been measured. For the northern hemisphere the location is usually taken as Verhoyansk (67° 33′N, 133° 23′E, altitude 122 m) where a temperature of −68 °C (−90 °F) was measured on 5 and 7 February 1892. The same temperature was measured at Oymyakon (63° 16′N, 143° 15′E, altitude 800 m) on 6 February 1933. During the period (from 1957) of regular measurement of air temperature in the interior of the Antarctic continent, several values between −85 and −90 °C (−121 and −130 °F) have already been reported at and near the Soviet base Vostok (78° 27′S, 106° 52′E).

The cold pole is sometimes alternatively defined as the location of lowest mean monthly temperature, or of lowest mean annual temperature, or of coldest air in the troposphere. In the last case it is usually indicated by the area of lowest THICKNESS on the chart of 1000–500 mb thickness or other representative thickness chart.

cold pool. A closed centre of low THICKNESS on a thickness chart, so called because it represents a region of low mean temperature in the isobaric layer concerned.

cold sector. That part of a DEPRESSION occupied by cold air on the earth's surface; it usually comprises about half to three-quarters of a recently formed DEPRESSION, and the whole of an old one.

cold trough. A pressure TROUGH (or trough on an isobaric contour chart) in which temperature is generally lower than in adjacent areas.

collection efficiency. That fraction of the total number of water drops (cloud or rain) lying within the geometrical volume swept out by a large drop which actually collide and unite with such a drop to form a larger drop. It is given by the product of the COLLISION EFFICIENCY and COALESCENCE EFFICIENCY.

collision efficiency. That fraction of the total number of water drops (cloud or rain) lying within the geometrical volume swept out by a large drop which actually collide with the large drop. The efficiency is a function mainly of the relative motion and the particle sizes involved. In general, it is less than unity because of the deflection of the streamlines of the particles on their near approach, in such a way as to avoid collision. An efficiency greater than unity occurs in some conditions, owing to the sucking of particles into the wakes of others.

The collision efficiency is an important factor in the COALESCENCE process of PRECIPITATION. See also COALESCENCE EFFICIENCY, COLLECTION EFFICIENCY.

collision frequency. The collision frequency (v) of a particle (e.g. molecule or electron) is the number of collisions it makes per second. Its inverse is the 'collision interval' (τ), which is the interval (in seconds) between collisions. Collision frequency is a function of gas pressure and density and is related to the MEAN FREE PATH (l) and the mean particle speed (v) by the relation $v = v/l$.

coloured rain. Rain which leaves a coloured stain on the ground, often red or rusty in hue ('blood rain'). The coloration is due to dust particles contained by the drops; the particles are carried from a sandy region by upper-air winds, sometimes for long distances. The phenomenon has been observed frequently in Italy, for example, but has also been known to occur in the British Isles, notably on 1 July 1968.

colour temperature. The colour temperature of a radiating body is that temperature obtained by insertion of the observed wavelength of peak radiation into Wien's formula. See RADIATION.

comfort zone. In relation to EFFECTIVE TEMPERATURE, the range of effective temperature within which most people can work with maximum efficiency. For acclimatized persons the comfort zone is considered to be from 19 to 24 °C (66–75 °F) with an optimum of 21 °C (70 °F).

comma cloud. A MESOSCALE cloud system, appearing in a satellite picture to have the shape of a comma. Such systems are associated with baroclinicity through at least some of the TROPOSPHERE and with deep CONDITIONAL INSTABILITY. They include polar lows that form in cold northerly outbreaks over warm oceans where CONDITIONAL INSTABILITY OF THE SECOND KIND appears to be the driving force, and in polar troughs in westerly flows behind major cold fronts where baroclinic slantwise ascent is the principal mechanism.

communications, meteorological. The rapid and accurate transmission of meteorological information including regular observations made at synoptic reporting stations, results of radiosonde ascents, and data obtained by meteorological satellites, is vital for the pursuance of meteorology as a practical applied science. Weather forecasting and synoptic meteorology started only after the invention of the electric telegraph, and meteorologists have been quick to make use of every subsequent advance in communications from wireless telegraphy to the latest communication satellites. Forecast products, as well as observational data, need the speed of modern communications for effective dissemination and exchange. The modern Global Telecommunication System is illustrated schematically in Figure 12. The WMO arranges the co-ordination of methods of operation and transmission techniques to be used. Further communication channels are used by civil aviation authorities, and most if not all countries have additional facilities dedicated to national defence. The switching and channelling of information is today largely controlled by computers with little human intervention.

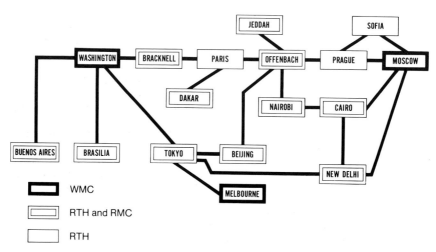

FIGURE 12. Routeing of the Main Telecommunication Network between the World Meteorological Centres (WMCs), Regional Telecommunication Hubs (RTHs) and Regional Meteorological Centres (RMCs) of the Global Telecommunication System.

compass. The magnetic or mariner's compass consists in its simplest form of a graduated card at the centre of which a magnetized steel needle is pivoted so that it may turn freely in a horizontal plane. The card is divided into 32 equal parts of $11\frac{1}{4}°$ each, these being N, N by E, NNE, NE by N, NE and so on (compass 'points'). The compass needle points to magnetic north which does not, in general, coincide with geographical north. See DECLINATION, MAGNETIC and GEOMAGNETISM.

Magnetic compasses have now largely given way to the gyrostatic compass which can be used in aircraft and on ships. This compass is one of several devices which utilize the property, possessed by a spinning flywheel, of maintaining its axis in space. The rotation of the flywheel is maintained electrically; it has the advantage of being independent of both magnetic and electrical disturbances.

compensation of instruments. An instrument designed to measure changes in a particular physical quantity (e.g. pressure) may be affected also by some other influence (e.g. temperature). To eliminate or minimize the influence of the disturbing element, a device may be introduced for the purpose of rendering the instrument insensitive to changes in the latter, in which case it is said to be 'compensated'. Thus, chronometers and aneroid barometers are ordinarily compensated for temperature.

composite forecast chart. A forecast chart so constructed as to depict, for a specified aircraft route, time of departure and air speed, the meteorological conditions which it is expected the aircraft will encounter at each point of the route.

computational instability. A phenomenon, which arises in numerical computations employing approximate methods, in which the error of the computation increases rapidly as the computation proceeds.

A particular effect of this kind, originally discovered by Courant, Friedrichs and Lewy, is of special importance in relation to NUMERICAL WEATHER PREDICTION. The effect is such that if the selected grid size is less than the distance travelled in the selected time interval by the fastest waves permitted by the equation, small errors of computation and of numerical approximation grow in successive time intervals to such an extent as to swamp the physical solution of the equation.

concrete minimum temperature. A reading obtained from a standard MINIMUM THERMOMETER which is exposed to the air with its bulb in contact with a concrete slab; the slab lies horizontally in an open situation and with its top almost flush with the ground.

Such readings, introduced at Meteorological Office stations on 1 December 1968, are less subject to very local influences than are GRASS MINIMUM TEMPERATURES. They are also more relevant to such operational problems as the incidence of ice on roads.

condensation. The process of formation of a liquid from its vapour; in meteorology, the formation of liquid WATER from WATER VAPOUR.

Since the capacity of air to hold water in the form of vapour decreases with temperature, cooling of air is the normal method by which first SATURATION, then condensation, is produced. Such cooling is effected by three main processes:

(i)   adiabatic expansion of ascending air,
(ii)   mixing with air at lower temperature, and
(iii) contact with the earth's surface at a lower temperature.

The water vapour condenses as cloud in (i), as fog or cloud in (ii), and as dew or hoar-frost in (iii).

Condensation in the atmosphere occurs at or near the temperature appropriate to the saturation VAPOUR PRESSURE, which is defined in terms of equilibrium between the vapour and a plane water surface, because of the presence in all parts of the TROPOSPHERE of an adequate supply of 'condensation nuclei', which are hygroscopic. In the absence of such nuclei a high degree of supersaturation (several hundred per cent) would be required to produce condensation — so-called 'homogeneous condensation'.

The main factors governing the rate of growth of a cloud droplet formed by direct condensation of water vapour on a condensation NUCLEUS (radius $> 0.1\ \mu$m, approximately) are the nuclei concentration and sizes, the degree of saturation or supersaturation, the excess saturation vapour pressure over a spherical droplet relative to a plane surface, the reduction of saturation vapour pressure associated with the solution which a hygroscopic salt forms with water, and the warming of the growing cloud droplets by the release of latent heat of condensation. Calculation shows that the drop grows quickly when small and increasingly slowly with increase of drop size. The process of direct condensation explains well the formation of normal cloud particles (diameter about 15 μm) but proceeds much too slowly to account for the size of drop often associated with PRECIPITATION (diameter about 1000 μm).

condensation level.   That level (geometric or pressure) at which CONDENSATION occurs in the atmosphere.

The term is normally used in relation to one or other of the processes of lifting, convection, or vertical mixing. The appropriate condensation level is found by means of an AEROLOGICAL DIAGRAM.

The 'lifting condensation level' is, for an air sample at any height, that (isobaric) level at which the dry ADIABATIC through the temperature intersects the HUMIDITY MIXING RATIO line drawn through the dew-point of the sample.

The 'convective condensation level', applied only to surface air, is the same as the lifting condensation level for surface air if the intersection of the dry adiabatic through the temperature and the humidity mixing ratio line through the dew-point lies to the right of the environment curve. If this is not the case, the convective condensation level is the level at which the humidity mixing ratio line through the surface dew-point intersects the environment curve.

The 'mixing condensation level' is the lowest level at which condensation occurs (if at all) as the result of complete vertical mixing throughout a given layer. Such mixing produces constant POTENTIAL TEMPERATURE and humidity mixing ratio throughout the layer; the mixing condensation level is the level of intersection of the dry adiabatic and humidity mixing ratio lines appropriate to these constant values.

condensation nucleus.   See NUCLEUS and CONDENSATION.

condensation trail.   An initially thin trail of water droplets or ice crystals produced by an aircraft engine exhaust when the humidifying effect of the water vapour exceeds the opposed heating effect of the exhaust air ('exhaust trail'). Such trails generally broaden rapidly and evaporate as mixing with drier air proceeds; when the atmosphere is already moist, however, they may be very persistent.

Theory indicates that for a given engine there is for a given pressure a critical air temperature above which a trail cannot form. A 'mintra' line corresponding to these critical pressure-temperature conditions is shown on the Meteorological Office TEPHIGRAM form. This line was calculated for a piston-engined aircraft (Spitfire). For modern jet engines the critical temperature is several degrees lower and in practice it is found that the appearance of trails is delayed until the temperature is several degrees below this. If the air is already saturated with respect to ice, very persistent trails are possible but trails persisting for several minutes are possible regardless of the humidity if it is cold enough. Over the British Isles it is usual to forecast transient trails if the temperature is between 11 and 14 °C below the mintra line and persistent trails for lower temperatures. The critical temperature is such that over the British Isles trails rarely form below about 8.5 km in summer and 6 km in winter. In tropical areas the heights are greater but in very cold conditions, such as winter in central Canada, trails can form at ground level.

Non-persistent condensation trails more rarely form, sometimes at low levels, in near-saturated air at aircraft wing-tips and propeller tips, because of the aerodynamically produced pressure falls which occur at these points ('aerodynamic trail').

conditional instability. See STABILITY.

conditional instability of the second kind (CISK). Conditional instability of the second kind, as distinct from ordinary conditional instability, is a suggested process whereby low-level convergence in the wind field produces convection and cumulus formation thereby releasing latent heat which enhances the convergence and increases convection; this 'positive feedback' may lead to the formation of a large-scale disturbance. The initial low-level convergence may arise either in the FRICTION LAYER or from dynamical effects induced by a tropospheric wave (wave CISK).

conditional symmetric instability. A type of mesoscale two-dimensional instability characteristic of organized ascending flow in a moist baroclinic zone, e.g. in a WARM CONVEYOR BELT. Ordinary convective instability of a parcel of air depends on gravitational forces arising from a deficit of density in the parcel compared to its environment, such forces being directed along the vertical. In inertial instability the body forces are due to departures of the motion from balanced flow and are directed horizontally. In conditional symmetric instability, a 'roll' or 'tube' of air, with axis parallel to the direction of flow, is subjected to both forces simultaneously and the resultant motion is in a slantwise direction. (It is assumed that there are no variations along the axis of the tube, i.e. the motion is 'two-dimensional'.) The instability is called conditional because the development depends on the release of latent heat, and symmetric because it is two-dimensional and hence symmetric about one of the axes; it is also sometimes referred to as 'slantwise convection'.

The potential energy available for this convective motion is called the slantwise convective available potential energy and may be estimated from the local vertical temperature and humidity structure as shown on an AEROLOGICAL DIAGRAM and knowledge of the horizontal wind and its shear.

A useful account, including the necessary mathematical analysis, has been given by Emanuel [13].

conduction (of heat). The process of heat transfer through matter by molecular impact from regions of high temperature to regions of low temperature without transfer of the matter itself. It is the process by which heat passes through solids; its effects in fluids are usually negligible in comparison with those of CONVECTION.

conductivity, thermal. That property of matter whereby thermal CONDUCTION occurs. Two quantities are defined:

(i)   thermal conductivity (k) by the equation $q = -k\, \partial T/\partial x$ (where q, the flux of heat per unit area, is proportional to $\partial T/\partial x$, the thermal gradient, and acts in the down-gradient direction, and

(ii) thermometric conductivity, or thermal diffusivity, (a), by the equation $a = k/c\rho$, where c is the specific heat and ρ the density of the substance.

For most soils the value of a is of the order 10^{-2} to $10^{-3}\,\text{cm}^{-2}\,\text{s}^{-1}$, though with appreciable dependence on density and water content. For still air a is about $0.2\,\text{cm}^{-2}\,\text{s}^{-1}$.

In a turbulent atmosphere (its normal state) the vertical heat transfer effected by molecular conduction is swamped by that effected by eddies. See EDDY CONDUCTIVITY.

confluence. The nearer approach to each other of adjacent STREAMLINES in the direction of flow. See DIVERGENCE.

confluent thermal ridge. A pattern of thickness lines which is concave towards high THICKNESS and in which the thickness lines crowd together in the direction of the THERMAL WIND. According to Sutcliffe's theory of development, CYCLOGENESIS (C) may be expected to occur behind and to the right, ANTICYCLOGENESIS (A) ahead and to the left of the pattern, as illustrated in Figure 13.

confluent thermal trough. A pattern of thickness lines which is concave towards low THICKNESS and in which the thickness lines crowd together in the direction of the THERMAL WIND. According to Sutcliffe's theory of development, CYCLOGENESIS (C) may be expected to occur ahead and to the right, ANTICYCLOGENESIS (A) behind and to the left of the pattern, as illustrated in Figure 14.

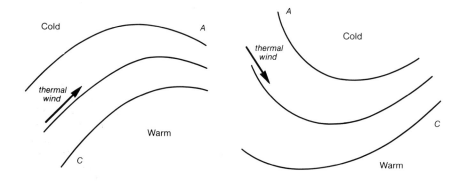

FIGURE 13.   Confluent thermal ridge   FIGURE 14.   Confluent thermal trough.

congestus (con).   A CLOUD SPECIES. (Latin for piled up.)

'CUMULUS clouds which are markedly sprouting and are often of great vertical extent; their bulging upper part frequently resembles a cauliflower'[2, p. 20]. See also CLOUD CLASSIFICATION.

Congress.   The Congress of the WORLD METEOROLOGICAL ORGANIZATION is the general assembly of delegates representing Members and as such is the supreme body of the Organization. It meets at intervals not exceeding four years.

conjunction, astronomical.   A planet or other heavenly body is said to be in conjunction when it is in line with the earth and sun, and on the same side of the earth as the sun.

conservation.   In statistics, a term which is sometimes used with the same meaning as PERSISTENCE or (time) COHERENCE.

conservative property.   A conservative air-mass property is one that remains unchanged, or almost unchanged (quasi-conservative), in a specified process or processes. Thus, for example, POTENTIAL TEMPERATURE is strictly conservative in a dry adiabatic process, is quasi-conservative in unsaturated vertical motion in the atmosphere, but is not conservative in a non-adiabatic process such as radiational cooling at constant pressure. See AIR-MASS ANALYSIS.

constancy of winds.   Owing to varying wind direction, the magnitude of the VECTOR mean (V_R) of a series of wind observations is less than that of the corresponding SCALAR mean (V_S). The quantity (q) defined by $q = 100\, V_R/V_S$, gives a measure of the constancy of wind direction. Values of q range between 0 (winds equally strong and frequent in all directions) and 100 (winds unvarying in direction but not necessarily in speed).

q is controlled in part by synoptic events and in part by the constraints placed on wind direction by local topography. Typical values of q for surface winds over and near the British Isles are within the range 15 to 50. Interpretation of a value of q is, however, unambiguous only on the assumption of a normal and CIRCULAR FREQUENCY DISTRIBUTION of the winds about the vector mean wind.

constant absolute vorticity trajectory (CAVT).   The TRAJECTORY of an air parcel, evaluated on the assumption that absolute VORTICITY is conserved during the motion in which there is no shear.

The basis of such trajectories is the POTENTIAL VORTICITY theorem, $(\zeta+f)/\Delta p =$ constant, which may also be written (see VORTICITY) as

$$\frac{\dfrac{V}{r} + \dfrac{\partial V}{\partial n} + f}{\Delta p} = \text{constant},$$

where r is a coordinate measured from the local centre of curvature of the trajectory. Under the restrictive conditions of a broad current of air moving horizontally at constant speed ($V = $ constant) with little lateral WIND SHEAR ($\partial V/\partial n \approx 0$) and in a thin layer at a level of non-DIVERGENCE ($\Delta p = $ constant), the above equation states that the air will move in such a way that latitudinal (ϕ) changes of vorticity are compensated by curvature (V/r) changes. A wave-like motion between two parallels of latitude results: the air passes through its mean latitude without curvature of path and acquires maximum anticyclonic curvature in its highest latitude and maximum cyclonic curvature in its lowest latitude. The wavelength and amplitude of the oscillation depend on the wind speed, the mean latitude, and the angle at which the trajectory intersects this mean latitude.

Tables for computing CAVT are contained in *Weather analysis and forecasting* [14].

constant-level balloon.   A balloon designed to float, usually for long periods, at a constant pressure or constant density level in the atmosphere. It carries instruments and reflectors which are monitored by satellites, thus providing information from parts of the earth remote from conventional upper-air stations. In one design, a pressure switch activates a valve controlling the release of ballast so as to maintain flight above a selected pressure level until the ballast is exhausted. In another, the super-pressure balloon, a tough non-extensible envelope capable of withstanding an excess of internal over external pressure is used; such a balloon will float at an isopycnic, or constant density, level. A tetroon is a small super-pressure balloon in the shape of a tetrahedron, usually used at low levels.

In the 'Global Horizontal Sounding Technique' (GHOST), large numbers of super-pressure balloons are released to float in isopycnic surfaces and are monitored for prolonged periods.

The 'Eole' balloon-satellite system was set up jointly by France and the USA to fly super-pressure balloons at about 200 mb in the southern hemisphere to provide reference-level data for the retrieval of temperatures from satellite soundings. (The name 'Eole' is the French form of Aiolos, the Greek god of the wind.)

constant-pressure chart. A SYNOPTIC CHART, relating to a surface of specified constant pressure (isobaric surface), on which contours of GEOPOTENTIAL height of the surface are drawn. Other elements, e.g. temperatures and winds observed at the given pressure, are usually entered on the same chart.

constituent body. The constituent bodies of the WORLD METEOROLOGICAL ORGANIZATION are CONGRESS, EXECUTIVE COMMITTEE, REGIONAL ASSOCIATIONS and TECHNICAL COMMISSIONS.

Contessa del Vento. A type of eddy cloud frequently found to the lee side of isolated mountains. It is often observed over Mt. Etna with a westerly wind. In its most characteristic form it consists of a rounded base or disc, sometimes with several vertically stacked discs, surmounted by a protuberance directed upwind.

continental climate. A type of CLIMATE, characteristic of the interior of large land masses of middle latitudes; the main distinguishing features are large annual and diurnal ranges of air temperature, with low rainfall a further characteristic feature. The most extreme form of continental climate is found in eastern Asia where there is an annual range of temperature (July mean minus January mean) of about 60 °C. See CONTINENTALITY.

continental drift. A hypothesis, generally linked with the names of F.B. Taylor and A. Wegener, of displacements, on a geological time-scale and by distances of the order of thousands of kilometres, of various parts of the earth's surface relative to other parts. According to this hypothesis continents now separated were once joined together as, for example, South America with Africa, and North America with Europe; in general, changes in both latitude and longitude are implied. Since the 1960s the hypothesis has become widely accepted and explained by the new science of plate tectonics. See also POLAR WANDERING.

continentality. In meteorology, a measure of the extent to which the climate at any place is subject to land, as opposed to maritime, influences. Because of the low specific heat of land relative to water and, more especially, the fact that heat exchanges by day and night, summer and winter, are confined to a shallow layer of a land surface, the diurnal and seasonal temperature changes of a land mass and its overlying air are great. Various measures of continentality have been suggested, most of them based on the observed annual range of temperature but depending also on latitude (the INSOLATION factor).

At a given latitude continentality increases, in general, with distance from the sea, but maximum continentality occurs at an appreciable distance downwind (relative to the PREVAILING WIND) from the centre of a land mass. A large range of mountains across the prevailing wind produces a marked increase of continentality to leeward of the mountains, mainly due to the clearer skies and consequently greater radiational heating in summer and cooling in winter that prevail there.

See also CONTINENTAL CLIMATE.

continuity, equation of. An equation which expresses the law of conservation of mass and states that the mass of air entering an elementary volume is equal to the increase of mass within it.

The equation has various forms, for example

$$\text{div } \mathbf{V} = -\frac{1}{\rho} \frac{d\rho}{dt},$$

$$-\frac{\partial w}{\partial z} = \text{div}_H \mathbf{V} + \frac{1}{\rho} \frac{d\rho}{dt}.$$

The latter equation states that if an air parcel contracts vertically it may expand horizontally and increase in density. Since local density change is small in the atmosphere the equation has the approximate form

$$-\frac{\partial w}{\partial z} \approx \text{div}_H \mathbf{V}.$$

Assuming that the vertical variation of pressure is governed by the HYDROSTATIC EQUATION, the equation of continuity in PRESSURE COORDINATES has the form

$$\text{div}_p \mathbf{V} = -\frac{\partial}{\partial p}\left(\frac{dp}{dt}\right).$$

See also DIVERGENCE, EULERIAN CHANGE, LAGRANGIAN CHANGE.

contour. In synoptic meteorology, a line of constant height of an ISOBARIC SURFACE. The terms 'absolute contour' and 'isohypse' are sometimes used. Contours represent STREAMLINES in strict GEOSTROPHIC WIND flow.

contrail. A common abbreviation for CONDENSATION TRAIL.

contrast threshold of the eye. A ratio (ϵ), important in measurements of VISIBILITY, defined by the relation

$$\epsilon = \frac{B-H}{H}$$

where B is the apparent brightness of an object which is just visible against a background of apparent brightness H.

The value of ϵ varies considerably with the observer and with viewing conditions. The value of 0.02 was originally assumed for normal observers in ordinary daylight illumination but experimental determinations have ranged from about 0.01 to 0.06. The WMO has recommended the use of the value 0.05 and this figure should be used for the purposes of conversion of attenuation measurements to corresponding visibility. See also VISIBILITY METER, TRANSMISSOMETER.

convection. A mode of heat transfer within a fluid, involving the movement of substantial volumes of the substance concerned. The convection process frequently operates in the atmosphere and is of fundamental importance in effecting vertical exchange of heat and other air-mass properties (water vapour, momentum, etc.) throughout the troposphere.

Two distinct, though not mutually exclusive, types of convection occur in the atmosphere. In 'free' or 'natural' convection the vertical motions are produced by BUOYANCY forces whereas in 'forced' convection they are produced by mechanical forces such as occur in SHEAR INSTABILITY.

convective atmosphere. See ADIABATIC ATMOSPHERE.

convective condensation level. See CONDENSATION LEVEL.

convective rain. Rain that is caused by the vertical motion of an ascending mass of air which is warmer than its environment; the horizontal dimension of such an air mass is generally of the order of 15 km or less and forms a typical CUMULONIMBUS cloud. Convective rain is typically of greater intensity than either of the other two main classes (CYCLONIC and OROGRAPHIC RAIN) and is sometimes accompanied by thunder.

The term is more particularly used in those cases in which the precipitation covers a large area as the result of the agglomeration of cumulonimbus masses.

convective region. An alternative for ADIABATIC REGION.

convergence. See DIVERGENCE.

conveyor belt. See COLD CONVEYOR BELT, WARM CONVEYOR BELT.

Coriolis acceleration. An apparent acceleration, which air possesses by virtue of the earth's rotation, with respect to axes fixed in the earth. It is sometimes termed 'geostrophic acceleration'.

The Coriolis acceleration is a three-dimensional vector which is given in vector notation by the expression $-2\Omega \times \mathbf{V}$ (see ACCELERATION). It is therefore everywhere at right angles to the earth's axis (in the plane of the equator) *and* to the air velocity; in the northern (southern) hemisphere it acts to the right (left) when viewed along the motion. The effect is often expressed in terms of the equal inertia force per unit mass termed the 'Coriolis force', or 'deviating force', or 'geostrophic force'.

The three components toward east, north and zenith are, respectively,

$$2\,\Omega\,(v \sin \phi - w \cos \phi),$$
$$-2\,\Omega\,u \sin \phi, \text{ and}$$
$$2\,\Omega\,u \cos \phi.$$

Only the component of the Coriolis acceleration which acts in the horizontal (xy) plane is significant in meteorological dynamics; the vertical component is of comparable magnitude, but is negligible compared with the large forces (gravity and pressure gradient) which act in this direction. For quasi-horizontal air motion $(w = 0)$ of velocity V, the horizontal Coriolis acceleration is $2\,\Omega\,V \sin \phi$.

The component of the earth's rotation in the horizontal plane is such that the earth simulates a flat disc rotating anticlockwise in the northern hemisphere and clockwise in the southern hemisphere (see ANGULAR VELOCITY OF THE EARTH). Air moving horizontally outwards from the centre of the disc appears to an observer stationed there to be deflected to the right in the northern hemisphere and to the left in the southern hemisphere, i.e. in a frame of reference fixed in the rotating earth the air has horizontal acceleration in these respective directions. The effect is negligible if the scale of the motion considered (radius of disc) is of the order of only a few kilometres.

Coriolis force. See CORIOLIS ACCELERATION.

Coriolis parameter. A quantity, denoted f (sometimes l), defined by the equation $f = 2\,\Omega \sin \phi$.

corona. A series of coloured rings surrounding the sun or moon. The space next to the luminary is bluish-white, while this region is bounded on the outside by a brownish-red ring, the two forming the 'aureole'. In most cases the aureole alone appears but a complete corona has a set of coloured rings surrounding the aureole — violet inside, followed by blue, green and on the outside yellow to red. The series may be repeated more than once, but the colours are usually merely greenish and pinkish tints.

The corona is produced by DIFFRACTION of the light by water drops. Pure colours indicate uniformity of drop size. The radius of the corona is inversely proportional to drop size; thus growth of a corona indicates decrease of drop size.

A corona is distinguished from a HALO (caused by REFRACTION) by its reversal of colour sequence, the red of the halo being inside, that of the corona outside; the dull red, which is the first notable colour in the aureole, ranks as outside the bluish tint near the luminary. An alternative criterion is that the colours of a halo are at the inner edge of a luminous area, while those of a corona are at the outer edge. For auroral corona see AURORA.

corona, solar. The outer atmosphere of the SUN, extending to great distances and comprising feebly luminous and highly ionized gases at a temperature of about 10^6 K.

corposant. See ST. ELMO'S FIRE.

correlation coefficient. The correlation coefficient, or more formally the product moment (r_{xy}), is the best known measure of association between pairs of variables. If $(x_1 y_1)$, ..., $(x_n y_n)$ are n pairs then

$$ r_{xy} = \frac{\sum\limits_{j=1}^{n} (x_j - \bar{x})(y_j - \bar{y})}{\left\{ \sum\limits_{j=1}^{n} (x_j - \bar{x})^2 \sum\limits_{j=1}^{n} (y_j - \bar{y})^2 \right\}^{1/2}} = \frac{\text{cov}(x,y)}{\sigma_x \sigma_y} $$

where $\bar{x}$, $\bar{y}$ are the MEANS and σ_x, σ_y the STANDARD DEVIATIONS of the variables and $\text{cov}(x,y)$ their COVARIANCE.

The value of the correlation coefficient can range from 1 (indicating complete functional dependence) through zero (independence) to -1 (implying complete dependence in opposing directions). Its significance can be judged from published tables provided that

(i)   the variables concerned are drawn from the NORMAL (FREQUENCY) DISTRIBUTION, and
(ii) each pair of variables is independent of the rest. See INDEPENDENCE.

When, as is usual with meteorological data, these conditions are not known to be satisfied, the correlation coefficient must be used with caution. It also suffers from the objection that it gives an exaggerated impression of the level of association between the variables; thus a value of $r = 0.5$ which appears to be half-way between independence and functional dependence corresponds to quite a low degree of association. In judging the value of a forecasting technique, for example, the percentage reduction of VARIANCE is a much more realistic measure of success.

A significant value of r is to be regarded as implying a measure of causal relationship only if this view is supported by strong physical reasoning.

correlogram. A diagram in which r_L, the AUTOCORRELATION at lag L, is plotted as ordinate, with $L = 1$, 2, 3, etc., for abscissa. The interpretation of a correlogram requires care, but it forms the basis of many of the advanced techniques of analysis of TIME SERIES.

cosmic radiation (or rays). Very high energy RADIATION which originates outside the earth and probably, in large part, outside the solar system.

The cosmic 'primary' radiation consists mainly of positively charged nuclei of hydrogen (protons) and heavier elements, which are deflected by the earth's magnetic field on near approach to the earth and enter the earth's atmosphere in numbers that increase with increasing geomagnetic latitude. Predominance of a positive charge in the primary radiation is inferred from the preponderance of entry of radiation from a westerly direction at all latitudes ('east–west asymmetry').

The primary radiation reacts to an appreciable extent with the nuclei of atmospheric atoms on penetrating to a level some 15–20 km above the ground, and so gives rise to a variety of secondary, tertiary, etc. products (especially various types of meson), some very penetrating ('hard'), others less penetrating ('soft'). Continuous recording of the intensity of the radiation which reaches the ground is made in many parts of the world. The intensity and average energy of cosmic radiation increase rapidly with height above mean sea level.

Before the era of nuclear explosions, measurement suggested that about one fourth of the conductivity of the air near the ground was due, on average, to the ionizing power of cosmic radiation; at low levels over the sea, and in all regions at a level greater than about 2 km, nearly all the conductivity of the air was due to cosmic radiation.

Among the time variations of cosmic radiation are: a solar-cycle variation more or less in phase with SUNSPOT number; a 27-day recurrence tendency; decreases associated with MAGNETIC STORMS and increases associated with intense SOLAR FLARES.

counterglow. The narrow coloured band or bands which are seen at TWILIGHT above the blue SHADOW OF THE EARTH thrown by the sun just above the horizon (eastern horizon at sunset, western at sunrise). The light from the bands is scattered back to the observer by dust particles and is predominantly soft red in colour, corresponding to the colour acquired by the incident solar beam as it passes through the atmosphere and suffers preferential scattering of the blue wavelengths by the air molecules. In some cases there are well marked higher bands of orange, yellow and green corresponding to a decrease in redness acquired by the incident light on passage through less dense regions of the atmosphere. The phenomenon is also termed 'countertwilight', or 'antitwilight', or 'antitwilight arch'.

The term 'counterglow' is also otherwise used as an alternative for GEGENSCHEIN.

counter sun. An alternative for ANTHELION.

countertwilight. An alternative for COUNTERGLOW.

Courant–Friedrichs–Lewy condition. See COMPUTATIONAL INSTABILITY.

covariance. If (x_1y_1), ..., (x_ny_n) are n pairs of variables, then

$$\text{cov}(x,y) = \frac{1}{n} \sum_{j=1}^{n} (x_j - \bar{x})(y_j - \bar{y}).$$

The covariance depends on the association between the variables, and has an expected value of zero if they are independent. When divided by the product of the STANDARD DEVIATIONS $(\sigma_x\sigma_y)$ it gives the CORRELATION COEFFICIENT.

crachin. Conditions of drizzle combined with low stratus, mist, or fog that occur at times between January and April in south China and in the coastal area between about Mui Bai Bung and Shanghai. The conditions are caused by an interaction between the TROPICAL MARITIME AIR and the POLAR MARITIME AIR circulating round the eastern side of the Asiatic anticyclone. Orographic and coastal lifting are also significant factors.

crepuscular rays.   These include three similar classes of phenomenon:

(i)   Sunbeams penetrating through gaps in a layer of low cloud and rendered luminous by water or dust particles in the air ('sun drawing water' or 'Jacob's ladder').

(ii)   Pale blue or whitish beams diverging upwards from the sun hidden behind cumulus or cumulonimbus clouds. The well defined beams are separated by darker streaks which are the shadows of parts of the irregular cloud.

(iii)   Red or rose-coloured beams, diverging upwards at twilight from the sun below the horizon. The light is scattered to the observer by atmospheric dust; the beams are separated by greenish-coloured regions which are the shadows of clouds or hills below the horizon.

In cases (ii) and (iii) the beams and shadows may persist across the sky before converging at the ANTISOLAR POINT (anti-crepuscular rays). The apparent divergence and convergence of the rays is an optical illusion produced by perspective. See Figure 15.

critical frequency.   In radio sounding of the IONOSPHERE, that radio frequency at which the wave just penetrates a specified layer. The square of the critical frequency is directly proportional to the maximum electron concentration in the layer.

critical height.   The minimum height above aerodrome level to which an approach to landing by an aircraft can safely be continued without visual reference to the ground.

critical temperature.   That temperature above which a specified gas cannot, and below which it can, be liquefied by pressure alone.

cross-section.   The representation in schematic form of conditions prevailing, or expected to prevail, in the atmosphere at a specified time in a vertical plane from the surface up to any desired height along a line from one point to another.

cryosphere.   That part of the earth's surface which at any one time is covered by ice or snow.

cumulonimbus (Cb).   One of the CLOUD GENERA. (Latin, *cumulus* heap, and *nimbus* rainy cloud.)
'Heavy and dense cloud, with a considerable vertical extent, in the form of a mountain or huge towers. At least part of its upper portion is usually smooth, or fibrous or striated, and nearly always flattened; this part often spreads out in the shape of an anvil or vast plume.
Under the base of this cloud which is often very dark, there are frequently low ragged clouds either merged with it or not, and precipitation sometimes in the form of VIRGA' [2, p. 48]. See Figure 16. See also CLOUD CLASSIFICATION.

cumulus (Cu).   One of the CLOUD GENERA. (Latin for heap.)
'Detached clouds, generally dense and with sharp outlines, developing vertically in the form of rising mounds, domes or towers, of which the bulging upper part often resembles a cauliflower. The sunlit parts of these clouds are mostly brilliant white; their base is relatively dark and nearly horizontal.
Sometimes cumulus is ragged' [2, p. 45]. See Figure 17. See also CLOUD CLASSIFICATION.

J.F.P. Galvin

D. Brown

A. Best

FIGURE 15. Crepuscular rays, depicting (top to bottom) the three classes (i), (ii) and (iii) referred to in the text.

FIGURE 16.   Cumulonimbus capillatus.

FIGURE 17.   Cumulus humilis with billow cloud above.

cup-contact anemometer. See ANEMOMETER, ANEMOGRAPH.

cup-counter anemometer. See ANEMOMETER.

cup-generator anemometer. See ANEMOMETER, ANEMOGRAPH.

curl. See VORTICITY.

current, ocean. A general movement, of a permanent or semi-permanent nature, of the surface water of the ocean. The term must not be used of tidal streams, which change direction and speed hour by hour. A 'drift current' is a drift of the surface water which is dragged along by a wind blowing over it. A drift current which is deflected by an obstruction, such as a shoal or land, forms a 'stream current'. A stream current may be formed as a counter current to a primary current, replacing the water displaced by the primary current. The 'set' of a current or tidal stream is the direction in which it is going.

Owing to the high specific heat of water, the main ocean currents are of great climatic importance. Ocean currents result from several causes, the most important being wind and differences of density resulting from differences of temperature and salinity.

Among the best known currents are those of the North Atlantic, notably the GULF STREAM, North Atlantic Drift, and the Labrador Current.

curve-fitting. The representation of experimental data or frequencies by a curve, generally an algebraic form determined by certain coefficients or parameters. The parameters are chosen to provide the 'best' or at least an acceptable fit to the data, using the LEAST SQUARES or other appropriate criterion. Curve-fitting is simplified if the data can be transformed so that curves of the appropriate form are represented by straight lines. For example, if the relationship is of the nature $y = Ax^n$, a plot on logarithmic paper yields the values of A and n; if of the nature $y = Ae^{bx}$, a plot on semi-logarithmic paper yields the values of A and b.

A simple curve form involving few parameters nearly always produces a more stable fit than a more complicated form. It is also possible for a fit to be too good, as judged by the CHI-SQUARE TEST, and hence unlikely.

cut-off high. A warm ANTICYCLONE, separated from the subtropical anticyclones, which is usually associated with a marked interruption of the zonal westerlies (BLOCKING). This feature forms initially as a ridge in the upper westerlies, intensifies into a closed circulation, and extends downwards towards the surface.

cut-off low. A DEPRESSION which lies equatorwards of the main mid-latitude belt of westerly winds. The feature forms initially as a trough in the upper westerlies, deepens into a closed circulation and extends downwards towards the surface.

cyanometer. An instrument for measuring the blueness of the sky.

cycle. A term for a recurrent phenomenon which is best reserved for changes of strictly periodic origin; the diurnal cycle of cloud or the annual temperature cycle are acceptable usages. The use of a neutral term such as OSCILLATION is preferable for roughly periodic changes which are not directly dependent on astronomical processes.

cyclogenesis. The initiation or strengthening of cyclonic circulation around an existing CYCLONE or DEPRESSION.

cyclolysis. The disappearance or weakening of cyclonic circulation around an existing CYCLONE or DEPRESSION.

cyclone. That atmospheric pressure distribution in which there is a low central pressure relative to the surroundings. It is characterized on a synoptic chart by a system of closed isobars, generally approximately circular or oval in form, enclosing a central low pressure (see Figure 18 under DEPRESSION). 'Cyclonic circulation' is anticlockwise round the centre in the northern hemisphere, clockwise in the southern hemisphere; in either case the sense of rotation about the local vertical is the same as that of the earth's rotation. See also DEPRESSION and TROPICAL CYCLONE.

cyclonic rain. Rain that is caused by the large-scale vertical motion associated with synoptic features such as DEPRESSIONS and FRONTS. It is one of a broad threefold classification, the other classes being OROGRAPHIC RAIN and CONVECTIVE RAIN.

cyclostrophic wind. A class of winds, of markedly curved flow, in which the CORIOLIS ACCELERATION is negligible in comparison with the CENTRIPETAL ACCELERATION. In such winds the cyclostrophic term is predominant in the expression for the GRADIENT WIND. A tropical cyclone provides an example in which the wind is mainly of this type.

D

daily variation. An alternative for DIURNAL VARIATION.

Dalton's law. See PARTIAL PRESSURE.

dawn. The time when light appears in the sky in the morning or the interval between the first appearance of light and the rising of the sun. See TWILIGHT.

dawn chorus. A type of radio disturbance which consists of a chorus of overlapping, rising tones, mainly in the middle audio-frequency range, and is most intense at local dawn. The disturbance is rarely heard at geomagnetic latitudes less than about 50°. Since it is correlated with geomagnetic disturbance it is considered to be initiated by extraterrestrial charged particles. It is also, however, sometimes related to the occurrence of ATMOSPHERICS and WHISTLERS.

day. A 'solar day' is the SYNODIC PERIOD of time between successive occasions on which the sun is in the MERIDIAN of any fixed place (sun 'transits'). A 'sidereal day' is the corresponding interval between successive transits of a distant fixed star. Since the earth moves in an orbit round the sun with the same sense of rotation as that about its own axis, the length of the solar day is slightly greater than that of the sidereal day.

Owing to the eccentricity of the earth's orbit and to the inclination of the equator to the ECLIPTIC, solar days are of slightly unequal lengths at different times of the year. The average length of the solar day is 86 400 s. This is called the 'mean solar day' and is taken as the length of the 'civil day', or the 'day' of ordinary parlance. The sidereal day contains 86 164.1 s.

Other small measured variations of length of day, on various time-scales, are of geophysical interest:

   (i)   A secular lengthening (about 10^{-3} seconds per 100 years) is attributed mainly to tidal friction.
   (ii)   Fluctuations of the order of decades are related to slow changes in the earth's magnetic field and are attributed to transfers of angular momentum between the core and mantle of the earth.
   (iii) An annual variation (length of day 2×10^{-3} s greater in February than in August) is mainly attributed to an annual variation of wind velocity on a global scale, involving angular momentum changes of the air opposite to those of the surface of the earth and so maintaining the total angular momentum of the earth–atmosphere system.

As used in geophysics, a 'lunar day' is the interval between successive transits of the moon; it varies in length from about 24 h 40 m to 25 h.

dayglow. The daytime AIRGLOW emission, thought to be rather more intense than the NIGHTGLOW but not observable at low levels against the intense background of scattered solar radiation.

daylight. The intensity of daylight illumination on a horizontal surface is recorded at some stations in the British Isles. The method consists of the measurement of the current emitted by a photocell on which radiation from the sun and sky falls after passing through a protective hemispherical dome of clear glass, a horizontal diffusing surface of translucent material, and a filter. The instrument, commonly called an illuminometer, is so designed that its spectral sensitivity is similar to that of the human eye.

The 'duration of daylight', or 'length of day', is the interval between SUNRISE and the following SUNSET. Tabulated values are contained in the *Smithsonian meteorological tables* [15].

débâcle. The breaking up in spring of the ice in rivers. The term is chiefly applied to the great rivers of Russia and Siberia and of the North American continent. Débâcle lasts from 2 to 6 weeks; during the period the rivers often overflow their banks, inundating the surrounding country. In southern Russia débâcle begins about the middle of March, in latitudes 55–60° N it begins early in April, but in the north it does not begin until May and in the extreme north of Siberia not until June. In Canada the débâcle in Ontario takes place in March and the water is free by April; in the St. Lawrence it is a little later, the river being free of ice in May.

debriefing, meteorological. Post-flight interview of aircrew in order to ascertain the meteorological conditions experienced during a flight.

decad. A period of 10 consecutive days. See also PENTAD.

decibel. Unit of relative measure of two flux densities, increasingly employed in various contexts such as electric power density, sound or light intensity — see, for example, NEBULE.

The decibel is one-tenth of a bel. Two flux densities (I_1, I_2) are said to differ by n bels when

$$\frac{I_1}{I_2} = 10^n, \text{ i.e. when } n = \log_{10}\frac{I_1}{I_2}.$$

Thus, flux densities differ by N decibels (the unit normally employed in preference to bel) when

$$N = 10 \log_{10}\frac{I_1}{I_2}.$$

declination. Angular distance of a heavenly body north or south of the celestial EQUATOR (the projection of the geographical equator upon the celestial sphere) measured along a GREAT CIRCLE passing through the celestial poles. In meteorology, SEASONS are controlled by the sun's declination which varies from about 23° 27′N at the June solstice, through 0° at the March and September EQUINOXES, to about 23° 27′S at the December SOLSTICE.

declination, magnetic. The departure (degrees east or west) of a compass needle from true (geographical) north. This angle, termed by mariners 'magnetic variation', varies in space and time.

deepening. In synoptic meteorology, 'deepening' of a depression signifies a decrease of pressure at the centre of the system with time. The converse term is 'filling'.

deformation. In hydrodynamics, the change of shape of a small element of fluid produced by space variations of the fluid velocity.

In meteorology, the term is used mainly in respect of the kinematical development of FRONTOGENESIS and FRONTOLYSIS. Thus, flow associated with a COL represents a deformation field of motion. If persistent, such a flow produces a relative crowding of isotherms along the axis of outflow from the neutral point of the col, and a displacement of isotherms away from the axis of inflow. Mathematically, the deformation involves two terms:

$$A = \frac{\partial u}{\partial x} - \frac{\partial v}{\partial y},$$

$$B = \frac{\partial v}{\partial x} + \frac{\partial u}{\partial y}.$$

Individually, these terms are not invariant with respect to a rotation of axes; the quantity

$$F^2 = A^2 + B^2$$

is, however, invariant and is usually taken as a measure of deformation in meteorology.

degree-day.   See ACCUMULATED TEMPERATURE.

degrees of freedom.   When a result is being tested for statistical SIGNIFICANCE, the number of degrees of freedom determines the standard of comparison, and hence the proper place to enter the statistical table of significance. The number of degrees of freedom is generally the number of items (or the number of cells in the CHI-SQUARE TEST) reduced by the number of constants calculated from the data in setting up the hypothesis to be tested.

dendrochronology.   The dating of past events, not necessarily climatic, by means of growth rings in trees.

dendroclimatology.   The interpretation of the varying width or density of the annual growth rings of certain trees ('tree rings') in terms of the corresponding year-to-year climatic fluctuations. This method of inferring past climatic fluctuations appears to have rather strict limitations owing to the various possible combinations of weather element values which may be associated with a given rate of growth. The most important of the elements involved are probably temperature and rainfall.

density.   The mass of unit volume of a substance, at a specified temperature and pressure. The dimensions are ML^{-3}.
   The density of the air is not measured directly but may be calculated in terms of the normally observed meteorological elements of pressure, temperature, and humidity, from the GAS EQUATION:

$$\rho = \frac{p}{RT} = \frac{Mp}{R^*T}.$$

The molecular weight of the air sample decreases with increasing percentage weight of water vapour in the air (regarded as a mixture of dry air and water vapour); relatively moist air is therefore less dense than relatively dry air at the same pressure and temperature.
   The value of M appropriate to a given sample is less readily obtained than the VAPOUR PRESSURE. Air density is therefore more readily obtained from the formula

$$\rho = \rho_0 \frac{p - \tfrac{3}{8} e}{p_0} \cdot \frac{T_0}{T}$$

where ρ_0 is the density of dry air at pressure p_0 and absolute temperature T_0.

The density of dry air at a pressure of 1000 mb and temperature of 290 K is 1.201 kg m^{-3}. Thus

$$\rho \,(\text{kg m}^{-3}) = 1.201 \times \frac{p - \tfrac{3}{8}\,e}{1000} \cdot \frac{290}{T}$$

where p and e are in millibars.

Air density in the British Isles is about 1.2 kg m^{-3} at the surface; mean values at 5 and 10 km, for example, are about 0.7 and 0.4 kg m^{-3}, respectively. At heights beyond those attainable by balloons (about 40 km) air density has been inferred, for example, from FALLING SPHERE experiments, ROCKETSONDES, GRENADES SOUNDING and SATELLITE orbits. Systematic diurnal and annual variations of density have been confirmed, the amplitude of variation generally increasing with height. At satellite heights (above about 150 km) there are also large variations of air density which are associated with solar and geomagnetic activity.

departure.   The amount, positive or negative, by which the value of a meteorological element departs from the normal, average or expected value for the place and time. It is also termed 'anomaly'.

deposit gauge.   A gauge designed for the measurement of the deposited products of ATMOSPHERIC POLLUTION. The solid and liquid products which enter the collecting bowl of such a gauge within a specified period (generally a calendar month) are subjected to volumetric, gravimetric and chemical analyses in order to determine their amount and constitution.

depression.   The term commonly applied to CYCLONES, of various intensities, in extratropical latitudes (see Figure 18). It is also used of a weak TROPICAL CYCLONE.

Central pressure of an extratropical depression varies from about 950 to 1020 mb. It is described as 'shallow' or 'deep' if encircled by few or many isobars, respectively. It is said to 'deepen' or 'fill up' if the central pressure decreases or increases, respectively, with time. Its diameter varies from about 200 to over 4000 km. The associated weather is unsettled, often with much precipitation and strong winds or gales. It is, in general, a highly mobile feature of the synoptic chart, with speeds ranging up to about 50 kn; large depressions may, however, remain almost stationary for several days. The general direction of movement is eastwards, though any direction of movement may occur. The extratropical depressions largely control the rainfall distribution of these latitudes and are responsible for much of the interchange of air between high and low latitudes, which is an essential feature of the GENERAL CIRCULATION of the atmosphere.

The typical extratropical depression is associated with FRONTS in the manner illustrated in Figure 19, in which three distinct stages in the 'life cycle' — the 'wave depression', 'warm-sector depression', 'occluded depression' — are shown. Such a cycle usually lasts some 3 to 4 days, sometimes appreciably longer. Deepening and rapid movement of the depression are common up to the start of the occlusion process, with subsequent filling up and slowing down. The formation of a fresh wave depression on the trailing cold front of a depression which is in a later stage of development is a very common occurrence. Typically there are four or five such members of a 'depression family'; the series normally ends with an incursion of cold air to unusually low latitudes.

In the early stages of a depression, the TROPOSPHERE is relatively warm near the centre (except in a shallow layer under the frontal surface) and the cyclonic circulation is shallow; in late stages the air is relatively cold throughout the troposphere and the cyclonic circulation is deep. Horizontal convergence and associated upward motion in the lower troposphere and horizontal divergence in the upper troposphere prevail near

FIGURE 18.   Depression centred over the Irish Sea at 1200 GMT on 21 December 1980.

FIGURE 19. The likely patterns at three stages in the life of a model low showing (above) horizontal distribution of frontal cloud and precipitation, and (below) vertical cross-sections along the straight lines shown in the top row. In the bottom row precipitation is shown by the vertical shading.

and in advance of the depression centre. The structure of horizontal divergence and vertical motion is reversed in the rear of the depression.

Important but less common types of extratropical depression which are not associated with fronts are the POLAR-AIR DEPRESSION, THERMAL LOW and OROGRAPHIC DEPRESSION.

depression, angle of. The angular depression of an object measured by an observer, with reference to the horizontal plane through the observer.

descendant. See GRADIENT.

desert. A region in which rainfall is insufficient, in relation to rate of evaporation, to support vegetation. The main desert regions are in latitudes lower than 50° and are marked by relatively large diurnal and seasonal ranges of temperature. In terms of synoptic meteorology, the main causes of desert conditions are either the presence of a persistent anticyclone, as in northern Africa where the Sahara coincides with the average position of the subtropical belt of high pressure, or a configuration of ground which shuts out the moisture-bearing winds as, for example, Gobi in central Asia.

For a formula used by W. Köppen and R. Geiger for the limit of rainfall which constitutes a desert climate see ARID CLIMATE.

desertification. The transformation of a vegetated zone on the boundary of a DESERT into an extension of that desert so that the vegetation disappears and is replaced by drifting sand dunes. Possible causes include CLIMATIC CHANGE AND VARIABILITY and the effects of human population pressure such as deforestation and over-grazing by domestic animals. It is possible that removal of vegetation by human agency may produce positive feedback owing to increase in the ALBEDO as has been suggested by Charney [16]. See also Warren [17].

desiccation. The permanent disappearance of water from an area due to a change of climate and especially a decrease of rainfall. Large areas in central Asia, Africa and western USA have been desiccated since the last GLACIAL PHASE, and Lake Chad (central Africa) appears to have shrunk by 80% during this century. See also EXSICCATION.

detachment. In meteorological literature, usually refers to the process of ejection of ELECTRONS from negative IONS which occurs in the high atmosphere by various types of particle collision and by the action of light photons ('photo-detachment').

development.   In synoptic meteorology, the intensification of circulation, CYCLONIC or ANTICYCLONIC.

A strong measure of 'compensation' exists in the atmosphere such that, for example, the surface pressure fall associated with strong surface cyclonic development is a small residual between horizontal convergence at low levels and horizontal DIVERGENCE at high levels (with upward motion through much of the troposphere). The converse situation exists in strong anticyclonic development. In R.C. Sutcliffe's 'development theorem', which is based on these conceptions, the relative divergence between selected low and high levels, e.g. surface and tropopause, is expressed, on various simplifying assumptions, in terms of the THERMAL WIND and VORTICITY, both of which may be measured from charts, i.e.

$$\mathrm{div}_p \, \mathbf{V}_1 \; - \; \mathrm{div}_p \, \mathbf{V}_0 = -\frac{1}{f}\left(\mathbf{V}_\mathrm{T}\,\frac{\partial f}{\partial s} \; + \; \mathbf{V}_\mathrm{T}\,\frac{\partial \zeta}{\partial s} \; + \; 2\mathbf{V}_\mathrm{T}\,\frac{\partial \zeta_0}{\partial s}\right)$$

where div_p is the divergence in an isobaric surface, $\mathbf{V}_1$ and $\mathbf{V}_0$ are the wind vectors at the upper and lower pressure levels, ζ_0 the vorticity at the lower level $\partial/\partial s$ denotes differentiation in the direction of the thermal wind. A positive (negative) result implies cyclonic (anticyclonic) development.

Of the three terms on the right-hand side of the above development equation, the first involves the variation of f (i.e. of latitude) in the direction of the thermal wind and is generally small. The second is the so-called 'development term' and is proportional to the strength of the thermal wind and to its variation of vorticity along its own direction. Contributions to the latter vorticity are made by the curvature and shear of the thermal or THICKNESS pattern, as illustrated, for example, in the CONFLUENT THERMAL RIDGE. The third term is the THERMAL STEERING term which is proportional to the thermal wind and to the variation of surface vorticity in the direction of the thermal wind ($\partial \zeta_0/\partial s$). This term predominates when the pattern of surface vorticity is well marked and the thermal wind almost zonal and without horizontal shear, the first two terms then being small.

deviation.   A general term for the difference between an observed value and what is in some sense the true, expected or average value for the place and time, for example STANDARD DEVIATION, deviation of a light ray from a straight path, etc. The angle between the surface wind and the isobars is sometimes termed the 'deviation of the wind'.

dew.   Condensation of water vapour on a surface whose temperature is reduced by radiational cooling to below the DEW-POINT of the air in contact with it. Of two recognized processes of dew formation, the more common occurs in conditions of calm (wind at 10 m less than 1 kn) when water vapour diffuses from the soil upwards to the exposed cooling surface in contact with it (e.g. grass) and there condenses. The second of the processes is one of 'dewfall' when, in conditions of light wind, downwind turbulent transfer of water vapour from the atmosphere to the cooled surface occurs. A night of clear skies, moist air and sufficient but not excessive wind is such as gives maximum dewfall rates (about 3 mg cm^{-2} h^{-1} in the British Isles). Such nights are relatively uncommon and are estimated by J.L. Monteith [18] to produce a total of only about 2.5 to 5.0 mm (0.1 to 0.2 in) dewfall per year in the British Isles. See also DROSOMETER, GUTTATION.

dewbow.   A RAINBOW formed on the ground, usually in the shape of a hyperbola, by the REFRACTION and REFLECTION of the sun's rays in dew drops.

dew-point.   The dew-point temperature (T_d) of a moist air sample, commonly termed simply the dew-point, is that temperature to which the air must be cooled in order that it shall be saturated with respect to water at its existing pressure and HUMIDITY MIXING RATIO.

T_d is that temperature for which the saturation VAPOUR PRESSURE with respect to water (e'_w) is identical with the existing vapour pressure (e) of the air, i.e.

$$e = (e'_w) \text{ at } T_d.$$

Dew-point may be measured indirectly from wet-bulb and dry-bulb temperature readings with the aid of a humidity slide-rule or humidity tables (see PSYCHROMETER), or directly with a 'dew-point HYGROMETER'.

dew-pond. A pond on high ground on chalk downs, artificially constructed with a watertight bottom. It is found that, despite the watering of cattle, such ponds retain water during all but the most prolonged droughts, after ponds at lower levels have dried up.

Observation does not support the theory, which gave rise to the name, that such ponds are replenished to a significant extent by night dewfall. The explanation for their persistence probably lies in the relatively large amounts of precipitation that fall at high levels and in the occurrence of FOG PRECIPITATION.

diabatic. A diabatic thermodynamic process is one in which heat enters or leaves the system. Meteorological examples are EVAPORATION, CONDENSATION, and emission and absorption of RADIATION. The established equivalent term 'non-adiabatic' is generally preferred because it better emphasizes the nature of the process involved. See also ADIABATIC.

diamond dust. Precipitation of very small unbranched ice crystals forming in air supersaturated with respect to ice at temperatures below $-30\,°C$. Diamond dust accounts for much of the annual average accumulated 'snowfall' in the interior of Antarctica.

diathermancy. The ability of a substance to transmit radiant energy. Oxygen and nitrogen are diathermanous, while water vapour, ozone and carbon dioxide absorb heat radiation of certain wavelengths; the atmosphere is therefore only partially diathermanous.

dielectric constant. If the capacity of a condenser with a given substance as dielectric is c_1, and that of the same condenser with a vacuum as dielectric is c_2, then the ratio c_1/c_2 ($=\epsilon$) is the dielectric constant of the substance.

The dielectric constant is a property of a medium which determines the curvature of the path of an electromagnetic wave through it and so is closely related to the REFRACTIVE INDEX (n) of the medium. For a given substance, ϵ is a function of temperature and frequency. At frequencies and for a substance in which there is no absorption of the wave energy the relationship between them is $\epsilon = n^2$. For air, the relatively small departures of ϵ from unity are determined mainly by the amount of water vapour present.

differential thermal advection. Differential thermal advection signifies a change in the vertical temperature structure, and so in the static STABILITY, of the air at a particular place, brought about by horizontal ADVECTION of air. Thus, the advection of relatively warm air at lower levels, or of cold air at higher levels, or both, is differential thermal advection of the type that leads to a decrease of static stability. In general, in northern hemisphere temperate latitudes veering (backing) of wind with increase of height signifies warm (cold) air advection in the atmospheric layer concerned.

diffluence. The separation of adjacent STREAMLINES in the direction of flow. See DIVERGENCE.

diffluent thermal ridge. A pattern of THICKNESS lines which is concave towards high thickness and in which the thickness lines separate in the direction of the THERMAL WIND. According to the theorem of DEVELOPMENT, cyclogenesis (*C*) may be expected to occur behind and to the left, anticyclogenesis (*A*) ahead and to the right, of the pattern, as illustrated in Figure 20.

diffluent thermal trough. A pattern of THICKNESS lines which is concave towards low thickness and in which the thickness lines separate in the direction of the THERMAL WIND. According to the theorem of DEVELOPMENT, cyclogenesis (*C*) may be expected to occur ahead and to the left, anticyclogenesis (*A*) behind and to the right, of the pattern, as illustrated in Figure 21.

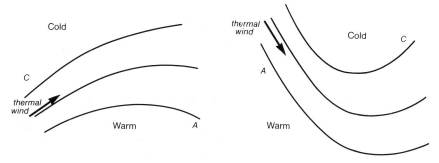

FIGURE 20.   Diffluent thermal ridge.   FIGURE 21.   Diffluent thermal trough.

diffraction. The bending of light rays produced by an obstacle in the path of the radiation, such as to produce an 'interference' pattern within the geometrical shadow region of the obstacle. The amount of the bending varies with the wavelength; resolution into spectral components therefore occurs in the case of visible light.

The phenomenon is responsible for a number of effects in atmospheric optics — see, for example, BISHOP'S RING, CORONA, GLORY. See also BABINET'S PRINCIPLE.

diffuse front. A front at which the changes in air-mass properties (temperature, humidity, wind velocity) from one side of the front to the other are not concentrated in a narrow belt but are spread through a wide 'frontal zone', which may be up to several hundred kilometres wide.

diffuse radiation. RADIATION which is received simultaneously from very many directions. In meteorology, it often signifies the solar radiation received (in particular, at the earth's surface) after SCATTERING by atmospheric molecules, cloud, dust, etc., as opposed to the parallel radiation received by direct incidence. It is also termed 'sky radiation'.

diffusion. Molecular diffusion is the process by which contiguous fluids mix slowly. The process follows laws similar to those of thermal CONDUCTION (heat diffusion). It is, however, so slow a process in the atmosphere as to be negligible in comparison with the mixing effected by turbulent eddies — 'EDDY DIFFUSION'. See EDDY.

diffusive equilibrium. That state of equilibrium, also termed 'gravitational equilibrium', in which the concentrations of different gases in a mixture of gases change with height according to the molecular weights of the gases involved. Efficient mixing prevents this state from being reached in the atmosphere below a level of about 80 km. See also GRAVITATIONAL SEPARATION.

diffusive separation. See GRAVITATIONAL SEPARATION.

diffusivity. A coefficient (K), of dimensions L^2T^{-1}, which measures the rate of DIFFUSION of a property (E).

In meteorology, the term refers to a measure of the rate of diffusion effected by atmospheric eddies, that effected by molecular processes being negligible in comparison. The coefficients involved are termed EXCHANGE COEFFICIENTS. That pertaining to the diffusion of matter is the 'eddy diffusivity', while the exchange coefficients for momentum and heat are the 'eddy viscosity' (see VISCOSITY) and EDDY CONDUCTIVITY, respectively. If q is the flux of E per unit area and $\partial E / \partial l$ is the spatial gradient of E, then K is defined by

$$q = -K\frac{\partial E}{\partial l}.$$

If K is constant in space and time the diffusion is called 'Fickian' and it follows that

$$\frac{\partial E}{\partial t} = K \nabla^2 E.$$

dimensions (of units). The powers to which the FUNDAMENTAL UNITS of mass (M), length (L) and time (T) have to be raised to express fully the units of a physical quantity; e.g. pressure = force per unit area = mass × acceleration per unit area and so has the dimensions $MLT^{-2}L^{-2}$, i.e. $ML^{-1}T^{-2}$. A pure number is assigned the dimension unity; temperature is allotted the special dimension θ.

Dines compensation. That property of the atmosphere whereby the sign of the DIVERGENCE is reversed at least once in a vertical column. This implies that the integrated divergence from the earth's surface to the top of the atmosphere, and the associated surface pressure tendency, are small residuals of much larger contributions. In general, a single change of sign occurs in the troposphere, and regions of convergence and divergence are separated by a level of zero divergence which is usually at around 600 mb.

This fundamental atmospheric property was inferred by W.H. Dines from his studies of the correlation between atmospheric variables at low and high atmospheric levels. See DEVELOPMENT, TENDENCY EQUATION.

direct circulation. A circulation in which the POTENTIAL ENERGY represented by the juxtaposition of relatively dense and light air masses is converted into KINETIC ENERGY as the lighter air rises and the denser air sinks. Land- and sea-breezes are examples of such a circulation. In an 'indirect circulation' the converse holds and the potential energy increases as kinetic energy decreases.

discontinuity. As a rule, the fundamental atmospheric variables of wind velocity, temperature and humidity are continuous functions of space and time, while pressure is invariably so. Occasionally, the space rate of change at a fixed time (or time rate of change at a fixed place) of wind velocity, temperature, humidity and also of pressure tendency is so abnormally great that the distribution of these variables may be regarded as discontinuous. Examples occur at well marked FRONTS and surfaces of SUBSIDENCE, more particularly, however, in association with SHOCK WAVES.

discriminant analysis. A method of distinguishing, by statistical analysis, between two or more sub-populations in a given sample of N items each of which is described by individual values of n random variables. It may be possible to discover a linear function (D) of the n variables such that the N values of D fall into two clearly separated groups, in which case D is described as a discriminant. It may be extended to the case of more than two groups; in general, p groups require $(p-1)$ discriminants. Discriminant analysis is a systematic method of searching for such functions and was developed largely in the field of biology and taxonomy. It has been applied to climatology, and to forecasting special types of weather from the output of numerical forecast models.

disdrometer.   Any apparatus for measuring the sizes of individual raindrops or cloud droplets and hence deriving the drop-size spectrum. The word seems to be derived from 'distribution of drop-size meter' and covers all sorts of devices from filter papers and coated glass slides to instruments incorporating laser beams and microelectronics.

dishpan experiment.   See ROTATING FLUIDS — LABORATORY STUDIES.

dispersion.   In physics, the separation of RADIATION into its component wavelengths, due to wavelength dependence of such processes as REFRACTION, DIFFRACTION, and SCATTERING.
   In statistics, the 'scatter' of a group of values. The most common measure of dispersion is the STANDARD DEVIATION: other measures are the mean deviation, the RANGE of the values, and the 'coefficient of variation' (standard deviation divided by mean).

dissipation trail.   An effect, opposite to the CONDENSATION TRAIL, in which the passage of an aircraft through cloud is marked by the appearance of a clear lane. The phenomenon, which is relatively rare, is found under conditions when the effect of the heat of combustion of the fuel released by the aircraft exhaust is sufficient to outweigh that of the released water vapour to the extent of causing tenuous cloud in the wake of the aircraft to evaporate. Occasionally, the effect may be produced by an aircraft flying in relatively dry air, just above a thin cloud layer, by the dragging down and mixing of the dry air with the cloud in the wake of the aircraft. A similar (spurious) effect is sometimes produced when the shadow of a condensation trail is cast on a thin cloud layer.

dissociation.   In geophysics, the breakdown of atmospheric molecules into component atoms by the action of ultraviolet radiation in the high atmosphere. The minimum energy required to effect dissociation of a molecule is its 'dissociation energy'.
   Dissociation in the atmosphere becomes increasingly important with increase in height above 80 km, mainly in respect of oxygen (molecular to atomic oxygen).

distrail.   A common abbreviation for DISSIPATION TRAIL.

distribution function.   The function $F(x)$ that gives the total frequency of members of a given POPULATION with variate values less than or equal to x. It is mathematically equivalent to the integral of the frequency function (see FREQUENCY DISTRIBUTION).

disturbance.   Sometimes used instead of DEPRESSION or TROUGH of low pressure, especially in broadcast GENERAL INFERENCES.

diurnal variation.   The changes of value, for example of a meteorological element, within the course of a (solar) day, more especially the systematic changes that occur within the average day.
   A systematic diurnal variation (or 'daily variation') may be revealed by determining the average values, in a large number of days, of the selected element at 0h, 1h, 2h,...,

24h, and removing the non-cyclic change (if any) which is given by the difference between the average values obtained for 0h and 24h. Irregular fluctuations present on individual days are eliminated by averaging over a sufficient number of days.

The systematic diurnal variation of atmospheric pressure is dominated in most latitudes by a 12-hourly oscillation which proceeds according to LOCAL TIME (maxima at 10h and 22h, minima at 4h and 16h) and whose amplitude decreases polewards from the equator; at 0° and 50°, for example, approximate values of the swing either side of the mean are 2.0 and 0.7 mb, respectively. In low latitudes the variation is prominent enough to be easily visible on a barogram, even with the near approach of a tropical storm. In middle latitudes the variation is normally obscured by non-periodic pressure changes, but is easily revealed by averaging. In high latitudes the 12-hourly local-time oscillation is very small and the main feature of the diurnal variation is a small 24-hourly UNIVERSAL TIME oscillation.

Temperature and relative humidity have systematic local-time diurnal variations, nearly opposite in phase, and of greater amplitude in summer than in winter. Minimum temperature, for example, occurs at about sunrise, and maximum temperature some 2 hours in winter, 3 hours in summer, after noon.

Systematic diurnal variations within the lunar day — 'lunar diurnal variations' — are obtained by averaging the magnitudes of elements arranged according to lunar time. Pressure variations of this type have been determined for many parts of the world, and an associated temperature variation in low latitudes has been discovered.

divergence. The divergence of the flux of a quantity (e.g. radiation or momentum) expresses the time rate of depletion of the quantity per unit volume. Negative divergence is termed 'convergence' and relates to the rate of accumulation.

In meteorology, divergence (or convergence) is mostly used in relation to the velocity vector and so refers to the flux of air particles themselves. The 'divergence of velocity' is a three-dimensional property which expresses the time rate of expansion of the air per unit volume. The following relation holds:

$$\text{div } \mathbf{V} \text{ (or } \nabla \cdot \mathbf{V}) = \frac{\partial u}{\partial x} + \frac{\partial v}{\partial y} + \frac{\partial w}{\partial z} \, .$$

In the atmosphere, div $\mathbf{V}$ is small and is of little direct interest. The main concern is with the 'horizontal divergence of velocity' (often referred to as the 'horizontal divergence' or even, misleadingly, the 'divergence') which, denoted div$_H\mathbf{V}$ or div$_2\mathbf{V}$, etc., is identified with the sum $(\partial u/\partial x + \partial v/\partial y)$, and expresses the time rate of horizontal expansion of the air per unit area.

Since div $\mathbf{V} \approx 0$,

$$\text{div}_H\mathbf{V} \equiv \frac{\partial u}{\partial x} + \frac{\partial v}{\partial y} = -\frac{\partial w}{\partial z} \, ,$$

i.e. horizontal divergence of air is closely associated with vertical contraction (usually termed 'vertical convergence') of an air column centred at the level concerned. Similarly, horizontal convergence is linked with 'vertical divergence' $(\partial w/\partial z)$.

Contributions to div$_H\mathbf{V}$ may be made by DIFFLUENCE and by increase of wind downstream. Since, however, diffluence is generally associated with decrease of wind downstream (and 'confluence' with increase of wind downstream), the respective contributions to div$_H\mathbf{V}$ are generally of opposite sign and are usually much larger than div$_H\mathbf{V}$ itself. Therefore, div$_H\mathbf{V}$ is a small remainder which is difficult to determine, even as to sign, in the free atmosphere from the substitution of wind observations in the expression $(\partial u/\partial x + \partial v/\partial y)$. It is more accurately determined from the VORTICITY EQUATION and from the distribution of vertical motion.

The dimension of div $\mathbf{V}$ is T^{-1}. Values of div$_H\mathbf{V}$ in large-scale motions in the free atmosphere generally range from zero up to about 10^{-5} s^{-1}. (It can be considerably greater locally in frontal zones.) On average, divergence (or convergence) is least at a level of about 600 mb — the so-called 'level of non-divergence'.

D-layer. That part of the IONOSPHERE, situated between about 65 and 80 km, which is mainly responsible for the absorption of radio energy reflected from higher levels, but from which radio waves of very low frequency may also be reflected. It is sometimes referred to only as the 'D-region' because it has apparently no peak value of electron concentration. The extra ionization caused at this level by SOLAR FLARE radiation gives rise to various types of 'sudden ionospheric disturbance'.

Dobson spectrophotometer. An instrument used in the routine measurement of atmospheric OZONE. The method consists of isolating, in the solar radiation spectrum, two wavelengths in the region of partial ozone absorption in the near ultraviolet (0.30 to 0.33 μm), allowing them to fall in rapid succession on a photomultiplier connected through an a.c. amplifier and rectifier to a galvanometer, and reducing, by means of a calibrated optical wedge, the intensity of the longer, less-absorbed wave until the photomultiplier outputs are equal and the galvanometer records no current. The amount of atmospheric ozone in a vertical column is obtained from the position of the wedge and the solar zenith angle at the time of observation.

Incorporated in the calibration of the instrument is the ratio of the intensities of the selected wavelengths outside the atmosphere. This is obtained, once for all, from a series of instrumental observations at various solar zenith angles and extrapolation to zero path length through the atmosphere. The extrapolation is based on the assumption, supported by photochemical theory, that there is little or no systematic diurnal variation of total ozone.

The standard measurements are those made against direct sunlight. Measurements are also made against the zenith sky, clear or cloudy, and against moonlight, using empirical relations with the direct sunlight observations.

The instrument may be used to obtain the vertical distribution of ozone, using the UMKEHR EFFECT.

doctor. A popular name given to the HARMATTAN of north-west Africa. The term is also used for the KATABATIC WIND that blows most evenings from the Blue Mountains over the Liquanea Plain in Jamaica.

doldrums. The equatorial oceanic regions of light variable (mainly westerly) winds, accompanied by heavy rains, thunderstorms and squalls. These belts, which are variable in position and extent, have a systematic north and south movement some 5° either side of a mean position, following the sun with a lag of 1 or 2 months.

Doppler effect. The observed frequency of a source moving with velocity v towards an observer, and emitting waves (e.g. radio, sound or light) of wavelength λ, is greater than the emitted frequency f by the amount $\delta f = v/\lambda$ — the 'Doppler effect'. For a receding source the observed frequency is less than f by the same amount. If, on average, emitting sources approach and recede from an observer in equal numbers, 'Doppler broadening', without shift, is observed. Frequency and spectral measurements may thus be employed to measure the velocity of fast-moving sources.

The Doppler-shift technique may be used to obtain a measure of the vertical velocities of precipitation elements. The frequency shift suffered by a radio wave on reflection from moving objects ($f = v/\lambda$) gives a measure of the mean 'line of sight' velocity of the elements, while variations of the intensity of the radar echo may be used to infer motions possessed by the various elements relative to one another.

Doppler radar. If a RADAR target is moving towards or away from the radio receiver, the frequency of a reflected signal is, because of the DOPPLER EFFECT, slightly different from the transmitted frequency. Doppler radar is designed to interpret this effect in terms of the radial velocities of targets. It is employed in meteorology to deduce various types of information, for example: air motion at various heights, the speeds of fall of precipitation particles and, on certain assumptions, vertical air motion within precipitation regions. See also WIND PROFILING.

Douglas sea and swell scale. Devised in the 1920s by Captain H.P. Douglas CMG, RN, Hydrographer of the Royal Navy, and recommended for international use in 1929 [19]. The scale consists of pairs of digits, each on a scale from 0 to 9, the first representing 'sea' and the second 'swell'.

downburst. A strong DOWNDRAUGHT causing an outward burst of damaging winds at or near the ground. Downbursts are often associated with severe local storms and hence with a great depth of instability and strong wind shear. They present a hazard to aviation, especially in the mid-western United States.

downdraught. A descending current of air, often moving with considerable speed and usually associated with strong convective motions in CUMULONIMBUS clouds and the development of heavy precipitation. The falling raindrops cool the surrounding air by evaporation making it denser than the environment, and viscous drag may accelerate the air still further. When the downdraught reaches the ground a SQUALL is often produced.

drag coefficient. A non-dimensional coefficient (C_D), also termed the 'skin-friction coefficient', which is defined by the equation $\tau = C_D \rho \bar{U}_s^2$, where $\bar{U}_s$ is the wind speed observed near the surface.
  C_D is a conventional, as opposed to a unique, parameter of surface roughness. Its value depends on the height at which wind speed is observed and also on the stability of the air. Wind speed observed at a height of 10 m is generally employed or, alternatively, the geostrophic wind, giving respectively the 10 m and geostrophic drag coefficients.

drainage area. The area whose surface directs water towards a stream above a given point on that stream. See also CATCHMENT AREA and GATHERING GROUND.

drainage gauge. See PERCOLATION.

drainage wind. An alternative for KATABATIC WIND.

drift. In oceanography, the velocity of an ocean current.

drifting dust or sand. 'Dust or sand, raised by the wind to small heights above the ground. The visibility is not sensibly diminished at eye level' [2, p. 120].

drizzle. Liquid precipitation in the form of water drops of very small size (by convention, with diameter between about 200 and 500 μm.
  Drizzle forms by COALESCENCE of droplets of STRATUS cloud. It falls from stratus cloud of low base, in which the widespread upcurrents which form the cloud have a velocity smaller than the terminal velocity of the drops concerned. High relative humidity below the cloud base is also required to prevent the drops from evaporating before they reach the earth's surface.
  For synoptic purposes, drizzle is classified as 'slight', 'moderate', or 'heavy': slight drizzle corresponds to negligible runoff from roofs, heavy drizzle to a rate of accumulation greater than 1 mm h^{-1}.

drop, droplet. Terms generally applied in meteorology to the particles of water which constitute liquid PRECIPITATION and CLOUD elements, respectively. The differentiation, whilst not precise, is essentially one of size, the limiting diameter being about 200 μm.

dropsonde. A RADIOSONDE which, instead of being carried upwards by a balloon, is dropped from a rocket or aircraft and is suspended from a parachute.

drosometer. An instrument for measuring the amount of DEW deposit. In the Duvdevani dew-gauge, the size, form and distribution of dew drops deposited overnight on a rectangular block, carrying a special paint and exposed at a standard height, are compared with standard photographs relating to a dew deposit of known amount. In other gauges, some of them recording, dew deposited (e.g. on a hygroscopic surface) is obtained by weighing. A basic uncertainty in all cases is the relationship between the amount of dew deposited on the gauge and that deposited on natural surfaces.

drought. Dryness due to lack of RAINFALL. Drought is a relative term and any definition in terms of rainfall deficit must refer to the particular rainfall-related activity that is under discussion. For example, there may be a shortage of rainfall during the growing season of a particular crop (agricultural drought) or during the winter runoff and percolation season affecting the supply of water for domestic and industrial use (hydrological drought); such shortages need to be compared with the climatological expectation for the place and season. For these reasons, tabulations of precisely defined 'absolute' and 'partial' droughts have been discontinued.

dry adiabatic (or dry adiabat). Line on an AEROLOGICAL DIAGRAM representing the dry adiabatic lapse rate. See ADIABATIC.

dry air. In physical meteorology this term generally signifies air which is completely dry, i.e. which contains no water vapour. In synoptic meteorology and climatology the term usually refers to air of low RELATIVE HUMIDITY. In the BEAUFORT NOTATION the letter 'y' signifies 'dry air', precisely defined as having a relative humidity of less than 60 per cent.

dry- and wet-bulb hygrometer. An alternative for PSYCHROMETER.

dry ice. A common term for solid carbon dioxide, frequently used in CLOUD SEEDING experiments. The substance is generally discharged from an aircraft into supercooled cloud in particles of about 10 mm diameter at the rate of a few kilograms per kilometre. The particles vaporize at a temperature of -78.5 °C. The local cooling of the ambient air effected by the sublimation process produces myriads of minute ice crystals.

dry season. A period of a month or more, recurring every year, which is marked in a given region (generally tropical or subtropical) by the complete or almost complete absence of precipitation. Winter is the dry season in most tropical regions; summer is, however, the dry season in the MEDITERRANEAN-TYPE CLIMATE.

dry spell. A period with RAINFALL below a specified amount. The period and amount of rainfall vary depending on the particular activity under discussion.

duplicatus (du). One of the CLOUD VARIETIES. (Latin for doubled.)
'Superposed cloud patches, sheets or layers, at slightly different levels, sometimes partly merged. This term applies mainly to CIRRUS, CIRROSTRATUS, ALTOCUMULUS, ALTOSTRATUS and STRATOCUMULUS' [2, p. 22]. See also CLOUD CLASSIFICATION.

dust. The atmosphere carries in suspension, often for long distances, solid particles of varying character and size. The chief sources of these particles are volcanic eruptions, meteors, dust and sand raised by winds, and the smoke produced in industrial and domestic combustion processes and in forest fires.

An important meteorological effect of atmospheric dust is its depletion of solar radiation by scattering and, to a smaller degree, by absorption. When present in appreciable quantity it gives atmospheric HAZE. Most dust particles are of a size sufficiently small to cause differential SCATTERING of sunlight and so produce, for example, highly coloured sunsets — see SUNRISE AND SUNSET COLOURS.

The concentration and size distribution of solid particles suspended in the atmosphere, and their effectiveness as condensation nuclei, are discussed under NUCLEUS.

dust counter. An instrument for counting the dust particles in a known volume of air. In Aitken's dust counter, condensation is made to occur on the nuclei present by ADIABATIC expansion of air, and the number of drops is ascertained.

dust (sand) haze. A suspension in the air of dust or small sand particles, raised from the ground prior to the time of observation by a DUSTSTORM (SANDSTORM).

duststorm. See SANDSTORM.

dust whirl (or **dust devil**). A WHIRLWIND, in which dust and sand are carried aloft from the ground by very strong convection from a hot, sandy region. The rotation may be in either direction round the centre, which is itself often free from dust. Heights of about 1 km have been reported but are generally less than 30 m. Speeds at which dust whirls move vary from less than 5 kn to over 25 kn. The phenomenon is sometimes termed 'sand pillar'.

D-value. The difference (D) between the actual height (Z) above mean sea level of a particular pressure surface and the PRESSURE ALTITUDE (Z_p), i.e. $D = Z - Z_p$.

A system of analysis based on the cross-section presentation of D-values is a method which well illustrates the vertical structure of pressure systems.

dynamical meteorology. The study of the causes and nature of motion of the atmosphere.

dynamic heating. The frictional heating experienced by a body due to its rapid motion through the air.

dynamic metre. See GEOPOTENTIAL.

dynamic pressure. In meteorology, the force exerted by the wind on unit area of the windward face of a surface. At points where the air is brought to rest relative to the surface it is given by the quantity $\frac{1}{2}\rho V^2$, where V is the relative wind speed.

dynamic stability. A term sometimes used, in a restricted sense, as a synonym of INERTIAL STABILITY. More generally this term, or the alternative 'hydrodynamic stability', signifies an atmospheric state characterized by there being no tendency for small wave-like perturbations of the flow to grow. Various influences operate in the atmosphere to cause the contrary state of (hydro)dynamic instability, e.g. GRAVITY, WIND SHEAR, or a strongly BAROCLINIC atmosphere. See also STABILITY.

dynamic temperature change. A change associated with the compression (warming) or expansion (cooling) of a gas. See ADIABATIC.

dynamo theory. The hypothesis, first proposed by Stewart, which explains the regular daily variations in the earth's magnetic field in terms of electrical currents in the lower IONOSPHERE.

dyne. The unit of force in the C.G.S. SYSTEM of units. It is the force that when applied to a mass of 1 gram produces an acceleration of 1 centimetre per second squared. The dimensions are MLT^{-2}.

$$1 \text{ dyne } = 10^{-5} \text{ NEWTONS.}$$

E

earth. The earth is a spheroid which is somewhat flattened at the poles relative to the equator because of rotation about its axis — the so-called 'equatorial bulge'. It is of mean radius 6371.229 km, of angular velocity 7.292×10^{-5} rad s^{-1}, of mean distance from the sun 1.4968×10^8 km, of mass 5.975×10^{24} kg and of mean density relative to water 5.5.

The respective percentages of water and land on the earth's surface are: in the northern hemisphere 60.7 and 39.3; in the southern hemisphere 80.9 and 19.1; for the earth as a whole 70.8 and 29.2.

earth-light. See MOON.

earthquake. A natural movement of the ground, originating below the surface, due to the build-up and sudden release of strain energy in a small underground region, termed the 'focus' or 'hypocentre' of the earthquake. The point of the earth's surface vertically above the focus is termed the 'epicentre'. The focus of most earthquakes is at a depth of less than 20 km. Earthquakes of much deeper focus (up to 700 km depth) occur in some areas, more especially in the Pacific Ocean area.

earth temperature. If the temperature of the earth's surface were constant, heat would be conducted continually upwards from the earth's very hot interior to the surface and there dissipated, mainly by radiation; a steady state would be reached, with temperature increasing with depth at an almost steady rate. This simple regime is much changed in the top layer of the earth by the fact that the surface temperature is raised, by absorption of solar radiation, to a higher value than that appropriate to the above conditions. The diurnal and annual temperature waves observed at the surface proceed downwards by heat conduction, but with rapid reduction of amplitude and progressive increase of time-lag relative to the surface waves. The diurnal wave extends to a depth of about 0.5 m and the annual wave extends to a depth of about 10 m; these values are approximately as the square root of the ratio of the periods (1 and 365 days, respectively), in accordance with the theory of heat conduction in solids.

earth thermometer. Obsolete term for SOIL THERMOMETER.

easterly wave. A shallow trough disturbance in the easterly current of the tropics, more in evidence in the upper-level winds than in surface pressure, whose passage westwards is followed by a marked intensification of cloudy, showery weather.

ecliptic. That GREAT CIRCLE on the CELESTIAL SPHERE which the sun appears to describe, with the earth as centre, in the course of a year; the plane in which the apparent orbit of the sun lies is the 'plane of the ecliptic'.

In terms of actual motion, the ecliptic is the path described by the earth round the sun in the course of a year.

The 'obliquity of the ecliptic' is the inclination of the plane of the ecliptic to that of the earth's equator. This value controls the limits of variation of the sun's DECLINATION in the course of the year and gives rise to the meteorological SEASONS. The obliquity of the ecliptic was close to 23° 26′ 26″ in 1990 and was then decreasing at the rate of about 0.47″ per year. The limits of variation of the value of the obliquity and the period or periods involved are as yet uncertain but are very probably significant factors in long-period CLIMATIC CHANGE AND VARIABILITY.

ecoclimatology. That branch of ecology which is concerned with the study of living matter (animal and plant) in relationship to its climatic environment.

eddy. A term in fluid motion for which, like that of the closely associated term TURBULENCE a brief comprehensive definition is impossible. Its essential characteristics are, however, covered by its definition as a mass of fluid which retains its identity for a limited time while moving within the surrounding fluid.

Eddies are often defined as departures from the main flow, the appropriate mean being taken either with respect to time or with respect to one or more dimensions of space. A common notation is to write [ ] for a space mean with departures indicated by *, and ‾ for a time mean with departures indicated by ' . Thus, for some variable q, we have $q = [q] + q* = \bar{q} + q'$, so that $[q*] \equiv 0 \equiv \bar{q}'$. However, $[q']$ and $\bar{q}*$ are not necessarily, or even usually, zero.

Eddies defined by $\bar{q}*$, where $\bar{q}* = \bar{q} - [\bar{q}]$, are called 'standing eddies', whereas those involving departures in time are called 'transient eddies', a method of classification much used in studies of the flow of energy or momentum induced by the GENERAL CIRCULATION.

Atmospheric eddies range in size from less than one centimetre (microscale turbulence) to some hundreds of kilometres (large depression or anticyclone) or more. The smaller-scale eddies play a vital part in effecting vertical mixing of such atmospheric properties as momentum, heat and water vapour ('eddy flux'), while the large-scale eddies are responsible for much of the meridional transport implicit in the general circulation of the atmosphere. See, for example, TURBULENCE, VISCOSITY, DIFFUSIVITY, EDDY SPECTRUM.

eddy conductivity. EXCHANGE COEFFICIENT (K), of dimensions $L^2 T^{-1}$, relating to the transfer of heat effected by eddies. For transfer in the vertical, for example, the coefficient may be regarded as being defined by the equation

$$ - \overline{w'T'} = K_z \frac{\partial \bar{T}}{\partial z} $$

where w' and T' are corresponding fluctuations of vertical velocity and temperature from mean values. The overbars denote time averages. K varies greatly in space and time.

See, for example, EDDY, DIFFUSIVITY, VISCOSITY, K-THEORY.

eddy diffusion. The mixing of atmospheric matter and properties which is effected by eddies. It is also termed 'turbulent diffusion'. See EDDY, DIFFUSION, DIFFUSIVITY.

eddy flux. The rate of transport across a unit surface area of atmospheric properties and matter (heat, momentum, water vapour, etc.) which is effected by atmospheric eddies. Frequently the vertical transport is implied. It is also termed 'turbulent flux'. See EDDY.

eddy shearing stress. An alternative for REYNOLDS STRESS.

eddy spectrum. Specification of the character of TURBULENCE in terms of the partition of kinetic energy between eddies of various sizes. It is normally obtained by CORRELOGRAM or FOURIER ANALYSIS of the time fluctuations of motion observed at a particular point. It is also termed 'turbulence spectrum', 'energy spectrum' (of turbulence), or 'power spectrum' (of turbulence).

effective height of an anemometer. The height over open level terrain in the vicinity of an anemometer which, it is estimated, would have the same mean wind speeds as those actually recorded by the anemometer.

No definite rules are laid down for obtaining the effective height of an anemometer. The nature, extent, height and distance of local obstructions and the height of the anemometer itself must be taken into account. The effective height may be different for different wind directions. See also EXPOSURE.

effective radiation. An alternative for NOCTURNAL RADIATION.

effective temperature. As defined by the American Society of Heating and Ventilating Engineers, the temperature of saturated motionless air which would produce the same sensation of warmth or coolness as that produced by the temperature, humidity and air motion under observation.

The definition implies that effective temperature is dependent on the amount of clothing worn. Nomograms which relate effective temperature to dry-bulb and wet-bulb temperatures and wind speed, for two categories of clothing, have been produced by the former Air Ministry [20]. See also COMFORT ZONE.

Ekman pumping. Because the motion in an EKMAN SPIRAL is convergent, with air flowing from high to low pressure, air is drawn into the lower layers of a cyclonic atmospheric vortex, rises vertically, and is expelled at the top of the boundary layer. This effect is known as Ekman pumping. The vertical motion at the top of the boundary layer leads to mean divergence over the whole atmospheric column above the boundary layer and hence to a decrease of cyclonic vorticity. (With anticyclonic circulations the effects are reversed in sign.) Ekman pumping is a much more efficient way of dissipating concentrations of VORTICITY than is EDDY DIFFUSION by itself.

Ekman spiral. An equiangular spiral which is the locus of the end points of the (idealized) wind vectors, starting from a common origin, within the FRICTION LAYER. The centre of the spiral corresponds to the GEOSTROPHIC WIND, while the surface wind is backed 45° from this direction (see Figure 22). Such a spiral, first derived by V.W. Ekman for the ocean currents produced by surface stress, is, however, obtained under assumed conditions that are rarely realized in the atmosphere — namely, no variation of eddy viscosity with height and no horizontal temperature gradient, within the friction layer.

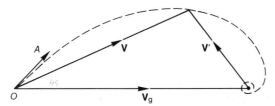

FIGURE 22. Equiangular spiral such that the ageostrophic wind component (**V′**) makes an angle of $\pi/4$ with the tangent at all points of the curve. **V**$_g$ is the geostrophic wind. **V′** decreases exponentially to zero with increasing height, and OA gives the direction of the resultant wind **V** at the surface.

E-layer. The lowest regularly observed layer of the IONOSPHERE occurring at a height of about 100 to 120 km and observable in normal ionospheric sounding mainly during the day. The maximum electron density is of the order 10^5 cm^{-3}. The structure of the E-region is at times complicated by the appearance of various 'ledges' of ionization and of intense 'sporadic E' ionization which exists in extensive patches and gives rise to anomalous reception of high-frequency radio waves.

electricity.   See ATMOSPHERIC ELECTRICITY.

electricity of precipitation.   The main features of surface measurements of the electric charge carried by solid and liquid PRECIPITATION are as follows:

(i)   Charges of both signs are carried by all kinds of precipitation.
(ii)   Rapid fluctuations of sign occur within short intervals of time.
(iii)   A net positive charge is, on average, carried to ground by all types of precipitation.
(iv)   The charges carried per cubic centimetre of precipitation and by individual drops are of the general order 3×10^{-10} and 3×10^{-7} coulombs, respectively.
(v)   Thunderstorm-type rain and snow carry, on average, larger charges than does steady rain.

These observations, together with the results of similar measurements made from aircraft, show that the processes involved are highly complex. As with the generation and separation of charges associated with a THUNDERSTORM, various theories have been advanced to explain the charges carried by precipitation. They include: effects associated with a mixture of water in the solid and liquid phases; the separation of charge on the breakup of large raindrops; and the selective capture by precipitation particles of positive IONS which stream upwards by BRUSH DISCHARGE from the ground in the disturbed surface POTENTIAL GRADIENT conditions which occur in the vicinity of rain clouds. The complexity of the observed effects makes it probable that several processes are important.

electric storm.   Little-used alternative for a THUNDERSTORM.

electrojet.   The term applied to the narrow belt of intense 'dynamo' electric currents which flow in the low ionosphere near the equator and which produce at the earth's surface in low latitudes a large augmentation of the solar and lunar diurnal variations of the magnetic elements, compared with other latitudes. See DYNAMO THEORY.

electromagnetic radiation.   The propagation of energy in which, according to classical 'electromagnetic theory', a transverse wave motion exists in the form of periodic fluctuations of the strengths of electric and magnetic fields which act at angles to the direction of propagation of the energy. The energy may be propagated through a medium or vacuum; the speed of propagation in the latter, and to a close approximation in air, is about 3×108 m s^{-1}.
Electromagnetic radiation is divided into various classes which differ only in wavelength. In order of increasing wavelength (decreasing frequency) they comprise GAMMA RADIATION, X-RAYS, ULTRAVIOLET RADIATION, VISIBLE SPECTRUM, INFRA-RED RADIATION, and RADIO WAVES. See also RADIATION.

electrometeor.   A little-used generic term for a visible or audible manifestation of atmospheric electricity, e.g. THUNDER, LIGHTNING, ST. ELMO'S FIRE, AURORA.

electrometer.   An instrument with a high input impedance (of the order of 10^9 to 10^{11} ohms) for measuring electric potential difference.

electron.   The elementary particle that carries the indivisible (unit) negative electric charge of 4.803×10^{-10} electrostatic units (1.6×10^{-19} coulombs).

electron-volt.   The electron-volt (eV) is the unit of energy employed in discussing processes of ION formation; it is the energy acquired by an electron on passing through a potential rise of 1 volt. $1 \text{ eV} = 1.6 \times 10^{-19}$ J.

elevation. In meteorology and in aviation, term used to denote the height of the ground above mean sea level.

Eliassen–Palm (EP) flux. A vector such that its divergence is equal to the zonal-mean poleward flux of the quasi-geostrophic POTENTIAL VORTICITY due to the EDDY component of atmospheric flow. The divergence of the EP flux provides a measure of zonal momentum forcing by the eddies. It is very useful in theoretical studies of the propagation of planetary waves; in fact, the GROUP VELOCITY of these waves as projected on the meridional plane is in the direction of the EP flux.

El Niño southern oscillation (ENSO). El Niño — The Child — is the name originally given by the local inhabitants to a weak warm ocean current flowing southwards along the coast of Ecuador and Peru around Christmas. The El Niño Southern Oscillation is the term now applied to a more extensive, intense, and prolonged warming of the eastern tropical Pacific Ocean occurring every few years which is associated with major anomalies in the patterns of atmospheric circulation and rainfall. Extreme ENSO events have serious effects on fisheries, bird life, and mainland weather. Noteworthy events were recorded in December of the following years: 1951, 1957, 1965, 1968, 1972, 1976, 1982, and 1986; of these, that of 1982 was outstanding, lasting well into the following year.

Elsasser's diagram. See RADIATION CHART.

emagram. See AEROLOGICAL DIAGRAM.

energy. A quantity, of dimensions ML^2T^{-2}, defined as the capacity for doing work. The unit of energy in the C.G.S. SYSTEM is the ERG, in the M.K.S. SYSTEM and SI UNITS the JOULE.
  The various forms of energy include, for example, POTENTIAL ENERGY, KINETIC ENERGY, HEAT, and radiant, electric, magnetic and chemical energy.

energy balance. In meteorology, a concept generally applied in the form of an 'energy-balance equation' which relates the net radiation flux at a portion of the earth's surface to the heat lost or gained by conduction to or from below, the heat lost or gained to or from the atmosphere by molecular and eddy processes, and the heat lost or gained at the surface by evaporation or condensation, respectively. See also RADIATION BALANCE.

energy cascade. In all forms of turbulent and unsteady fluid motion (see TURBULENCE), the energy is partitioned between 'eddies' of different sizes (see EDDY SPECTRUM). There is a constant tendency for energy to be passed from eddies of one particular size to those of another size forming what is termed an 'energy cascade'; in three-dimensional turbulence the flow of energy is from large to small eddies until the energy is ultimately dissipated by molecular viscosity as described in L.F. Richardson's well known rhyme:

> Big whirls have little whirls that feed on their velocity,
> And little whirls have lesser whirls and so on to viscosity.

(See KOLMOGOROFF SIMILARITY HYPOTHESES.)
  Two-dimensional turbulence, of which the general circulation of the atmosphere is to the first approximation an example, has very different properties. The stretching of vortex lines and increases of vorticity on ever smaller scales typical of three-dimensional turbulence cannot occur; in fact, the mean-square vorticity, or

ENSTROPHY, is an invariant of the motion (in the absence of viscous dissipation) just as much as the total energy, and this has important consequences.

It may be shown [21] by expansions in ORTHOGONAL FUNCTIONS over a sphere that the total energy (E) and enstrophy (F) may be expressed as

$$E = \sum_{1}^{\infty} b_n = \text{constant}$$

$$\text{and} \quad F = \sum_{1}^{\infty} \lambda_n b_n = \text{constant}$$

where λ_n and b_n are derived from the expansions, and n is the wave number.

For large n, $\lambda_n \approx n^2$, so that we can interpret the two relations above by a mechanical analogy. Suppose we have a weightless beam balanced on a fulcrum with a weight E on one side and weights $b_1, b_2, ..$, etc. on the other at distances $\lambda_1^{1/2}, \lambda_2^{1/2}$ from the fulcrum on the other. It follows that any reallocation of energy among different eddies is equivalent to a shifting of the weights, and this shifting must preserve the moment of inertia (i.e. F). Hence at least three weights (or eddies) must take part in any energy transformation and the fraction that can flow to higher wave numbers is strictly limited. Deeper analysis shows that there is in fact a strong tendency for energy to flow 'backwards', i.e. to ever decreasing wave numbers until it all piles up at the lowest frequency. This has been called the 'Infra-red catastrophe in Flatland fluid dynamics' by Lilly [22], a catastrophe from which we escape in practice by the action of surface drag and dissipation in the boundary layer.

energy spectrum. See EDDY SPECTRUM.

ENSO. Abbreviation for EL NIÑO SOUTHERN OSCILLATION.

enstrophy. Half the square of the VORTICITY of two-dimensional fluid flow. Mean enstrophy is conserved over a closed region provided the motion is adiabatic and there is no viscous or other dissipation of energy.

enthalpy. A thermodynamic quantity (H) which represents the 'total' heat content per unit mass of a substance; the units normally employed are joules per kilogram.

A change in enthalpy (dH) of a mass of gas is the heat gained or lost by the gas in an isobaric process. It is given by $dH = c_p dT$. The isobaric transport of enthalpy is an important item in the overall heat balance of the atmosphere.

In recent meteorological literature enthalpy is frequently referred to as 'sensible' heat as opposed to latent ('hidden') heat. See also HEAT.

entrainment. A term applied to the mixing of environment air into the updraught of a cumuliform cloud. The effects of such mixing (and also of the mixing into the cloud of environment air trapped by the rising cloud turrets) on the temperature, moisture and momentum properties of the cloud are progressively spread through the cloud. This general mixing process is a significant factor which is not taken into account in the PARCEL METHOD and SLICE METHOD of deducing the kinetic energy available in an atmosphere in which there is static instability. Entrainment is thought to influence the growth of droplets in small cumuliform clouds.

entropy. If, in a reversible thermodynamic process, a substance absorbs a quantity of heat dQ at absolute temperature T, the ratio dQ/T represents the increase of entropy of the substance. Entropy per unit mass is normally measured in joules per kilogram per kelvin and has the dimensions $L^2 T^{-2} \theta^{-1}$.

Entropy is a function of the pressure, volume, and temperature of the substance but requires for its evaluation the arbitrary choice of a state of zero entropy. Interest in the atmosphere is mainly confined to the changes of entropy to which air is subject in the course of a specified process. Of chief interest is the ADIABATIC process in which no heat is supplied to, or withdrawn from, the air ($\mathrm{d}Q = 0$) and in which, therefore, the entropy (S) remains constant; such a process is called 'isentropic' and is characterized by a constant POTENTIAL TEMPERATURE (θ) since S and θ are related by

$$S = c_p \log \theta + \text{constant}.$$

Eole.   See CONSTANT-LEVEL BALLOON.

equation of continuity.   See CONTINUITY, EQUATION OF.

equation of state.   See GAS EQUATION.

equation of time.   See TIME.

equations of motion.   See MOTION, EQUATIONS OF.

equator.   The earth's equator is that GREAT CIRCLE whose plane is perpendicular to the earth's axis of rotation (polar axis).

equatorial air.   The AIR MASS participating in the equatorial low-latitude circulation. Originating in the subtropical anticyclones, the air generally becomes moist and unstable when it stagnates in the DOLDRUMS region. See also TRADE WINDS.

equatorial bulge.   The slight bulging of the EARTH in low relative to high latitudes. The earth's equatorial diameter is about 43 km greater than its polar diameter. As a result of the bulging the force of GRAVITY varies with latitude.

equatorial stratospheric waves.   Internal gravity waves on the planetary scale which are affected by the earth's rotation as well as buoyancy and are often referred to as 'inertia-gravity oscillations'. There is an extensive mathematical theory of such waves; introductory accounts are provided by Holton [9] and by Haltiner and Williams [23]. The theory shows that planetary-scale waves are unable to propagate energy vertically unless their frequency exceeds the CORIOLIS PARAMETER; because planetary waves have periods of several days they are therefore vertically 'trapped' except comparatively near the equator.

The two most important types are the eastward-travelling Kelvin wave and the westward-travelling mixed Rossby-gravity wave. The horizontal distributions of pressure and velocity characteristic of these waves are shown in Figure 23. The Kelvin wave has distributions of pressure and zonal velocity that are symmetric about the equator and has no meridional velocity component. (Originally, the term 'Kelvin wave'

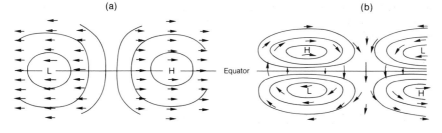

FIGURE 23.   Velocity and pressure distribution in the horizontal plane for (a) Kelvin waves, and (b) mixed Rossby-gravity waves. (After Matsuno [24].)

105

was applied to the propagation of an oceanic disturbance parallel to a shelving coastline; in the atmosphere, the part of the coastline is played by the equator where the Coriolis parameter changes sign.) The mixed Rossby-gravity wave, on the other hand, has pressure and zonal velocity distributions that are antisymmetric about the equator, and has a symmetric distribution of meridional velocity. The dynamics of these waves may be studied by the PERTURBATION METHOD. They have been identified in observational data from the equatorial stratosphere, and are probably excited by the large-scale pattern of convective heating in the equatorial troposphere. They transmit energy upwards through the stratosphere with the appropriate GROUP VELOCITY although the vertical phase velocity is in the opposite direction. See also QUASI-BIENNIAL OSCILLATION.

equatorial trough.   The shallow TROUGH of low pressure, generally situated on or near the equator, marking the convergence zone of air which moves equatorwards from the subtropical anticyclones of either hemisphere. The trough over the oceans lies in the belt of the DOLDRUMS and has a north and south movement which follows the sun with a time-lag of 1 or 2 months. The trough is not, however, a permanent feature of the synoptic chart in all longitudes at all times of the year; in particular, it is generally absent over most land areas in the northern hemisphere summer; there is a trough, however, over the southern Sahara and one over northern India.

equatorial westerlies.   The north-east or south-east TRADE WINDS are deflected on crossing the equator and acquire a westerly component to become the 'equatorial westerlies' when the INTERTROPICAL CONVERGENCE ZONE is farther than some 5° from the equator.

Over the eastern Indian Ocean and the Malaysian region westerly winds are normally found near the equator throughout the year. These are called equatorial westerlies although the turning of the wind usually begins before crossing, or without crossing, the equator.

equinoctial gales.   The implication contained in this term, which is in fairly wide popular use, that GALES are either more frequent or more severe near the EQUINOXES than at other times, is not supported by observations. In all parts of the British Isles, for example, the peak frequency of moderate or severe gales is near the winter solstice and the minimum frequency near the summer solstice.

equinox.   One of two periods of the year, which occur on about 21 March (spring or vernal equinox) and about 22 September (autumnal equinox), when the astronomical day and night are equal, each lasting 12 hours. (The sun is visible by refraction for a little longer than the duration of the astronomical day.)

At an equinox, when the sun is said to be 'on the equator', time of sunrise (or of sunset) is the same all along a MERIDIAN. Apart from the small refraction effect, the sun then rises exactly in the east and sets exactly in the west all over the world.

equiscalar surface.   In meteorology, a surface along which a specified scalar quantity, e.g. pressure or temperature, is constant.

equivalent constant wind.   An alternative for BALLISTIC WIND.

equivalent head wind.   That uniform wind which, directed at all points along the track of an aircraft, results in the same average ground speed as that actually attained. A positive head wind is such as to oppose the flight of an aircraft. A negative head wind is a 'tail wind', i.e. ground speed in excess of air speed. The term 'equivalent tail wind' is also used.

equivalent potential temperature. See EQUIVALENT TEMPERATURE.

equivalent temperature. The (isobaric) equivalent temperature (T_e) of a moist air sample is the temperature that would be attained on the assumption of condensation at constant pressure of all the water vapour in the sample, all the latent heat released in the condensation being used to raise the temperature of the sample. T_e is given by the equation

$$T_e = T + Lr/c_p$$

where T is the dry-bulb temperature and r the mixing ratio of the sample, L the latent heat of condensation, and c_p the specific heat at constant pressure. Where r is expressed in grams per kilogram,

$$T_e \approx T + 2.5r.$$

The 'equivalent potential temperature' (θ_e) is found on an AEROLOGICAL DIAGRAM by progressing along the dry adiabatic line from T_e to the 1000 mb level. See also PSEUDO-EQUIVALENT TEMPERATURE.

erg. The unit of work or energy in the C.G.S. SYSTEM of units. It is the work done by a force of 1 DYNE in moving its point of application 1 centimetre in the direction of the force. The dimensions are ML^2T^{-2}.

$$1 \text{ erg} = 10^{-7} \text{ JOULE.}$$

error. An error of observation is the departure of a measured quantity from its true value. Such an error is in general partly 'systematic' and partly casual or 'random' in nature. The systematic component which may arise, for example, through errors of the scale or zero setting of the instrument, or from the PERSONAL EQUATION of the observer, can often be recognized (and corrected) by comparing averaged measurements made with the instrument in question with similar measurements made with an instrument of known accuracy, while the random component of error can be reduced, in proportion to the square root of the number of observations, by averaging several independent measurements of the same quantity.

error function. See NORMAL (FREQUENCY) DISTRIBUTION.

Ertel potential vorticity. See POTENTIAL VORTICITY.

escape velocity. The minimum velocity which a molecular or atomic gas particle must attain in order that it may escape from the gravitational field of a specified planet and so from the planetary atmosphere. Such escape can take place only from the outer fringe (EXOSPHERE) of the atmosphere, where the particle MEAN FREE PATH is large and the probability of its collision with another particle correspondingly small.
The escape velocity, v (km s^{-1}), is given by

$$v = \left(\frac{2GM}{a}\right)^{1/2}$$

where G is the constant of gravitation $= 6.67 \times 10^{-11}$,
   M the mass (kg) of planet and
   a the distance (m) of level of escape from centre of planet.
   For the earth, $v \approx 11.2$ km s^{-1}; for the moon, $v \approx 2.4$ km s^{-1}.

etesian winds. A Greek term for the winds which blow at times in summer (May to September) from a direction between north-east and north-west in the eastern Mediterranean, more especially in the Aegean Sea. The winds are termed 'meltemi' in Turkey.

Eulerian change. The time rate of change of an element at a fixed point in an Eulerian system of coordinates, i.e. the 'local change', designated $\partial/\partial t$. It is to be contrasted with the LAGRANGIAN CHANGE to which it is related by the velocity components and the gradients of the elements in the various component directions.

Eulerian wind. That class of winds, in the classification of H. Jeffreys, in which the earth's rotational term (involving the CORIOLIS FORCE) and the frictional term are unimportant relative to the acceleration term. The EQUATION OF MOTION reduces to

$$\frac{d\mathbf{V}}{dt} = -\frac{1}{\rho}\nabla_H P$$

where $\nabla_H P$ is the horizontal gradient of pressure.

The CYCLOSTROPHIC WIND of the tornado or tropical cyclone is an example of an Eulerian wind.

eustasy, glacial. The release or absorption of water in ice-caps, with consequent rise or fall of mean SEA LEVEL.

evaporation. In meteorology, the change of liquid water or ice into water vapour. In certain usages the term signifies only the liquid to vapour phase change, as distinct from SUBLIMATION which signifies the solid to vapour phase change. The rate of evaporation is controlled by the water and energy (mainly solar radiation) supplies and by the ability of the air to take up more water.

Since large amounts of energy are required to effect the above change of state, the evaporation that proceeds continuously from the earth's free water surfaces, soil, snow and ice fields, and vegetation, and from ice and liquid water within the atmosphere, is a fundamental item in the energy balance of the earth–atmosphere system. Evaporation also plays a basic role in the earth HYDROLOGICAL CYCLE and HYDROLOGICAL BALANCE.

The interpretation of direct measurements of evaporation presents certain problems. It is difficult to reconcile the results obtained from the different forms of EVAPORIMETER, or to relate them precisely to evaporation which occurs from a free natural water surface. Measurements of the changes in weight of a sample of soil are also used to a limited extent as a measure of evaporation, and conform better to natural conditions, but this type of apparatus presents many experimental difficulties.

Indirect measurements often use the relationship: evaporation equals measured RAINFALL minus measured RUNOFF. In such cases the water storage by the soil is either experimentally held constant, or is assumed, by the use of long-period mean values, to be constant.

Indirect assessment of evaporation can also be made using a theoretical formula based on incoming and outgoing radiation, wind, and humidity conditions. Work on these lines has been done by H.L. Penman, C.W. Thornthwaite and others. Calculations made in this way are said to give the potential evaporation.

The extension of this work into the realms of agriculture and horticulture has led to the concept of potential TRANSPIRATION and EVAPOTRANSPIRATION.

evaporation fog. FOG which is formed by evaporation of relatively warm water into cool air. Examples are ARCTIC SEA SMOKE and FRONTAL FOG.

evaporimeter.   An instrument for determining the rate of EVAPORATION of water into the atmosphere. Evaporimeters fall into two classes:

(i)   those employing a free water surface, and
(ii) those in which evaporation takes place from a continuously wetted porous surface of blotting-paper, fabric or ceramic material.

Class (i) was exemplified by the Meteorological Office standard evaporation tank (no longer in general use). It consisted of a square tank with sides 1.83 m long and 610 mm deep, and was constructed of wrought-steel plates, overlapped and rivetted together. The tank was sunk into the soil with its top 76 mm above the surrounding ground. The evaporation, expressed in millimetres and tenths, was deduced from daily readings of level made with a micrometer gauge, allowance being made for rainfall. The level of the water was kept at 50 to 75 mm below the top of the tank by adding or removing water as required. Evaporimeters of class (ii) have been designed by Piche, Livingston, Owens and others, but they have not come into general use in this country as meteorological instruments.

evapotranspiration.   The combined processes of EVAPORATION from the earth's surface and TRANSPIRATION from vegetation. 'Potential evapotranspiration' is the addition of water vapour to the atmosphere which would take place by these processes from a surface covered by green vegetation if there were no lack of available water.
   The meteorological variables used in the estimation of potential evapotranspiration are air temperature, solar radiation or duration of sunshine, vapour pressure, wind speed and length of day. Mean annual values of estimated evapotranspiration range from about 37 to 43 cm in northern Scotland to 50 to 55 cm in southern England with, however, substantial regional and local variations.
   Evapotranspiration over a particular period may also be estimated from a formula which contains the above variables. It is then possible, by allowing for the rainfall of the period, to estimate the soil moisture deficit and assess the amount of artificial irrigation which may be required to keep a particular crop growing in the most productive manner. Such methods are now widely used by farmers and horticulturists.

exchange coefficient.   A general term for the coefficients of VISCOSITY (momentum), CONDUCTIVITY (heat), and DIFFUSIVITY (matter), which are defined, on analogy with molecular exchange processes, in respect of the vertical transfer effected by atmospheric eddies. See also K-THEORY.

Executive Committee.   The Executive Committee of the WORLD METEOROLOGICAL ORGANIZATION is the executive body of the Organization. Its primary functions include implementing decisions taken by Members, administering the finances, considering and taking action on resolutions and recommendations of Regional Associations and Technical Commissions, and studying and making recommendations on any matter affecting international meteorology and the operation of Meteorological Services. All Members act as representatives of the Organization and not as representatives of particular Member States or Territories. The Executive Committee normally meets at least once a year.

Exner function.   Denoted by Π and defined by

$$\Pi = (p/p_0)^\kappa$$

where p_0 is the pressure at the ground, sea level, or other standard height. (Some writers include a multiplicative factor (c_p) on the right-hand side.) Use of Π instead of p

is sometimes made in the mathematical treatment of atmospheric dynamics when it provides a convenient method of accounting for compressibility; the role of density is taken by (potential temperature)$^{-1}$, and if terms involving the reciprocal of the speed of sound are neglected the equations become identical with those for an incompressible fluid. See, for example, the paper by Tapp and White [25].

exosphere. The outermost 'fringe region' of a planetary atmosphere — in particular, that of the earth — in which the gas density is very small, MEAN FREE PATH large, and COLLISION FREQUENCY so small that particles moving upwards with sufficient velocity may escape from the atmosphere. For the earth, the exosphere is considered to extend upwards from about 700 km. See ESCAPE VELOCITY and ATMOSPHERE.

expansion. An increase in size of a sample of material; such an increase may be due to heat, or to the release of mechanical strain, or the absorption of moisture, or some other physical or chemical change. The 'size' may be a length, an area, or a volume.

expert system. A computer program which, within some specified field, simulates the performance of a human expert. It does much more than carry out a repetitive task, and can in fact assess evidence, exercise judgement, and make decisions. Some more elaborate expert systems can 'learn' from their successes and failures. In the general field of weather forecasting, expert systems are being applied to special areas such as NOWCASTING where the human forecaster finds it difficult or impossible to keep up with all the new information that is being made continuously available.

exponential atmosphere. An atmosphere in which the pressure scale-height (h) and the density scale-height (H) are constant and equal, so that the pressure (p) and density (ρ) decrease exponentially with height (z). In such an atmosphere

$$p_z/p_0 = \rho_z/\rho_0 = e^{-z/h} = e^{-z/H}$$

where subscript '0' denotes a reference level. See SCALE HEIGHT.

exposure. In meteorology, the method of presentation of an instrument to that element which it is designed to measure or record, or the situation of the station with regard to the phenomena to be observed. If meteorological observations are to be of full value, attention must be paid to the exposure of the instruments. Details are to be found in the *Observer's handbook* [1, pp. 183–199]. Uniformity of exposure is of the greatest importance, and for that reason the pattern of the THERMOMETER SCREEN has been standardized in most countries, while in the British Isles a standard height of 1 ft (30 cm) above ground for the rim of the RAIN-GAUGE has been fixed.

It is important, too, that the sites of the thermometer screen and rain-gauge should not be unduly shut in; on the other hand, a very open exposure, as on a bare moor, is undesirable for a rain-gauge, as is also a position on a slope, a roof, or near a steep bank. In these cases the catch is reduced by the effect of wind eddies due to the obstruction of the gauge itself. An ideal exposure for a SUNSHINE RECORDER requires that there should be no horizon obstruction in the direct line from sun to recorder at any time of the year.

The question of the exposure of ANEMOMETERS is one of great difficulty. The effect of the ground on a uniform current of air blowing above it is to reduce the velocity by an amount which increases as the ground is approached and, at the same time, to introduce into the motion unsteadiness which is manifested by the creation of eddies in the air. The motion is then said to be turbulent (see TURBULENCE). The record from an ANEMOGRAPH erected in a turbulent wind shows a large number of gusts and lulls corresponding to various parts of the eddy motion. The degree of GUSTINESS recorded by an anemograph is thus an indication of the exposure, the breadth of trace increasing

110

with roughness of terrain. The ideal exposure is at the top of a mast 10 m (33 ft) high erected on a flat plain with no obstructions.

exsiccation. Drying by the draining away or driving away of moisture. The term implies some change, frequently the result of human agency, which decreases the quantity of moisture available without any appreciable change in the average rainfall. It is used in contrast with DESICCATION, which implies an actual drying up due to a change of climate. Examples of exsiccation are the washing away of the soil due to the cutting down of forests, with consequent conversion of a fertile region into the semblance of a desert, the advance of sand dunes across cultivated ground, and the draining of swampy ground.

extinction coefficient. A term synonymous with 'attenuation coefficient' (see ATTENUATION) but often reserved as a measure of the combined effects of ABSORPTION and SCATTERING of wavelengths within the VISIBLE SPECTRUM. The extinction coefficient has the dimension L^{-1}; its value in the atmosphere varies from about 10 km^{-1} in thick fog to 0.01 km^{-1} in air of very good visibility.

In relation to VISIBILITY, scattering is much the more important of the two extinction processes.

extremes. The highest and lowest values of meteorological elements in a specified period. The following definitions illustrate, with reference to maximum temperature, the nomenclature employed:

(i)   Daily maximum temperature: highest temperature reached between two fixed times 24 hours apart.

(ii)   Mean daily maximum temperature: mean of the 'daily maximum temperatures' observed during a given calendar month, either in a specified year or over a specified period of years.

(iii)   Monthly maximum temperature: highest of the 'daily maximum temperatures' observed in the course of a given calendar month in a specified year.

(iv)   Mean monthly maximum temperature: mean of the 'monthly maximum temperatures' observed during a given calendar month over a specified period of years.

(v)   Absolute monthly maximum temperature: highest of the 'monthly maximum temperatures' observed during a given calendar month over a specified period of years.

(vi)   Annual maximum temperature: highest of the 'daily maximum temperatures' observed during a given calendar year.

(vii)   Mean annual maximum temperature: mean of the 'annual maximum temperatures' observed over a specified period of years.

(viii) Absolute maximum temperature: highest temperature observed during the whole period of observation ('absolute extreme').

extreme value analysis. A branch of STATISTICS that deals with the estimation of the greatest (or smallest) value of some physical quantity likely to occur in a specified period which is usually longer than the period for which observations are available.

Both the theory and practice of extreme value analysis are difficult and uncertain, and any conclusions arrived at by the statistician must be regarded as tentative and treated with caution. A major difficulty is the frequent presence of more than one distinct statistical population in the data. For example, winds over a subtropical island may consist mainly of the trades, but will also occasionally be caused by tropical storms; in the British Isles, the strongest winds and gusts are due to tornadoes and will need separate treatment from the gales due to depressions.

Two characteristic parameters sought by extreme value analysis are defined as follows:

(i) The return period (T_v) of a value (V_1) is defined as the mean (or expected) number of years between successive occurrences of a value equal to or greater than V_1.

(ii) The once-in-N-year value (V_N) is defined as the extreme value expected to occur in a randomly chosen period of N consecutive years.

Because of the skewness of the distribution of extreme values, the return period of the once-in-N-year value is not in general equal to N years.

Under certain strict assumptions, not always true in practice, it is possible to derive three different mathematical distributions of extreme values known as Fisher–Tippett distributions; the difference arises from the three ways in which the tails of the parent populations can behave, i.e. whether the population is bounded below only, bounded above only, or is unbounded above and below.

eye of storm.   The central region of an intense TROPICAL CYCLONE, usually some 10–50 km in diameter and fairly symmetrical, though subject to time fluctuations. The main features are absence of rain, small horizontal pressure gradient, light winds, high and turbulent sea, and layered clouds, often well broken.

eye of the wind.   A nautical expression indicating the direction from which the wind blows.

F

facsimile transmission. A system of telegraphy providing reproduction, in the form of fixed images (photographic or otherwise), of the form and possibly of the depth of tone, or of the colours, of an original document, whether hand written, printed, or pictorial.

Fahrenheit scale. A scale of temperature introduced about 1709 by the German physicist Fahrenheit, who was the first to use mercury as the thermometric substance. Primary fixed points were the temperature of a mixture of common salt and ice and the temperature of the human body; with reference to these, the freezing-point of water was marked 32° and the boiling-point of water was marked 212°. See TEMPERATURE SCALES.

fair-weather cumulus. A common alternative for cumulus HUMILIS.

falling sphere. A device for measuring air density and winds in the stratosphere and mesosphere. The acceleration of a rigid sphere is measured during its descent after ejection from a METEOROLOGICAL ROCKET. The density of the air is a function of the acceleration and drag coefficient of the sphere.

Two distinct forms exist. The first form consists of an inextensible balloon, 1–2 m in diameter. This is inflated to a pressure considerably in excess of the ambient pressure. The acceleration of the sphere is measured by tracking it very accurately with a high-precision radar, or optically with a high-precision theodolite. The second form is a rigid instrumented sphere, about 15 cm in diameter, which contains an accelerometer. The acceleration of the sphere is measured directly and is transmitted by radio to a ground station.

Air density measurements can be made by this method from a height of 100 km downwards. If the sphere is tracked by radar its horizontal drift can be used to give winds below a height of 65 km.

fall-out. A term applied both to the process of deposition of solid material on the earth's surface and to the deposited material itself. It may be used in such a sense as to signify only 'dry deposition' (mainly the result of gravitational settling); in such a sense it is used in contrast to the term WASHOUT. The term fall-out is, however, used mainly in respect of the radioactive debris associated with a nuclear explosion. The process of washout is then often important in the manner of deposition of the material, especially when the fall-out is other than 'close-in'. See RADIOACTIVE FALL-OUT.

fallstreak. A term often used for the supplementary cloud feature VIRGA.

false cirrus. A popular expression for cirrus SPISSATUS or anvil cirrus — see ANVIL CLOUD.

fata morgana. A rare and complex form of MIRAGE in which horizontal and vertical distortion, inversion and elevation of objects occur in changing patterns. The phenomenon occurs over a water surface and is produced by the superposition of several layers of air of different REFRACTIVE INDEX.

113

Ferrel cell. A mid-latitude mean atmospheric circulation cell proposed by Ferrel in the 19th century in which air flows poleward and eastward near the surface and equatorward and westward aloft. This disagrees with reality. However, the term is sometimes used nowadays to describe any mid-latitude circulation identifiable in mean meridional wind patterns.

Ferrel's law. That freely moving objects swing to the right in the northern hemisphere and to the left in the southern hemisphere. The term is little used by meteorologists. See CORIOLIS ACCELERATION.

fetch. The fetch of an air-stream is the length of its traverse across a sea or ocean area.

fibratus (fib). A CLOUD SPECIES. (Latin for fibrous.)
'Detached clouds or a thin cloud veil, consisting of nearly straight or more or less irregularly curved filaments which do not terminate in hooks or tufts.
This term applies mainly to CIRRUS and CIRROSTRATUS' [2, p. 18]. See also CLOUD CLASSIFICATION.

Fickian diffusion. See DIFFUSIVITY.

FIDO. The term, abbreviated from 'Fog Investigation and Dispersal Operations', signifying the system, developed in Great Britain during the Second World War, of effecting temporary and local clearance of fog by the burning of petrol at intervals alongside an airfield runway.

fiducial point. A reference point of an instrument or scale, as, for example, the ivory tip of the FORTIN BAROMETER to which the mercury in the cistern must be brought before a reading is taken.

fiducial temperature. An alternative for STANDARD TEMPERATURE.

field capacity. The mass of water retained by a previously saturated soil when free drainage has ceased, expressed as a percentage of the mass of dry soil. It varies from about 7 per cent in light sand to about 60 per cent in heavy clay and corresponds to a CAPILLARY POTENTIAL of about 500 cm of water or a pF (see ATMOSPHERIC CHEMISTRY) of 2.7. See SOIL MOISTURE.

field mill. A type of ELECTROMETER employed, in various forms, more especially in field measurement or recording of atmospheric POTENTIAL GRADIENT. The charge generated in a conductor which is alternately exposed to and sheltered from the atmospheric electric field is conveyed to a meter; d.c. or a.c. amplification is often used.

filling. See DEEPENING.

filtering. The processing of values of a variable to emphasize certain patterns of variation while suppressing others. Filtering of the basic hydrodynamic equations of motion may, for example, involve the assumption of QUASI-GEOSTROPHIC MOTION, while filtering of TIME SERIES, or of measurements at equal intervals in space, is carried out by means of weighted MOVING AVERAGES. The object is generally to isolate oscillations lying within a certain waveband, or range of frequencies, while reducing NOISE and oscillations in other wavebands.
The filtering of a time series is carried out as follows: if $x_1, x_2, ..., x_N$ is a time series, and $W_0, W_1, W_2, ..., W_K$ are the constants of a symmetrical filter of order K, then corresponding to any term x_n we find the value of

$$y_n = \sum_{j=0}^{K} \tfrac{1}{2} W_j (x_{n-j} + x_{n+j}).$$

For $K \leq n \leq N-K$ the terms y_n, y_{n+1}, y_{n+2}, etc., can be found from the x series.

If

$$x_n = A \cos(2\pi fn + B)$$

then

$$x_n = A \cos(2\pi fn + B). \sum_{j=0}^{K} W_j \cos 2\pi fj,$$

so the effect of filtering a harmonic wave by a symmetric filter is to leave the frequency and phase unchanged, and to multiply the amplitude by the magnification factor

$$M(f) = \sum_{j=0}^{K} W_j \cos 2\pi fj.$$

The problem of designing a filter to pass a particular waveband thus reduces to that of choosing the weights $W_0, W_1, ..., W_K$ to make $M(f)$ take values near unity in a certain range of frequencies and values near zero for other frequencies.

finite-difference method. A method of approximation to space or time derivatives of a mathematical function, widely used in meteorology, in which the derivative is represented as the difference between values of the function at two points in space or time. The appropriate separation between the points depends on the scale of the phenomenon being studied.

fireball. This term is used to signify BALL LIGHTNING. It is also used of the larger type of METEOR, with the brightness of a first magnitude star or greater.

The rapidly expanding, white-hot ball of gas which is produced on explosion of a nuclear weapon is also termed a 'fireball'; the term is applied up to the time at which the volume of gas becomes, through adiabatic and radiational cooling, no longer incandescent.

firn. A German word meaning old snow which is in process of being transformed into GLACIER ice. The word is also used to denote an accumulation area of old snow above a glacier and is synonymous with the French word 'névé'. There is no corresponding word in English.

firn wind. An alternative for GLACIER WIND.

fitness figures (numbers). A scale of figures, developed in Great Britain during the Second World War, used as a measure of the meteorological fitness of a particular airfield for the landing of aircraft. The scale is governed mainly by visibility and observed cloud base in relation to the maximum height of obstructions near the airfield.

F-layer. A two-part layer of the IONOSPHERE. The lower (F_1) part is best observed in daytime in summer and has a maximum electron concentration, at about 160 km, of the order 3×10^5 cm^{-3} with overhead sun. The higher (F_2) part, also called the *Appleton layer*, has a maximum electron concentration at a height which is very variable in space and time within the approximate range 220 to 400 km. The F_2-layer is, in terms of CHAPMAN LAYER theory, highly anomalous and is also the seat of 'ionospheric storm' phenomena.

flight level. A surface of constant atmospheric pressure which is related to a specific pressure datum (1013.25 mb) and is separated from other such surfaces by specific pressure intervals (WMO and ICAO definition).

It is conventionally given a numerical value to the nearest 1000 feet in units of 100 feet in accordance with the structure of the ICAO STANDARD ATMOSPHERE; for example, the 500 mb level is written as FL 180, the ICAO standard height being 18 289 ft.

float barograph. A (seldom-used) type of recording SIPHON BAROMETER. See BAROGRAPH.

floccus (flo). A CLOUD SPECIES. (Latin for tuft.)

'A species in which each cloud unit is a small tuft with a cumuliform appearance, the lower part of which is more or less ragged and often accompanied by VIRGA.

This term applies to CIRRUS, CIRROCUMULUS and ALTOCUMULUS'[2, p. 18]. See also CLOUD CLASSIFICATION.

flux density. The rate of transport (flux) of a specified quantity (e.g. RADIATION) across unit area of a surface.

fluxplate. An instrument of thermo-electric design for recording heat flux electrically.

foehn. An alternative for FÖHN.

fog. Obscurity in the surface layers of the atmosphere, which is caused by a suspension of water droplets, with or without smoke particles, and which is defined, by international agreement, as being associated with VISIBILITY less than 1 km. In British practice, use of the term in forecasts for the general public relates to visibility less than 200 yards (180 m). 'Ice fog' is an obscurity produced by a suspension of numerous minute ice crystals.

Although fogs which are entirely composed of smoke or dust particles do occur, the more persistent and thick fogs of industrial areas contain also water droplets. The term SMOG is sometimes used of such fogs. The frequent occurrence of fog in industrial areas is due in large measure to the plentiful supply of hygroscopic particles which are able to act as condensation NUCLEI when the relative humidity is less than 100 per cent.

Fogs which are composed entirely or mainly of water droplets are generally classified according to the physical process which produces saturation or near-saturation of the air; examples are RADIATION FOG, ADVECTION FOG, UPSLOPE FOG, and EVAPORATION FOG, this last including FRONTAL FOG and ARCTIC SEA SMOKE which is also known as 'steam fog' and has various other synonyms. Natural fogs are frequently the result of the combined action of two or more such physical processes.

fogbow. A white RAINBOW of about 40° radius seen opposite the sun in fog. Its outer margin has a reddish tinge, its inner a bluish tinge, but the middle of the band is quite white. A supernumerary bow is sometimes seen inside the first bow and with the colours reversed. The bows are produced in the same way as the ordinary rainbow but owing to the smallness of the drops, the diameters of which are about 50 μm, the colours overlap and the bow appears white.

fog-point. The air (screen) temperature at which FOG forms.

fog precipitation. Fog precipitation, sometimes also termed 'fog drip', signifies the precipitation of liquid water from non-rain-bearing clouds due to the interception of the cloud particles by trees and other vegetation.

The relatively few measurements of this phenomenon suggest that it is common in places which have a high frequency of orographic cloud and suitable vegetation. In measurements over the plateau of Table Mountain, Cape Town, for example, one day in three yielded fog precipitation in the absence of rainfall measured in the normal way, and the total amount of fog precipitation measured in the year was nearly twice that of the measured rainfall.

föhn. A warm dry wind that occurs to leeward of a range of mountains. The name originated in the Alps but is now used as a general term for this type of wind.

Conspicuous föhn winds tend to occur in conditions stable enough for LEE WAVES, i.e. when the atmosphere is stable and the POTENTIAL TEMPERATURE increases markedly with height. In such conditions the air reaching the ground on the lee side may have descended from levels well above the mountain tops and, having suffered adiabatic compression, will be much warmer than would otherwise be expected.

The classical explanation of the warmth of the föhn was that it was due to the condensation and loss through precipitation of water vapour caused by forced ascent on the windward side. The WET-BULB POTENTIAL TEMPERATURE would remain constant, but the air to leeward, having lost much of its vapour, would have a much higher dry-bulb temperature. This process is now thought to play at most a subsidiary role.

föhn wall. Associated with the FÖHN effect, a mass of precipitating clouds often forms over the windward slopes of the hills. Cloud continues some way down the lee slope but evaporates in the descending current, terminating along a line parallel to the main ridge of hills.

foot-candle. An obsolescent unit of illumination, equal to 1 LUMEN per square foot. See LUX.

force. That which alters or tends to alter the state of rest or motion of a body. Force is a vector quantity with both magnitude and direction and has dimensions MLT^{-2}. In meteorology, 'specific force' (force per unit mass), with dimensions LT^{-2}, is used in place of total force. See also ACCELERATION.

forecast. The term, first applied in meteorology by Admiral FitzRoy, which signifies a statement of anticipated (meteorological) conditions for a specified place (or area, route, etc.) and period of time.

A threefold classification of forecasts, in terms of the period covered, is recognized:

(i) 'Short-period' forecast for part or whole of a 24-hour period, often with a 'further outlook' for the following 24 hours.
(ii) 'Medium-range' forecast for some 2 to 5 days, and
(iii) 'Long-range' forecast for a period longer than about 5 days ahead, for example a month or season.

The short-period forecast generally contains information concerning wind velocity, weather (state of sky, precipitation, fog, frost, thunder, etc.) and temperature (relative to seasonal normal); for special purposes, additional information is given — for example, upper winds for aviation. In medium-range and more especially long-range forecasts the information given is in more general terms and is often confined to precipitation and temperature.

In short-period forecasting, the methods of SYNOPTIC METEOROLOGY are normally used to anticipate changes in surface pressure distribution and positions of fronts, on which, together with an appeal to physical reasoning and to precedent, the forecast is based. Alternatively, the changes in pressure distribution are derived by numerical methods (see NUMERICAL WEATHER PREDICTION). Purely statistical methods of short-period forecasting are sometimes applied. In medium-range forecasting, the synoptic method predominates, with concentration of attention on the dominant circulation features. In long-range forecasting, significant but limited success has been achieved with synoptic methods in which time-averaging of the pressure and circulation patterns is employed, and with a synoptic ANALOGUE method.

forked lightning. LIGHTNING in which many luminous branches from the main discharge channel are visible. Such branching occurs in response to local variations of SPACE CHARGE close to the main channel.

Fortin barometer. A form of mercury BAROMETER, the zero of whose scale is fixed by a pointer inside the cistern which is made partly of leather. By adjustments of a screw, the level of mercury in the cistern is brought to the scale zero ('FIDUCIAL POINT') before each reading is taken.

Foucault pendulum. A pendulum designed by J. Foucault in Paris in 1851 to give experimental proof of the earth's rotation. The direction of swing of an oscillating large iron ball, suspended just above a tray of fine sand by a wire over 60 m in length, was observed to change clockwise (northern hemisphere) at a rate corresponding to $15°$ per sidereal hour $\times \sin \phi$. The observation is consistent with an apparent deviating force arising from the rotation of the earth about a polar axis.

Fourier analysis. An alternative for HARMONIC ANALYSIS.

Fourier series. A representation of any function (f) of an independent variable x in terms of sines and cosines of multiples of that variable. Such a series was first developed by J. Fourier in 1822. In symbols:

$$f(x) + A_0 + A_1\sin x + A_2\sin 2x + \dots$$
$$+ B_1\cos x + B_2\cos 2x + \dots .$$

See HARMONIC ANALYSIS.

Fourier transform. A mathematical method of expressing an arbitrary function, $f(x)$, say, as a combination of an infinite number of harmonic oscillations, such that

$$f(x) = (2\pi)^{-\frac{1}{2}} \int_{-\infty}^{+\infty} g(t)e^{ixt}dt.$$

The function $g(t)$ is given by

$$g(t) = (2\pi)^{-\frac{1}{2}} \int_{-\infty}^{+\infty} f(x)e^{-ixt}dx.$$

The pair of functions $f(x)$ and $g(t)$ are said to be Fourier transforms of each other; i is $\sqrt{-1}$.

One may also consider real or imaginary parts of the above integrals, e.g.

$$f(x) = (2\pi)^{-\frac{1}{2}} \int_{0}^{\infty} g(t) \cos (xt)dt$$

and $$g(t) = (2\pi)^{-\frac{1}{2}} \int_{0}^{\infty} f(x) \cos (xt)dx.$$

foyer.   The place of origin of a group of ATMOSPHERICS.

f-plane approximation.   The assumption that the earth may be treated as flat for the purpose of numerical calculation with the Coriolis parameter (f) put equal to a constant (f_0) appropriate to the latitude under consideration.

f.p.s. system.   A system of units, seldom used in meteorology, which is based on the foot, the pound, and the second as FUNDAMENTAL UNITS. In this system the unit of force is the *poundal*, and the unit of work is the *foot-pound*.

fractional volume abundance.   A synonym for VOLUME FRACTION.

fractus (fra).   A CLOUD SPECIES. (Latin for broken.)
'Clouds in the form of irregular shreds, which have a clearly ragged appearance.
This term applies only to STRATUS and CUMULUS' [2, p. 19]. See also CLOUD CLASSIFICATION.

Fraunhofer lines.   See SUN.

frazil ice.   ICE which forms in spicules or small plates in rapidly flowing rivers, and at times in the sea, the movement of the water preventing the ice crystals from forming a solid sheet of ice. The formation has been best observed in the rivers of Canada; the word is from the French-Canadian *frasil*, meaning cinder, the frazil crystals being supposed to resemble cinders from a forge.

free atmosphere.   The atmosphere above the BOUNDARY LAYER, where the influence of surface drag and heat flux on air motion is assumed negligible.

free lift.   The free lift of a balloon is given by the excess load that would be required to make the balloon float at a constant level; it is the excess of the BUOYANCY force over the gross weight.

free period.   That period of vibration of a system, determined by its physical characteristics, which the system adopts when set in motion by the application of an external force which is then removed; the corresponding frequency of vibration is termed its 'natural frequency'. See ATMOSPHERIC TIDES, RESONANCE.

freeze, freezing.   With reference to the weather, the term 'freezing' is used when the temperature of the air is below 0 °C. Freezing conditions for an appreciable time over a widespread area are in the USA termed a 'freeze'.

freezing drizzle.   Supercooled water drops of drizzle which freeze on impact with the ground or other objects to form GLAZE. See SUPERCOOLING.

freezing fog.   Supercooled water drops of fog which freeze on impact with the ground or other objects to form RIME. See SUPERCOOLING.

freezing level.   Commonly the lowest height above mean sea level at which, for a given place and time, the air temperature is 0 °C; it is now generally termed the height above mean sea level, or if appropriate the PRESSURE ALTITUDE, of the 0 °C isotherm.
Over the British Isles the freezing level has average values of about 600 m in winter and 3000 m in summer with, however, large day-to-day variations from these values. The term 'melting level' is to be preferred from the physical point of view since water in the atmosphere is not necessarily frozen at temperatures below 0 °C — see SUPERCOOLING.

freezing nucleus. See NUCLEUS.

freezing-point. That constant temperature at which the solid and liquid forms of a given pure substance are in equilibrium at STANDARD ATMOSPHERIC PRESSURE. For pure-water substance the temperature is 0 °C and is termed the 'ice-point'. In meteorology, the term 'freezing-point' is often used to signify 'ice-point'.

In practice, a cooling liquid may not freeze at the freezing-point because of a pressure variation from standard atmospheric pressure, or the presence of impurities, or the phenomenon of SUPERCOOLING.

freezing rain. Supercooled water drops of rain which freeze on impact with the ground or other objects to form GLAZE. See SUPERCOOLING.

frequency. The statistical term for the number of occasions a variable takes a certain value or lies in a certain range of values.

In a wave vibration the frequency is the number of complete vibrations (cycles) per unit time. It is numerically equal to the velocity divided by the wavelength and is usually expressed in cycles per second, kilocycles per second, or megacycles. The unit of frequency is termed the HERTZ (Hz).

frequency distribution. A term used rather loosely to cover both a graph of the actual frequencies of occurrence of values of a statistical variable in a sample, and what is better called a PROBABILITY DISTRIBUTION. When the variable can take only discrete values, the frequency distribution must refer to these. When, however, the variable is continuous, and the data are to be analysed mathematically, or represented in a HISTOGRAM, it is nearly always desirable to group them into ranges of equal width, e.g. 0–1, 1–2, 2–3, etc. Ranges of unequal width are inconvenient mathematically, although an open range containing, say, all cases above a certain value may be used in tables. Data grouped in equal ranges can readily be processed to find their STANDARD DEVIATION and MEAN.

A frequency distribution is usually, when possible, given mathematical form as a *frequency function* which gives the frequency of a variate value x as a function of x or, for continuous variates, the frequency in an elemental range dx; the total frequency over the complete range of x is normally taken as unity. The frequency function may be regarded as the derivative of the DISTRIBUTION FUNCTION.

friction. The mechanical force of resistance which acts when there is relative motion of two bodies in contact, or of a body in contact with a medium, or of adjacent layers of a medium, or of adjacent media. Within a fluid, the friction that arises from molecular collisions is termed VISCOSITY.

In meteorology, the effects of friction are important in the flow of air over the earth's surface ('surface friction') and also when there is WIND SHEAR. Surface friction affects the wind velocity within the FRICTION LAYER and is important in all scales of motion up to and including that implied in the general circulation of the atmosphere, in which it plays a vital part in the overall momentum balance that is achieved. The effects of surface friction visible on the synoptic scale are a decrease of SURFACE WIND speed relative to that appropriate to the pressure gradient (GEOSTROPHIC WIND or GRADIENT WIND), and a 'frictional outflow' of surface air from higher to lower pressure. The magnitudes of these effects increase with the surface roughness and decrease with increasing height within the friction layer. Typical values for airflow over the ocean at surface wind level are a decrease of speed by about one-third and a 'cross-isobar' wind direction of about 10 to 20°.

Direct measurements of the surface friction effect at individual points have been made employing 'drag plates'. Estimates of the larger-scale effects have also been made using vertical wind profiles. See also EKMAN SPIRAL.

friction layer. The atmospheric layer, extending from the earth's surface to about 600 m (2000 ft) above ground, in which the influence of surface FRICTION on air motion is appreciable.

friction velocity. That reference velocity (u_*), employed in the study of fluid flow over a rough surface, which is defined by the equation

$$u_* = \sqrt{(\tau_0/\rho)}$$

where τ_0 is the surface drag per unit area and ρ is the fluid density.

Generally in meteorology, τ_0 is little different from τ, the REYNOLDS STRESS in the fluid, within a shallow layer near the surface.

u_* increases with roughness of surface and with mean wind speed ($\bar{u}$). In meteorology, u_* is of the general order ($\bar{u}/10$).

fringe region. See EXOSPHERE.

front. A term introduced into synoptic meteorology by Norwegian meteorologists in 1918. A 'frontal surface' is a sloping transition zone separating two air masses of different density and so of different temperature; a 'surface front' is the zone (usually represented on charts as a line) along which a frontal surface intersects the earth's surface.

Frontal surfaces have very gentle slopes, of about 1 in 100. Horizontal convergence and associated vertical motion are essential features of a well marked front; the upward motion results, especially within the warmer air mass, in the condensation and precipitation which are associated with a typical active front. Measurements in the free atmosphere show that the horizontal gradients of temperature within the separate air masses are not negligible (though, by definition, smaller than in the frontal zone), and that large local variations of temperature and humidity often occur near the edges of frontal cloud and precipitation.

A front necessarily lies in a trough of low pressure and is marked by discontinuities of wind velocity and, in general, of pressure tendency. The more important and extensive fronts, such as the POLAR FRONT, may be traced from the earth's surface to the tropopause.

frontal contour chart. A synoptic chart of the contours (usually expressed in pressure units) of a selected frontal surface, i.e. a plan view of the intersection of the frontal surface with selected isobaric surfaces.

frontal fog. FOG which forms at and near a FRONT. Such fog forms when raindrops, falling from relatively warm air above a frontal surface, evaporate into cooler air close to the earth's surface and cause it to become saturated.

frontogenesis. The development or marked intensification of a FRONT. This process — the intensification of the horizontal temperature gradient in a restricted zone — is effected mainly by horizontal CONFLUENCE and/or CONVERGENCE in conditions of suitably orientated isotherms.

frontolysis. The disappearance or marked weakening of a FRONT (converse of FRONTOGENESIS). This process is effected mainly by horizontal DIVERGENCE of air from the frontal zone, usually accompanied by SUBSIDENCE.

frost. Frost occurs when the temperature of the air in contact with the ground, or at thermometer-screen level, is below the freezing-point of water ('ground frost' or 'air frost', respectively). The term is also used of the icy deposits which may form on the ground and on objects in such temperature conditions (GLAZE, HOAR-FROST).

Since the sensation of cold depends not only on air temperature but also on the accompanying wind speed, the fourfold classification of frosts used in forecasts of this condition in the United Kingdom is varied with wind speed. Thus, frost is classified as 'slight', 'moderate', 'severe', or 'very severe' for screen temperature ranges of -0.1 to $-3.5\,°C$, -3.6 to $-6.4\,°C$, -6.5 to $-11.5\,°C$, or below $-11.5\,°C$, respectively, if the accompanying wind speed is less than 10 kn; and for screen temperature ranges of -0.1 to $-0.4\,°C$, -0.5 to $-2.4\,°C$, -2.5 to $-5.5\,°C$, or below $-5.5\,°C$, respectively, if the wind speed is 10 kn or more.

frost day.   A frost day is defined as a period of 24 hours ending at 0900 GMT (or at 2100 GMT, where observations are made at 2100 GMT) in which the minimum air temperatures in the screen is below $0\,°C$ ($32\,°F$). See also GROUND FROST.

frost heaving.   The uneven lifting and distortion of the ground close to the surface. It results from the expansion of water within the soil on freezing associated with the local formation of ice crystals, which accumulate into ice 'lenses'. The phenomenon may result in damage to road surfaces and loosening of the root hold of plants.

frost hollow.   A local hollow-shaped region in which, in suitable conditions, cold air accumulates by night as the result of katabatic flows. See KATABATIC WIND. Such regions are subject to a greater incidence of FROSTS, and to more severe frosts, than are the surrounding areas of non-concave shape.

frost-point.   The frost-point (T_f) of a moist air sample is that temperature to which the air must be cooled in order that it shall be saturated with respect to ice at its existing pressure and HUMIDITY MIXING RATIO.

T_f is that temperature for which the saturation VAPOUR PRESSURE with respect to ice (e'_i) is identical with the existing vapour pressure (e) of the air, i.e.

$$e + e'_i \text{ at } T_f.$$

Frost-point may be measured indirectly from wet-bulb and dry-bulb temperature readings with the aid of a humidity slide rule or humidity tables (see PSYCHROMETER), or directly with a FROST-POINT HYGROMETER.

frost-point hygrometer.   A development by G.M.B. Dobson and A.W. Brewer of the well known dew-point principle, for use in aircraft flying in air at sub-freezing temperatures. A suitable polished surface (see HYGROMETER) is provided by a hollow cylinder of anodized aluminium which is painted black and closed at the top; it is usually referred to as a 'thimble'. The thimble is cooled from below by a controlled flow of liquid air and is surrounded by a heating coil. Cooling and heating rates are adjusted till a deposit of ice on the surface is observed, either visually or photoelectrically, to remain constant; the temperature of the surface, measured by a resistance thermometer, is then the FROST-POINT of the air.

frost, protection against.   Various methods are practised of affording protection to crops in orchards, vineyards, etc., against damage by frost. Protective measures may be effective only on RADIATION NIGHTS characterized by calm or light wind conditions; no method is effective on occasions of frost in which there is substantial natural air movement.

The methods include: direct heating of the air near the ground; the production of a smoke screen over the crop (see SMUDGING); the flooding or sprinkling of crops, thus adding the thermal content of the water and increasing the effective specific heat of the soil; and the use of large fans designed to mix cold air near the surface with warmer air aloft.

frost smoke. An alternative for ARCTIC SEA SMOKE.

Froude number. In fluid flow, a non-dimensional parameter (Fr) defined by the relationship $Fr = \bar{u}/(\delta g)^{1/2}$, where $\bar{u}$ is the free stream velocity, δ the thickness of the BOUNDARY LAYER, and g the gravitational acceleration.
More useful in meteorology is the 'internal Froude number' defined by

$$Fr = \bar{u}/(\delta N)^{1/2}$$

where N is the BRUNT–VÄISÄLÄ FREQUENCY. (Occasionally the Froude number is defined as the square, or inverse square, of the quantities defined here.)

F-test. Also known as the 'variance-ratio test'. If two estimates of VARIANCE are available, from two independent samples based respectively on n_1 and n_2 DEGREES OF FREEDOM, then the CHANCE EXPECTATION of their ratio (arranged to have the larger variance in the numerator) may be found from the F-table published, for example, in Brooks and Carruthers [26, p. 384]. A value exceeding the chance expectation will be regarded with some confidence as implying that the samples have not been drawn from the same population.

fumigation. The bringing down to the ground of pollution, accumulated beneath a stable layer, owing to the onset of convective mixing. Usually, the pollution accumulates overnight and fumigation begins as the surface temperature rises during the morning.

fundamental units. The units of mass (M), length (L), time (T) and temperature (θ) on which the less fundamental, or 'derived' units, e.g. the units of pressure or viscosity, are based.

funnel cloud. The cloud formed at the core of a WATERSPOUT or TORNADO, sometimes extending right down to the earth's surface, attributed to the reduction of pressure at the centre of the vortex. Similar cloud formations are sometimes seen without a waterspout or tornado at the ground ('cloud pendants'). See also TUBA.

funnelling. The term 'funnelling' (sometimes 'canalization') is applied to the phenomenon in which the surface wind is constrained by topographical features to blow along a valley and is thereby increased in speed.

further outlook. A statement in brief and general terms appended to a detailed forecast and giving the conditions likely to be experienced in a period beyond that covered by the actual forecast.

fusion. An alternative for 'melting'. See MELTING-POINT.

G

Gaia. A term introduced by J.E. Lovelock [27] to convey the idea that the surface of the earth (including the oceans), the atmosphere, and the biosphere form a total system that has reached its present state by a process of natural selection dependent on the presence of living organisms. (The name is that of the Greek Earth Goddess.)

gale. A WIND with a mean speed in the range 34–40 kn (force 8 on the BEAUFORT SCALE of wind force, where it was originally described as 'fresh gale') and/or gusts reaching 43–51 kn, at a free exposure 10 m (33 ft) above ground. In general, a mean speed over a period of 10 consecutive minutes, as reported in synoptic code, is implied by the term 'gale'; where this is not intended, the specific threshold for the gusts is used.

While the term 'gale' applies strictly to the speed limits given above, and higher winds are referred to in other terms, e.g. severe gale, storm, etc., statistics of gales refer to the attainment of mean speeds of 34 kn or over. See also GALE, DAY OF.

gale, day of. A day on which the wind speed at the standard height of 10 m attains a mean value of 34 kn or more over any period of 10 consecutive minutes during the day. The average numbers of days per month with gale at coastal anemometer stations in Great Britain varies fairly smoothly through the year with a maximum in midwinter and a minimum in midsummer.

Galerkin methods. A class of numerical methods for solving partial differential equations in which the dependent variables are represented, not as values at points over a grid, but as sums of functions that have a prescribed spatial structure, the coefficient associated with each function normally being a function of time. They include spectral methods and finite element methods. Spectral methods employ orthogonal functions, e.g. FOURIER SERIES or surface harmonics, and finite element methods employ functions that are zero except in a limited region where they are low-order polynomials.

gale warning. The Meteorological Office issues notice of the probability of gales (mean winds of at least force 8 on the BEAUFORT SCALE in exposed situations or on the open sea, or with gusts to 43 kn or more) by broadcast warnings and by messages to ports and fishing stations recommended by the responsible local authorities.

gamma (or γ) **radiation.** ELECTROMAGNETIC RADIATION of wavelength less than that of X-rays and of great penetrative power. Gamma radiation (also called 'gamma rays') is produced during the disintegration of many radioactive elements. Emitted by radioactive material in the ground, it is responsible for part of the total IONIZATION of the air at lower levels over land. It also constitutes the chief hazard in 'close-in' RADIOACTIVE FALL-OUT. See also ALPHA PARTICLE, BETA PARTICLE.

gas. A fluid of unlimited capacity for expansion under diminishing pressure. The term is applied to any substance which obeys approximately BOYLE'S LAW and CHARLES'S LAW, and the combination of these two laws in the GAS EQUATION.

gas constant. See GAS EQUATION.

gaseous concentration, units of.   The earth's atmosphere, in addition to oxygen and nitrogen, contains small quantities of other gases including CARBON DIOXIDE, the INERT GASES, OZONE, and many others, either natural or artificially introduced like CHLOROFLUOROCARBONS.

Consider a mixture of gases occupying a volume V at total pressure p and temperature T. If the ith constituent has partial pressure p_i, mass m_i and molecular weight M_i, then its concentration relative to the total mixture may be expressed in several ways. The most useful are as follows:

(i)   Mass fraction $r_i = m_i/(\sum_i m_i)$.

(ii)   Volume fraction $f_i = v_i/V$ where v_i is the volume that would be occupied by m_i at (p,T). It is numerically equal to the SI UNIT MOLE FRACTION N_i where

$$N_i = \frac{m_i/M_i}{\sum_i (m_i/M_i)}$$

(iii)   Mass per unit volume $= m_i/V$.
(iv)   Number density, the number of molecules per unit volume.

Often conglomerations of letters such as 'ppbv' (meaning parts per billion by volume), are used instead of volume fraction; this is, however, not advisable because such conglomerations are undefined in lists of approved symbols issued by the WMO, the British Standards Institute, the Royal Society and the Royal Meteorological Society. Furthermore they do not follow the general laws of formation of unit symbols, nor is the word 'billion' scientifically defined (standing for either 10^9 or 10^{12}).

gas equation.   The pressure (p), specific volume (α), density (ρ), and temperature (T) of a PERFECT GAS are related by the 'gas equation' ('equation of state'):

$$p\alpha = RT \text{ or } p = R\rho T$$

where R the 'specific gas constant' $= R^*/M$. R^* is the universal gas constant (gas constant per gram mole of perfect gas) and M the MOLECULAR WEIGHT of the gas concerned.

Values quoted in Appendix A of WMO publication *Technical regulations* [28] are:
$R^* = 8.31432$ J mole^{-1} K^{-1}
For dry air up to about 25 km, $M = 28.9644$ and $R = 287.05$ J kg^{-1} K^{-1}.
For water vapour $M = 18.0153$ and $R = 461.51$ J kg^{-1} K^{-1}.

For moist air the value of M is smaller, and that of R is greater, than the respective values for dry air by amounts which depend on the percentage weight of water vapour in the air — see MOLECULAR WEIGHT.

gathering ground.   An area from which water is obtained by way of rainfall, drainage or percolation. See also CATCHMENT AREA and DRAINAGE AREA.

Gaussian (frequency) distribution.   An alternative for NORMAL (FREQUENCY) DISTRIBUTION.

GDPS.   Abbreviation for Global Data-Processing System.

gegenschein.   A faintly luminous patch of light on the line of the ZODIACAL BAND at the point of the sky opposite to the sun. It is sometimes termed COUNTERGLOW, which is, however, also used of a different phenomenon.

general circulation. The term 'general circulation' has different meanings in different contexts and there is no unique definition. In its widest sense it is used to imply all aspects of the three-dimensional global flow and energetics of the whole atmosphere. In this sense the general circulation is exceedingly complex, involving fluctuations on all time-scales and space-scales as well as many other considerations. For many purposes, therefore, study of the general circulation is rendered more specific by the application of some form of averaging, in time or in space, or in both. This gives rise to various kinds of average circulation in which their temporal variations of high frequency and/or the spatial variations of small scale are filtered out so as to reveal the character of the longer-period/larger-scale events. The term 'general circulation' is often used to denote one or other of the various kinds of average circulation which can be so generated, usually supported by appropriate statistical information concerning the temporal or spatial variability about that average.

As an example, a systematic annual variation of the circulation is revealed by averaging the flow or pressure fields, which are intimately related, over individual calendar months or seasons. The number of years' data required for adequate suppression of the fluctuations is small in low and high latitudes but is much greater in middle latitudes, especially in the northern hemisphere where occurrences of strikingly abnormal mean circulation over large areas for a single month are not uncommon.

An idealized mean surface circulation with associated pressure distribution, appropriate to equinox and a uniform surface of the earth, is considered to be:

(i) A narrow belt of light variable winds converging in a shallow belt of low pressure on the equator (DOLDRUMS).

(ii) TRADE WINDS (or 'tropical easterlies'), north-east in the northern hemisphere and south-east in the southern hemisphere, between about latitudes 30° N and S and the doldrums.

(iii) Light, variable winds associated with high-pressure belts (subtropical anticyclones) in latitudes 30–40° N and S.

(iv) Belts of 'westerlies', south-west in the northern hemisphere and north-west in the southern hemisphere, between about latitudes 40–60° N and S.

(v) Variable winds converging in low-pressure belts at about 60° N and S ('temperate storm belts').

(vi) Regions of outflowing winds with an easterly component, diverging from high pressure near the poles.

The actual mean surface circulations in January and July are shown in Figures 24 and 25. The appreciable departures from the idealized flow, more especially in the northern hemisphere, are due mainly to the non-uniform character of the earth's surface, with associated continental winter anticyclones and summer depressions.

The general circulation is maintained against dissipative processes (mainly friction with the earth's surface) by the heat energy from the sun. The earth and its atmosphere are continuously losing heat by radiation, mostly in the infra-red region of the spectrum, and in the long term there must be a global balance between this loss and the incoming solar energy. Because of the special radiative properties of its constituents, the earth's atmosphere prevents much of the long-wave radiation from being lost to space whilst at the same time being nearly transparent to solar radiation. In the absence of the atmosphere the average surface temperature of the globe would be around $-17\,°C$ instead of $+15\,°C$ and the diurnal and annual ranges of temperature would be very much larger than they are.

Although there is a long-term global balance between incoming and outgoing radiation, considerable imbalances exist locally. Indeed, there is a substantial excess of effective incoming radiation in low latitudes up to about 40° N and S and a deficit from there to the poles. Alternatively, one may say that the mean atmospheric temperature in low latitudes is lower, and in high latitudes higher, than that appropriate to local radiative balance between incoming solar and outgoing terrestrial radiation. This situation is possible because of the global-scale mixing operation performed by the general circulation. In particular the circulation mixes the air of higher and lower latitudes. In middle latitudes the mixing is mainly effected by the cyclones and anticyclones which *inter alia* transform into kinetic energy the available potential energy established by differential heating.

The existence of a latitudinal mixing of air implies a three-dimensional mean circulation. At any given latitude the continuity demands that average mass transport of air polewards, integrated over all heights, equals that equatorwards; mean meridional flow in the upper air (of about 1 m s^{-1}) is polewards in low (0–30°) and high (60–90°) latitudes and equatorwards in middle (40–60°) latitudes, balancing the opposite meridional components of the lowest layers. The (very small) mean vertical components of motion which are implied act upwards in the low-pressure belts near the equator and 60° N and S and downwards in the high-pressure belts at 30–40° N and S and near the poles.

At any given level above the surface the mean flow is more nearly along a parallel of latitude than is the case at the surface. The main feature of the upper-air mean circulation is an increase of the westerly wind component with height within the troposphere (increasing THERMAL WIND component) except in equatorial latitudes where easterly winds generally prevail. The strongest mean winds are the westerlies near the tropopause in about latitudes 30–40° N and S; in winter their mean speed is in excess of 80 kn, in summer some 30–40 kn. Above the tropopause the latitudinal gradient of temperature is reversed and the westerly wind component decreases with height except in high latitudes in winter — the 'polar-night JET STREAM'.

Apart from diagnostic analysis of the observations the large-scale motions of the atmosphere may be studied to some extent by means of laboratory experiments in which a fluid contained in a rotating vessel is differentially heated. Clearly the extent to which such a model can simulate realistically the behaviour of the actual atmosphere is severely limited; nevertheless useful insight into some of the properties of atmospheric motions can be gained. See ROTATING FLUIDS — LABORATORY STUDIES.

An alternative line of study is to set up the mathematical equations governing the motion, heat balances, etc. of the atmosphere and to solve the system numerically on an electronic computer. Essentially one attempts to study the time evolution of the atmosphere starting from prescribed initial conditions. Such studies have come to be known as numerical simulation experiments because in a sense the computer is analogous to a laboratory with the system of equations taking the place of a physical model and indeed often being described as a mathematical model of the atmosphere. Such experiments may be conducted for a variety of reasons. Thus one may integrate forward in time for a long period in order to monitor the behaviour of the model and, if this closely resembles that of the real atmosphere, to exploit events within the model to gain a deeper understanding of the way the circulation operates and fulfils its various functions. Alternatively one may wish to appraise the effect on the circulation and climate of variations in the external parameters or of possible methods of climate modification. Numerical simulation is thus a powerful method for studying the atmospheric general circulation. See also POLAR VORTEX, QUASI-BIENNIAL OSCILLATION, CONDITIONAL SYMMETRIC INSTABILITY.

general inference. A term used in weather forecasting for a description of the general pressure distribution and the changes of pressure distribution which are in progress, together with a general statement of the type of weather likely to be experienced. It usually precedes a series of more detailed forecasts for individual districts and gives the framework on which these forecasts are based.

genitus. See CLOUD CLASSIFICATION.

FIGURE 24.   Mean surface-wind directions in January.

128

FIGURE 25.   Mean surface-wind directions in July.

129

geodesy. The science concerned with the size and shape of the earth.

geodynamic metre. The standard unit of dynamic height defined as $10 \text{ m}^2 \text{ s}^{-2}$. See GEOPOTENTIAL.

geomagnetism. The study of the nature and causes of the earth's magnetic field and its variations.

Measurements of the magnetic field vector are made at a world-wide network of observatories. Continuous photographic recordings of the three independent components, horizontal force, angle of declination and vertical force, are made, and the variations recorded on such magnetograms are reduced to absolute measure by regular calibration.

More than 99 per cent of the earth's magnetic field is of internal origin. Observations since the 17th century have revealed the presence of slow, but cumulatively large, changes of strength and orientation of the field. Their cause is now thought to be large-scale vortices in the conducting molten core or the earth, slowly circulating across the earth's main field and so producing, by dynamo action, slowly changing regional magnetic fields. The origin of the main field itself is not yet satisfactorily explained.

Recent studies in PALAEOMAGNETISM indicate some very large changes of field orientation on the geological time-scale. World-wide extension of these studies has yielded evidence of CONTINENTAL DRIFT on this time-scale.

The small part of the earth's field which is of external origin is produced by the entry into the high atmosphere of electromagnetic waves and particles emitted by the sun. Each of these produces characteristic effects on the magnetograms which thus provide a valuable continuous measure of solar activity.

The magnetic 'disturbance' produced by solar particles is a highly complex phenomenon which is most frequent in the 'auroral zones', at about 70° geomagnetic latitude, where it is never entirely absent. At higher levels of disturbance, especially beyond the rather arbitrary lower level of a 'magnetic storm', rapid field variations are world-wide and are accompanied by ionospheric disturbances, which produce anomalous radio reception, and by large earth currents, which adversely affect cable telegraphy. At such times AURORA is visible far equatorwards of its normal position. Field-strength fluctuations during large storms amount to about 3–8 per cent of the undisturbed value, depending on latitude. Field direction changes are in general a few degrees, but are much larger in high latitudes.

The connection between magnetic disturbance and the physical state of the sun is shown, for example, by a 27-day recurrence tendency (solar rotation period) and an approximate 11-year (solar-cycle) variation of disturbance frequency and intensity. The correlation between individual magnetic disturbances and visual, radio or photographic evidence of solar activity is, however, relatively weak. Dependable prediction of magnetic storms has therefore not yet been achieved.

geometric mean. See MEAN.

geophysics. That branch of physics concerned with the earth and its atmosphere. The seven International Associations which at present constitute the International Union of Geodesy and Geophysics (IUGG) are Geodesy, Geomagnetism and Aeronomy, Scientific Hydrology, Meteorology and Atmospheric Physics, Physical Oceanography, Seismology and Physics of the Earth's Interior, and Vulcanology; other participating disciplines in the INTERNATIONAL GEOPHYSICAL YEAR project were Cosmic Rays, Ionosphere, and Solar Activity.

geopotential. The potential energy per unit mass of a body due to the earth's gravitational field referred to an arbitrary zero. The dimensions are $L^2 T^{-2}$.

Geopotential (Φ) depends on geometric height (z) and the acceleration of GRAVITY (g) in accordance with the equation

$$\Phi = \int_0^z g\,dz,$$

mean sea level being the selected level of zero potential. In general, a geometrically level surface is not one of constant geopotential because of changes of the value of g along it.

Geopotential is the potential energy acquired by unit mass on being raised through unit distance in a field of gravitational force of unit strength. It is, from the dynamical point of view, a better measure of height in the atmosphere than is geometric height; energy is in general lost or gained when air moves along a geometrically level surface but not when it moves along an equigeopotential surface. A 'dynamic height' unit, as defined above, may be used but, by international agreement, a 'geopotential height' unit is preferred because it has the advantage of giving an even better fit, on average over the world, with geometric height, while retaining the dimensions and physical significance of geopotential. The two units are related by the equation:

1 standard geopotential metre = 0.980665 geodynamic metre.

The equation which defines the relationship between geopotential height (Z) and geometric height (z) is $Z = gz/9.80665$. Thus, where g has its near average value of 9.80665 m s^{-2}, heights in geopotential metres and geometric metres are the same; for $g < 9.80665$ m s^{-2} the height in geopotential metres is the smaller, for $g > 9.80665$ m s^{-2} it is the bigger.

geopotential metre.   See GEOPOTENTIAL.

geosphere.   That part of the EARTH which is either solid or is composed of water, i.e. the LITHOSPHERE and HYDROSPHERE combined.

geostrophic approximation.   See QUASI-GEOSTROPHIC MOTION.

geostrophic coordinates.   These coordinates are defined on the ordinary (x,y) plane by

$$X = x + (v_g/f), \quad Y = y - (u_g/f).$$

They are the positions that particles would have had if they had moved with their geostrophic velocity at every instant provided that the GEOSTROPHIC MOMENTUM APPROXIMATION was valid.

geostrophic departure (or **deviation**).   See AGEOSTROPHIC WIND.

geostrophic momentum approximation.   The replacement of the true wind velocity ($\mathbf{V}$) in the momentum term ($d\mathbf{V}/dt$) of the EQUATION OF MOTION by the geostrophic wind ($\mathbf{V_g}$). The true wind is, however, retained in the advective part of d/dt so that, for example, the x component of $d\mathbf{V}/dt$ becomes

$$\frac{\partial u_g}{\partial t} + u\frac{\partial u_g}{\partial x} + v\frac{\partial u_g}{\partial y} + w\frac{\partial u_g}{\partial z};$$

the approximation is equivalent to neglecting

$$\frac{1}{f}\frac{d^2\mathbf{V}}{dt^2}$$

131

in comparison with

$$\frac{d\mathbf{V_g}}{dt^2}.$$

Using this approximation, a set of equations may be written down, superficially similar to the PRIMITIVE EQUATIONS, which form a balanced system, i.e. they cannot describe gravity waves. They can, however, describe regions of large shear vorticity (such as jet streams and fronts) provided that the *curvature* vorticity (see VORTICITY) is much less than the CORIOLIS PARAMETER, and are thus more useful in the study of such regions than equations derived from the GEOSTROPHIC APPROXIMATION. This set of equations is often called 'the semi-geostrophic equations', and is particularly useful in two-dimensional cases. Good accounts have been given by Hoskins [29, 30]. A valuable aid in the consideration and solution of the semi-geostrophic equations is the introduction of GEOSTROPHIC COORDINATES.

geostrophic vorticity.   Vorticity evaluated on the assumption of GEOSTROPHIC WIND conditions, as from the contours of an ISOBARIC SURFACE.

If z is the height at a point on an isobaric surface, then the 'relative geostrophic vorticity' is given by

$$\zeta_g = g\left\{\frac{\partial}{\partial x}\left(\frac{1}{f}\frac{\partial z}{\partial x}\right) + \frac{\partial}{\partial y}\left(\frac{1}{f}\frac{\partial z}{\partial y}\right)\right\} = (g/f)\nabla^2 z.$$

See also VORTICITY, LAPLACIAN.

geostrophic wind.   That horizontal equilibrium wind $(\mathbf{V_g})$, blowing at right angles to the pressure gradient, which represents an exact balance between the PRESSURE GRADIENT FORCE $(-(1/\rho)\nabla_H p)$ and the horizontal component of the CORIOLIS FORCE $(f\mathbf{V_g})$, where $\nabla_H p$ is the horizontal pressure gradient. Low pressure is to the left of the wind vector in the northern hemisphere, to the right in the southern hemisphere (see Figure 26). The magnitude of $\mathbf{V_g}$ is given by

$$V_g = |(1/f\rho)\nabla_H p|.$$

In terms of the height gradient of an isobaric surface $(\nabla_p z)$, $\mathbf{V_g}$ is given by

$$V_g = |(g/f)\nabla_p z|.$$

Geostrophic wind scales, based on the above relationships, are used and give a good approximation to the actual wind in the free atmosphere provided that isobaric curvature is not excessive. A scale based on the isobaric surface relationship has the advantage of being independent of air density and so may be used at any level in the atmosphere.

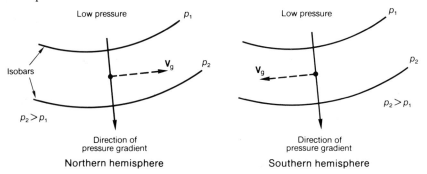

FIGURE 26.   Geostrophic wind.

132

GHOST. Abbreviation for Global Horizontal Sounding Technique. See CONSTANT-LEVEL BALLOON.

glacial phase. A period during an ICE AGE (alternating with INTERGLACIAL PHASES) when there is marked extension of ICE SHEETS equatorwards from polar regions; for the northern hemisphere average extension south of 75° N has been proposed (by G.C. Simpson) as a suitable definition of such a period.

glaciation. A term applied to the sudden transformation, near the top of a developing shower cloud, of supercooled cloud droplets into ice crystals; the transformation is often marked by the formation of ANVIL CLOUD.

glacier. A mass of ice which is formed over mountains where there is net accumulation of snow. It is considered that growth of glaciers is favoured by snowy winters and cool summers, while recession or disappearance of glaciers is favoured by relatively dry winters and warm summers. Observed changes in thickness and areal extent of glaciers are used as important climatic indicators.

glacier wind. A gravitational (downhill) flow which develops above a GLACIER in daytime, especially in summer, because of the low temperatures then attained close to the glacier, relative to the surroundings. Maximum wind speed occurs some 2 m (6 ft) above ground.

glaciology. The study of the distribution and behaviour of snow and ice on the earth's surface. There is appreciable meteorological significance in this study, particularly in relation to climatic changes — see GLACIER. See also EUSTASY, GLACIAL.

Glaisher stand. A form of stand devised by James Glaisher for the exposure of thermometers. The stand consists of a vertical portion, partially roofed, on which the thermometers are mounted, with a doubly roofed sloping rear portion designed to prevent the front portion becoming heated from the rear. The whole is capable of rotation about a vertical axis so that direct sunshine may be prevented from affecting the thermometers at all times. The stand has been superseded by the THERMOMETER SCREEN.

glaze (clear ice). A generally homogeneous and transparent deposit of ice formed by the freezing of supercooled drops of drizzle or rain on objects the surface temperature of which is slightly above 0 °C or below.

It may be produced by the freezing of non-supercooled drops of drizzle or rain immediately after impact with surfaces the temperature of which is well below 0 °C.

Glaze is observed at the earth's surface when raindrops fall through a layer of sub-freezing temperature of sufficient depth.

On rare occasions glaze may attain sufficient thickness to bring down telephone wires or present a serious hazard to ships by adding weight to the superstructure. There was a notable occurrence of glaze in late January 1940 over much of south and west England and in Wales [31]. See BLACK ICE, ICE FORMATION ON AIRCRAFT.

Glaze on the ground must not be confused with 'ground ice', which is formed when:

(i)   water from a precipitation of non-supercooled drops of drizzle or rain later freezes on the ground,
(ii) snow on the ground freezes again after having completely or partly melted, or
(iii) snow on the ground is made compact and hard by traffic.

Global Data-Processing System (GDPS). The co-ordinated global system of meteorological centres and arrangements for processing, storage and retrieval of meteorological information within the framework of the WORLD WEATHER WATCH.

Global Horizontal Sounding Technique (GHOST). See CONSTANT-LEVEL BALLOON.

global radiation. The sum of solar direct and diffuse radiation received by a (usually horizontal) surface. See RADIATION.

Global radiation is also measured on inclined surfaces in which case the orientation of the surface is specified.

glory. The system of coloured rings similar to those of a CORONA round sun or moon, surrounding the shadow of an observer's head on a bank of cloud or mist. The phenomenon is also termed 'anticorona'. A several-fold effect is sometimes observed, while a FOGBOW may be seen to surround a glory.

When light passes through circular holes in an opaque screen, colours are produced by DIFFRACTION. If little mirrors all facing the sun could be substituted for the droplets in a cloud, the light from each mirror would behave as if it came through a hole from the reflection of the sun and similar diffraction colours would occur. The action of the droplets is probably analogous. The mathematical theory developed by B. Ray is on these lines. Earlier writers had supposed that the phenomena were produced by the diffraction, by particles comparatively near the surface, of light reflected from deeper portions of the fog or cloud.

gold-beater's skin. The prepared outside membrane of the large intestine of an ox, used until recently by gold-beaters for separating leaves of gold foil. It is hygroscopic, and the variations in its physical dimensions allow it to be used as a sensitive indicator of RELATIVE HUMIDITY in the RADIOSONDE.

Gold slide. An attachment, devised by E. Gold, for a marine mercury BAROMETER to allow the rapid correction and reduction to sea level of the reading of such a barometer with sufficient accuracy.

The ATTACHED THERMOMETER is mounted on a brass stock and the corresponding 'barometer correction scale' of millibars (1 mb per 6 K difference from STANDARD TEMPERATURE) is mounted on a vertical slide. The position of the zero of this scale is altered according to the (closely approximate) corrections required for index error, latitude difference from 45°, and height of barometer above sea level; the required correction is then read from that part of the scale, in its adjusted position, opposite the end of the thermometer column.

goodness of fit. A term sometimes used for the measure of agreement between a set of observed frequencies and the frequencies expected according to some hypothesis which forms the subject of the CHI-SQUARE TEST.

gradient. The word 'gradient' is used in surveying and in common practice to indicate the slope of a hill, i.e. the change in height per unit distance along the hill. In mathematics, the gradient of a function (ϕ) is a vector, written grad ϕ or $\nabla\phi$ or del ϕ, whose direction is that in which ϕ increases most rapidly and whose magnitude is the rate of increase of ϕ with distance in this 'up-gradient' or 'ascendant' direction. In meteorology, the PRESSURE GRADIENT FORCE acts from high to low values of pressure, i.e. in the 'down-gradient' or 'descendant' direction, and hence approximately at right angles to the GEOSTROPHIC WIND.

For synoptic and other purposes, attention is often confined to gradients in the horizontal plane, as for example the pressure gradient at mean sea level, as defined by mean-sea-level isobars. In common but not universal usage, the term 'temperature gradient' is reserved for temperature change with horizontal distance, change of temperature with height being referred to as 'temperature lapse rate' (or simply 'LAPSE rate'). The same is true of humidity. Similar considerations apply to gradients in an isobaric surface.

gradient wind. That equilibrium horizontal wind (V) blowing parallel to curved isobars of radius of curvature r, whose CENTRIPETAL ACCELERATION (V^2/r) represents the net inward horizontal force acting per unit mass of air. The only forces considered to be acting are the horizontal components of the PRESSURE GRADIENT FORCE ($-(1/\rho)\nabla_H p$ and CORIOLIS FORCE ($2\Omega V \sin \phi$) (see Figure 27). The equations, for cyclonic and anticyclonic curvature of isobars, are:

cyclonic (acceleration in direction of pressure gradient force)

$$-(1/\rho)\nabla_H p + 2\Omega V \sin \phi = -\frac{V^2}{r} \, ,$$

anticyclonic (acceleration in direction opposite to pressure gradient force)

$$-(1/\rho)\nabla_H p + 2\Omega V \sin \phi = +\frac{V^2}{r} \, .$$

In middle latitudes the gradient wind speed is normally a rather closer approximation to the actual wind speed than is the GEOSTROPHIC WIND speed (V_g). V and V_g are related by the equation

$$V = V_g \pm \frac{V^2}{2r\Omega \sin \phi}$$

($+$ for anticyclonic, $-$ for cyclonic curvature)

where the term $V^2/2r\Omega \sin \phi$ is the so-called 'cyclostrophic component' of the wind. This latter is the AGEOSTROPHIC WIND component of the gradient wind; it is normal to and to the left of the acceleration.

FIGURE 27.   Gradient wind, drawn here for the northern hemisphere for (a) cyclonic motion, and (b) anticyclonic motion.

gram.   The unit of mass in the C.G.S. SYSTEM of units. It is one-thousandth part of the standard kilogram which was originally supposed to represent the weight of a cubic decimetre of pure water at 4 °C, but subsequent research has shown that the relationship is not exact.

gram-calorie.   See CALORIE.

Grashof number.   In heat transfer by free CONVECTION, a non-dimensional parameter (Gr) defined by the relationship $Gr = gl^3 \Delta T/v^2 T$ where l is a characteristic length, g the gravitational acceleration, v the kinematic VISCOSITY, and ΔT the difference between the temperatures of the surface and the air in contact with it.

grass minimum temperature. The minimum temperature indicated by a STANDARD MINIMUM THERMOMETER freely exposed in an open situation at night with its bulb in contact with the tips of the grass blades on an area covered with short turf. See also GROUND FROST, CONCRETE MINIMUM TEMPERATURE.

gravitational equilibrium. An alternative for DIFFUSIVE EQUILIBRIUM.

gravitational separation. The separation of particles, which are free to fall to earth, by virtue of their different TERMINAL VELOCITIES. For particles of a given substance, for example raindrops, the terminal velocities acquired increase with particle size.

Gravitational separation of the gases which form the atmosphere — usually then termed 'diffusive separation' — in accordance with the different weights of their molecules or (at high levels) atoms is negligible below a level of about 80 km, because of the efficient mixing of the air at the lower levels. See also DIFFUSIVE EQUILIBRIUM.

gravity. The force of attraction between material bodies. The law of universal gravitation is that every mass attracts every other mass with a force which varies directly as the product of the attracting masses and inversely as the square of the distance between their centres of mass. The gravitational acceleration (g), i.e. the acceleration produced in a body which is free to move by the downward pull of the earth, is the force acting per unit mass of the body and is about 9.8 m s^{-2} near the earth's surface. It decreases with increasing height in the atmosphere, being about 3 per cent less at a height of 100 km.

Because of the earth's rotation about its axis the observed force at the earth's surface is the vector resultant of the universal gravitational force and the centrifugal force which arises from the earth's rotation. (The force, acting at right angles to the earth's axis is $\Omega^2 r$, where r is the distance from the earth's axis.) Since the centrifugal force varies with latitude, being zero at the poles ($r = 0$) and a maximum at the equator (but of opposite sense there to the gravitational force), the force of apparent gravity also varies with latitude and has a maximum value at the poles and minimum value at the equator. The direction of the force of apparent gravity defines the direction of the local vertical, being normal to a *level* surface at the corresponding point. Because of the spheroidal shape of the earth (equatorial radius about 21.5 km longer than polar radius) — also the result of the earth's rotation — the force of gravity is not, in general, directed exactly towards the earth's centre. Small local anomalies of gravity occur and are associated with local topographical features and variations of mass distribution.

gravity wave. A type of wave which depends for its existence on the restoring force provided by BUOYANCY, and might be better referred to as a 'buoyancy wave'. So-called external gravity waves can propagate along the boundary separating two essentially homogeneous fluids; ocean surface waves are an example. In a stably stratified fluid, such as the atmosphere, gravity waves can propagate vertically as well as horizontally and are known as internal waves. As well as buoyancy forces yielding the natural BRUNT-VÄISÄLÄ FREQUENCY, internal gravity waves can be affected by wind shear and, on a large scale, the earth's rotation.

(The term 'gravity wave' as used in meteorology has no connection with the gravitational waves that have been postulated as a consequence of Einstein's General Theory of Relativity.)

gravity-wave drag. In certain conditions of atmospheric stability and wind shear, LEE WAVES set up by mountain ranges may attain large amplitudes and then 'break' in a way analogous to the way that ocean waves break on reaching the sea-shore. This means that much of the horizontal momentum of the air that it possessed upwind of the mountains is lost in a host of small-scale turbulent eddies over, and immediately to the lee of, the mountains so that, when reasonably smooth flow is resumed well downwind, the atmosphere will have been permanently deprived of this momentum over a considerable range of height. The net effect is that mountain ranges tend to act as sinks of momentum, and the process is known as gravity-wave drag. Gravity-wave drag is sub-grid-scale, highly non-linear and difficult to handle mathematically, but may be parametrized to a certain extent, e.g. by Palmer *et al.* [32].

Such approximate methods, when inserted in numerical forecast models, produce useful improvements in forecasts of upper winds and the depth of depressions.

grazing season. The period of the year during which animals can be put out to graze with the expectation of adequate fodder and without risk of POACHING. It does not coincide with the GROWING SEASON.

great circle. Any plane which passes through the centre of a sphere cuts the surface of the sphere in a 'great circle'.

green flash. On some rare occasions the last glimpse obtained of the sun at sunset, or the first glimpse at sunrise, is a brilliant green — the 'green segment' — lasting a few seconds. On still rarer occasions a 'green flash' or 'green ray', also lasting a few seconds, shoots above the horizon from the upper limb. The explanation is the greater REFRACTION of the short waves (violet, blue, green) than of the long waves (red) of white sunlight, coupled with the greater degree of RAYLEIGH SCATTERING experienced by the violet and blue rays. In a hazy atmosphere such differential scattering may not be appreciable and the flash may then appear blue or violet. It is probable, though not yet confirmed, that an unusual degree of refraction, such as occurs with a low-level inversion of temperature, is required for the phenomenon.

Differential refraction of white light is also the cause of the analogous very rare phenomenon of the 'red flash' which may occur when the sun's disc appears just below a bank of clouds near the horizon.

The green flash has been observed in association with the moon and planets on rare occasions.

greenhouse effect. The effect, analogous to that which has been supposed to operate in a greenhouse, whereby the earth's surface is maintained at a much higher temperature than the temperature (about 256 K) appropriate to balance conditions with the solar radiation reaching the earth's surface. The atmospheric gases are almost transparent to incoming solar radiation, but water vapour and carbon dioxide in the atmosphere strongly absorb LONG-WAVE RADIATION emitted from the earth's surface and re-emit the radiation, in part downwards. CHLOROFLUOROCARBONS (CFCs) and methane also make a contribution to the greenhouse effect which is becoming of increasing significance. See ABSORPTION.

gregale. A strong north-east wind occurring chiefly in the cool season in the south central Mediterranean, but used also of a strong north-east wind in other parts of the Mediterranean, for example the south coast of France (grégal) and in the Tyrrhenian Sea (grecale).

grenades sounding. A method of atmospheric SOUNDING in which wind velocities and temperatures are inferred, as functions of height up to about 90 km, from an analysis of the speed of travel of sound waves originating from a series of grenades fired at height intervals of a few kilometres from an ascending METEOROLOGICAL ROCKET. The sounds of detonation are received by an array of microphones on the ground. The times and positions of grenade bursts are determined by optical or radio means.

grey body. A 'grey-body radiator' is defined as one which emits, for every wavelength, the same proportion of the maximum or 'black-body' radiation at a given temperature. Treatment of the atmosphere as a grey-body radiator has been found to give results which are too inaccurate to be useful; consideration of the detailed ABSORPTION spectrum is required. See RADIATION.

gridding. In synoptic meteorology, a system of graphical addition or subtraction of two fields represented by isopleths, used for example in upper-air work and in obtaining 'anomaly patterns'.

Gridding may be performed on a single chart or, alternatively, two superimposed charts may be used, each with a set of isopleths, the points of intersection of which are made visible by illuminating the charts from below.

grid length. The distance between two consecutive points used for the solution of a set of equations by the FINITE-DIFFERENCE METHOD.

Grosswetter. A German term which is used to indicate, with special reference to long-range weather forecasting, the main features of weather over a specified region and period of time. The use of this term involves the concept that certain variable but non-random influences govern the large-scale pattern of weather development. 'Grosswetterlagen' are the synoptic situations corresponding to the types of Grosswetter.

ground discharge. Lightning flash from cloud to ground. See LIGHTNING.

ground frost. From 1906 to 1960, inclusive, the Meteorological Office practice was to record a 'ground frost' when the grass minimum temperature reached 30 °F (-1 °C) or below (30.4 °F for thermometers read to tenths) and to base the 'number of days of ground frost' on this criterion. The reason for this choice is rather obscure; it may have been based on the belief that a temperature appreciably below 32 °F (0 °C) is required before damage is caused to the tissues of growing plants.

From 1 January 1961, the statistics issued have referred to the 'number of days with grass minimum temperature below 0 °C, and no statistics have referred to 'ground frost'. The use of the term 'ground frost' in forecasts signifies a grass minimum temperature below 0 °C (32 °F).

ground, state of. Observations of the state of ground are made at selected stations. The code provides for ten different conditions of the ground in the vicinity of a station for each of two classes — one with and the other without snow or measurable ice cover. Details are given in the *Observer's handbook* [1, pp. 95–97].

ground water. In HYDROLOGY, the water retained at all levels below the WATER TABLE.

group velocity. The velocity with which wave energy is propagated. A fluid disturbance can often be regarded as a 'wave packet', composed of a very large number of superposed simple harmonic waves having appreciable amplitude only within a limited domain; the envelope of such a wave packet will, to a first approximation, move with the group velocity (v_G) where

$$v_G = c - \lambda \frac{dc}{d\lambda}$$

where c is the (variable) phase velocity of the simple harmonic wave and λ the corresponding wavelength.

Many disturbances correspond to wave packets rather than simple waves so that they, and their associated energy, travel at the group velocity, not the phase velocity. It is, in fact, possible for v_G and c to have opposite signs, for example in vertically propagating equatorial stratospheric waves.

growing season. That period of the year during which the growth of vegetation proceeds. Nowadays a mean screen temperature exceeding 0 °C is regarded as the most useful critical value for crops grown in temperate climates, but in the past 42 °F and 6 °C have both been used. For crops normally grown in tropical climates a higher value such as 10 °C is appropriate. The rough length of outdoor growing season for various localities may be obtained from the corresponding curve of annual variation of mean temperature.

Gulf Stream. Originating in the eastern area of the Gulf of Mexico, the ocean current known as the Gulf Stream flows through the Straits of Florida and up the eastern coast of the USA, following the edge of the continental shelf. Leaving the coast at about 40° N, it proceeds as a weaker and broader current across the Atlantic and reaches the British Isles at about 50° N. Its mean speed across the Atlantic is 5–8 km per day; near the American coast mean speeds of about 50 km per day are reached in spring and summer. Though one of the strongest and most constant of ocean currents, the Gulf Stream is subject to variability and even to reversals. While the term 'Gulf Stream' is popularly applied to the entire current system described above, more precise technical definition subdivides the system into the Florida Current to about 40° N, the Gulf Stream eastwards to 45° W, and the North Atlantic Drift farther eastwards and northwards.

It is popularly supposed that the temperate climate of the British Isles is due to the warmth conveyed by the Gulf Stream; a more accurate view is that the temperate, maritime nature of the climate is due to the prevalence of south-west to west winds, which are warmed by the water in the North Atlantic Drift. The ocean circulation is itself partly driven by the winds.

gust. A rapid increase in the strength of the wind relative to the mean strength at the time. It is a much shorter-lived feature than a SQUALL and is also different in nature, being due mainly to mechanical interference with the steady flow of air; it is, therefore, a pronounced feature of air flow near the ground, especially where there are large obstructions. Other factors, notably temperature lapse rate and vertical wind shear, are important, however, in determining the existence and magnitude of gusts. Such factors may produce gusts in circumstances where mechanical interference with the flow appears to be insignificant, as in CLEAR-AIR TURBULENCE.

The definition of the gust implies the existence of 'negative gusts' (lulls) of wind. The fluctuations of either sign are involved in the definition of 'gustiness'. The range between gusts and lulls increases with the mean wind strength; there is, however, also a dependence on anemometer exposure and, at most stations, on wind direction. See also GUSTINESS, TURBULENCE.

gustiness. The important characteristics of a fluctuating wind, characterized by GUSTS and lulls about a mean level, are the frequency and strength of the gusts. The former is usually expressed by the number of wind maxima occurring within a specified period of time. The strength is usually defined in one of the following ways:

(i) In normal surface observations by a 'gustiness factor', i.e. by the percentage ratio of the difference between the maximum and minimum horizontal wind speeds to the mean wind speed recorded in a given period. In a selection of eight British stations, values of the gustiness factor ranging from 25 to 100 per cent were obtained, large dependence of the factor on mean wind direction being found at some of the stations. A gustiness factor in ordinary observations may also be defined in terms of wind direction; the angular width (in radians) is a measure of lateral gustiness which for small values is nearly equivalent to the speed ratio.

(ii) In micrometeorology, if u', v', w' denote the fluctuations of the velocity components in three mutually perpendicular directions from the corresponding mean components, then the gustiness components may be represented by the standard deviations $\sigma_u = \{\overline{(u')^2}\}^{1/2}$ etc. If the resultant mean wind ($\overline{V}$) is horizontal and in the direction of the x- axis, then the longitudinal, lateral and vertical 'intensities' of gustiness (or turbulence) are defined by $\sigma_u/\overline{V}$, $\sigma_v/\overline{V}$, $\sigma_w/\overline{V}$, respectively.

For some purposes, especially the study of the effect of gusts on aircraft, the structure of individual gusts — as represented by the shape of the profile of gust velocity plotted against distance (or time) — is important. The most important 'gust shapes' are (a) the 'flat-topped' gust, in which the velocity increases uniformly, and (b) the 'sinusoidal' gust. In theoretical work, these are used as the standard gust in the United Kingdom and the USA, respectively. Other gust-shape classifications are the 'sharp-edged' gust and the 'triangular' gust. The 'gust length' is the distance occupied by the ascending part of the gust profile; it is typically about 30 m. See also TURBULENCE, EDDY, EDDY SPECTRUM.

guttation. The exuding of liquid water from the tips of plants, usually under conditions of a warm, moist soil. The phenomenon is often mistaken for one of DEW formation.

H

haar. A local name in eastern Scotland and parts of eastern England for a sea fog which at times invades coastal districts. Haars occur most frequently in spring and early summer months. See ADVECTION FOG, SEA FRET.

haboob. The name, derived from the Arabian *habb* meaning to blow, applied to a DUSTSTORM in the Sudan north of about 13°N. Such storms occur from about May to September and are most frequent in the afternoon and evening.

Hadley cell. A simple thermal circulation first suggested by George Hadley in the 18th century as part explanation of the TRADE WINDS and still thought to be approximated to in the troposphere between latitudes 0 and 30°. If the effects of the earth's rotation are neglected, the circulation comprises a high-level poleward flow from heat source to heat sink in response to a horizontal pressure gradient (at a high level, pressure is greatest above the heat source since pressure decreases less rapidly with height in the warmer air column), and a compensating low-level flow towards the heat source; upward and downward motions at the heat source and sink, respectively, complete the circulation — see Figure 28.

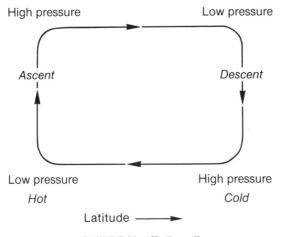

FIGURE 28. Hadley cell.

Haidinger's brush. A faint and transitory figure, in the shape of a yellowish brush with a small blue 'cloud' on either side, which appears when a source of strongly polarized light, such as the blue zenith sky with the sun close to the horizon, is closely observed. Because of physiological differences, the phenomenon is apparently not seen by all observers in quite the same form.

hail. Solid PRECIPITATION in the form of balls or pieces of ice (hailstones) with diameters ranging from 5 to 50 mm or even more. The stones fall from CUMULONIMBUS clouds and are commonly spherical or conical in shape although they sometimes form irregular lumps by agglomeration. The structure often consists of concentric layers of

141

alternately clear and opaque ice, and the density varies between 0.1 and 0.9 g cm^{-3} according to structure.

SMALL HAIL is generally classed as ICE PELLETS, but as hail if the diameter should equal 5 mm or more.

Because of the high liquid-water concentration of cumulonimbus clouds, the growth of ice particles, following their initial formation on ice nuclei at temperatures well below 0 °C and at saturation with respect to water, is mainly by collision and coalescence with supercooled water drops. The particles are supported within the cloud by strong updraughts (of about 10 m s^{-1}) and so are able to attain an appreciable size.

The opacity of the ice is caused by the trapping of air bubbles within the ice on freezing of supercooled water drops on the ice particles. Whether or not air bubbles are trapped depends on the surface temperature of the hailstone when the water freezes on to it. While the layer structure may reflect fluctuations in the height of the hailstone round the −5 °C level when it is supported by a varying updraught nearly equal to the terminal velocity of the stone, it is more likely that it reflects changes in the local supercooled water content in the air round the stone. Clear (opaque) rings reflect slow (rapid) freezing of supercooled water drops on the stone, associated with a high (low) water concentration in the cloud, and that the concentric rings result, therefore, from variations of liquid-water content. (The freezing of the drops involves the release of latent heat which must be dispersed before the freezing can be complete.)

hail, day of.   Up to and including 1959 a day on which HAIL, of whatever variety, was observed even though in so small a quantity as to yield no measurable amount in the rain-gauge.

A day of hail is now defined as one when pieces of ice, 5 mm or more in diameter, are observed to fall at any time between midnight and midnight. The hailstones can fall separately or be agglomerated into irregular lumps. Days with ICE PELLETS, SNOW PELLETS and ICE PRISMS are counted separately.

half-life.   In RADIOACTIVITY, the time taken for the activity of a given quantity of a radioactive element to decrease to one-half of its original value. The half-lives of radioactive elements range from a small fraction of a second to many thousands of years.

halo phenomena.   The term 'halo', which might be applied to any circle of light round a luminous body, is restricted by meteorologists to a circle produced by REFRACTION through ice crystals, in contrast to a CORONA which is produced by DIFFRACTION. All the optical phenomena produced by REFLECTION and refraction of light by ice crystals are sometimes grouped together as halo phenomena.

The most common halo is a luminous ring of 22° radius surrounding the sun or moon, the space within the ring appearing less bright than that just outside. The ring, if faint, is white but if more strongly developed the inner edge is a pure red, outside which yellow may be detected. The halo of 22° is very common. In England it can be seen on about one day in three. See Figure 29.

The halo of 22° is sometimes within a circumscribed nearly elliptical halo, the points of contact being at the highest and lowest points. The complete circumscribed halo is only seen when the elevation of the sun is 40° or more. With lower elevations separate tangent arcs are seen. These phenomena are explained by the presence of prismatic ice crystals floating with their axes horizontal.

The angle of 22° is the angle of MINIMUM DEVIATION for light passing through a prism of ice (refractive index 1.31) with faces inclined at 60°. Thus the occurrence of the halo of 22° radius indicates the presence of ice crystals with faces inclined at 60°. Alternative faces of a hexagonal prism are inclined at this angle, and as hexagonal

FIGURE 29.   Halo of 22° radius.

prisms are frequently found amongst ice crystals the halo is probably due to the refraction of light through such prisms.

A halo of 46° is to be seen occasionally, though seldom complete. This halo requires crystals with faces at right angles.

Another group of phenomena requires prismatic crystals with their axes vertical. In this group are PARHELION (mock sun) and the CIRCUMZENITHAL ARC. See Figure 30.

In weather lore, haloes are often spoken of as presaging storms. Haloes are, however, too common to be reliable signs of impending exceptional weather.

harmattan.   A dry wind blowing from a north-east or sometimes easterly direction over north-west Africa. Its average southern limit is about 5° N in January and 18° N in July. Beyond its surface limit it continues southwards as an upper current above the south-west monsoon. Being both dry and relatively cool, it forms a welcome relief from the steady damp heat of the tropics, and from its health-giving powers it is known locally as 'the doctor' in spite of the fact that it carries with it from the desert great quantities of dust. This dust is often carried in sufficient quantity to form a thick haze, which impedes navigation on the rivers.

harmonic analysis.   The resolution of a series of measurements (for example a TIME SERIES) into harmonic components, or sine waves, of which the periods are fixed in advance but the PHASE and AMPLITUDE are determined from the data. The periods chosen are usually sub-multiples of a particular period, the fundamental. Since the analysis involves the use of FOURIER SERIES it is also termed 'Fourier analysis'.

The reverse process, that of reconstituting the original function by addition of the harmonic components, is termed 'harmonic synthesis'.

Harmonic analysis is perhaps the simplest example of the general method of representing data by means of ORTHOGONAL FUNCTIONS. When the fundamental period can be determined from independent evidence, harmonic analysis is a powerful tool for separating harmonic components of variation from NOISE. It has been used effectively, for example, to investigate the semi-diurnal variation of pressure and the

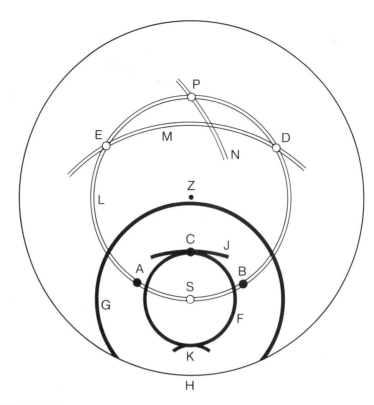

FIGURE 30. Diagram of an unusually complete halo display at about midday on 6 March 1941 in the west Midlands: composite diagram based on observations from various locations. Appearances due to refraction, which may be brilliantly coloured, as on this occasion, are shown black; appearances due to reflection, which are always white, are shown by the finer lines. The outer circle, H, is the complete horizon, with Z, the zenith, in the centre; S is the sun, P the anthelion, A, B, C, D and E are parhelia or mock suns, F is the 22° halo, G the 46° halo, J and K are upper and lower arcs of contact of F; L is the parhelic circle or mock-sun ring (parallel to the horizon), M the arc through the mock suns at 120° (usually a pair of arcs not joined in the middle), and N the oblique arc through P (one of a symmetrical pair which may sometimes be seen together).

annual variation of temperature. If, however, the fundamental period is chosen by guesswork, or by inspection of the data, it is often impossible to determine the statistical significance of the results. In these circumstances harmonic analysis is likely to be less fruitful than methods of analysis which start from the CORRELOGRAM.

harmonic dial. A representation, on a polar diagram, of the results of HARMONIC ANALYSIS, in which periodic components of a given frequency are compared. The components may, for example, refer to different stations at a given epoch, or to results for different seasons at a given station. Points are plotted on the diagram at positions corresponding to the amplitude, measured on a linear scale from the origin, and to the epoch of maximum of the component. The scale of angular measure employed in the analysis and the corresponding scale of time are generally both shown on such diagrams.

harmonic mean. See MEAN.

haze. A suspension in the air of extremely small, dry particles invisible to the naked eye and sufficiently numerous to give the air an opalescent appearance. There is no upper or lower limit to the horizontal visibility in the presence of which haze may be reported.

In most cases the particles composing haze are small enough (less than about 1 μm) to cause differential scattering of sunlight and to contribute, for example, to sunrise and sunset colours.

For the reduction of visibility by water droplets see FOG and MIST. See also DUST HAZE.

hazemeter. A term sometimes used synonymously with VISIBILITY METER. The 'loofah hazemeter' is an instrument in which the intensity of the light scattered at a particular angle to the original beam passing through an enclosed air sample is used as a measure of the SCATTERING coefficient of the air and hence its visibility (neglecting direct ABSORPTION of light).

head wind. See EQUIVALENT HEAD WIND.

health resort station. A CLIMATOLOGICAL STATION which participates in the Meteorological Office health-resort scheme whereby participating stations send daily a coded report or reports relating to observed temperature, rainfall, sunshine and weather for issue to the Press. See *Observer's handbook* [1, p. 8] for details.

heap clouds. CLOUDS of appreciable vertical development (CUMULUS and CUMULO-NIMBUS), as opposed to LAYER CLOUDS.

heat. A form of energy, normally measured in CALORIES or JOULES. The dimensions are ML^2T^{-2}.

The transfer of heat to or from a substance is effected by one or more of the processes CONDUCTION, CONVECTION, and RADIATION. The common effect of such a transfer is to alter either the temperature or the state of the substance (or both). Thus, a heated body may acquire a higher temperature ('sensible' heat) or may change to a higher state (thus acquiring latent or 'hidden' heat).

heat capacity. An alternative for THERMAL CAPACITY.

Heaviside layer. A layer of the IONOSPHERE at about 110 km height, now usually termed the E-layer; it has also been termed the 'Kennelly–Heaviside layer'.

heiligenschein. A diffuse, white ring of light cast on dewy grass surrounding the shadow of the head of an observer. The phenomenon occurs mainly when the sun's elevation is low and the observer's shadow long.

helicity. The scalar product of the velocity and the vorticity of a fluid, i.e. **V**.ζ. (Some authors refer to **V**.ζ as *helicity density*, reserving the term helicity for the integrated value over a finite volume of fluid.) The idea of helicity has been applied to various aspects of fluid mechanisms and dynamical meteorology, including the development of rotation in SUPERCELL thunderstorms.

heliostat. An instrument mounting designed to provide automatic orientation of the instrument towards the sun or automatic direction of the light from the sun on to the instrument. A familiar form is that used, in conjunction with a PYRHELIOMETER, to obtain continuous measurement of direct solar radiation.

helium. So named because of its original discovery in the sun's atmosphere, helium is one of the INERT GASES. It occurs in very low concentration in the atmosphere, with VOLUME FRACTION and MASS FRACTION, relative to dry air, of 5.2×10^{-6} and 7.2×10^{-7}, respectively. Being very light, with a molecular weight of only 4.003, it escapes continuously from the top of the atmosphere at a rate which is in approximate balance with the rate of production of ALPHA PARTICLES (helium nuclei) near the earth's surface.

Helmholtz instability. See KELVIN-HELMHOLTZ INSTABILITY.

helm wind. A strong, cold, north-easterly wind which blows down the western slope of the Cross Fell range in Cumbria. Its greatest frequency is in late winter and spring. When the helm wind blows, a heavy bank of cloud (the 'helm') rests along or just above the Cross Fell range and a slender, nearly stationary roll of whirling cloud (the 'helm bar'), parallel with the 'helm', appears above a point 1 to 6 km from the foot of the fell. The helm wind is very gusty and often violent as it blows down the steep fell sides but ceases under the helm bar cloud. To the west of this point a light westerly may prevail over a short distance. The helm wind has a marked drying effect. The phenomenon is an example of a LEE WAVE.

The term 'helm wind' is applied to similar winds with associated cloud elsewhere, e.g. in the Lake District, but the full development is confined to the strip of country east of the River Eden, particularly near Cross Fell itself.

hertz (Hz). The unit of frequency of a periodic function, equal to one cycle per second.

heterosphere. Term for that region of the ATMOSPHERE, upwards of about 85 km, in which the composition of the atmosphere changes, hence also the mean molecular weight of the gases, mainly because of the partial DISSOCIATION of oxygen and DIFFUSION. If forms a contrast with the underlying HOMOSPHERE.

high. A term sometimes used in synoptic meteorology to indicate a high-pressure system. See ANTICYCLONE.

hill fog. A term generally used of low cloud which envelops high ground. The production of saturation and condensation by forced uplift of the air is not necessarily implied in the use of this term as it is in the case of UPSLOPE FOG.

histogram. A graphical representation of the frequencies of occurrence of a variable in certain ranges. The variable is taken as abscissa, and on each range is drawn a rectangle of area proportional to the number of cases in that range. If the ranges are all made equal then the heights of the rectangles are proportional to the frequencies.

The histogram allows the general character of a FREQUENCY DISTRIBUTION to be taken in at a glance; it is especially useful when, as for example winter temperatures in Alaska, a description in terms of the MEAN and STANDARD DEVIATION would be misleading.

hoar-frost. Ice crystals in the form of scales, needles, feathers or fans deposited on surfaces cooled by radiation or otherwise. The deposit is frequently composed in part of drops of DEW frozen after deposition and in part of ice formed directly from water vapour at a temperature below 0 °C (sublimation). The presence of fog, in so far as it checks the radiational cooling of surfaces, tends to prevent the formation of hoar-frost. See also RIME.

hodogram. See HODOGRAPH ANALYSIS.

hodograph analysis. A method of analysis of a wind sounding. By convention, the individual wind vectors at selected levels are plotted on a polar coordinate diagram from the origin of the diagram (representing station position) towards the direction from which the wind blows (see Figure 31). The method has the advantages of being easy to apply in practice and of being in accordance with the normal method of representing wind direction at a station. The lengths of the vectors are proportional to the corresponding wind speeds. The sense of each vector is towards the origin.

The lines joining the 'starting points' of airflow at successive levels form a hodograph (or hodogram). Each such line corresponds to the WIND SHEAR vector in the layer concerned. When taken in the appropriate sense (higher to lower level) it represents, on the assumption of geostrophic flow, the corresponding THERMAL WIND.

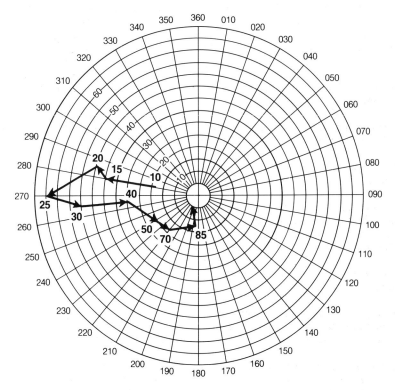

FIGURE 31. Hodograph relating to winds observed at Stornoway (58° 13'N, 6° 19'W) at 2300 GMT on 16 July 1980. Successive arrows represent the vector difference between the wind at the top of a layer (denoted by figures, in tens of millibars at arrow tail) and the wind at the bottom of a layer (figures at arrow head). The bottom of the lowest layer is the earth's surface.

Hollerith system. A now little-used mechanical system for the processing of climatological data which are represented by the position of holes punched into cards of standard size.

homogeneous atmosphere. A hypothetical atmosphere in which air density is constant with height; the lapse rate is, by definition, the AUTOCONVECTIVE LAPSE RATE (g/R). The height (z) of such an atmosphere is from the hydrostatic and gas equations given by

$$z = p_0/g\rho_0 = RT_0/g$$

147

where p_0, ρ_0 and T_0 are the surface pressure, density and temperature. By ignoring the small variation of g with height, z is seen to be proportional to the absolute air temperature at the surface. For $T_0 = 273$ K, z is about 8 km. See also SCALE HEIGHT.

homogeneous condensation, freezing.   See HOMOGENEOUS NUCLEATION.

homogeneous nucleation.   Homogeneous or 'spontaneous' nucleation signifies the initiation of either CONDENSATION or FREEZING in the absence of either condensation or freezing nuclei, the processes are also termed *homogeneous* or *spontaneous* condensation and freezing, respectively.

Aggregates of water molecules continuously form and evaporate in supersaturated air which is free from condensation nuclei. The probability that such aggregates will attain a critical size big enough for them to become more stable than the vapour and so act as centres on which further rapid growth (spontaneous condensation) will occur is small for degrees of supersaturation less than about 400 per cent; such an order of supersaturation does not occur in the atmosphere.

In air which is deficient in freezing nuclei, lowering the temperature below 0 °C increases the probability that aggregates of water molecules may take up an ice-like configuration and grow to a size sufficient for them to act as centres on which ice crystals may rapidly form. The variation with temperature of the probability of homogeneous nucleation is such that, in practice, there appears to be a critical temperature below which it occurs and above which it is absent. For water it is about -40 °C.

homosphere.   That region of the ATMOSPHERE, extending from the earth's surface to about 85 km, in which, neglecting water vapour, the composition of the atmosphere is constant (apart from some gases in very small concentration, e.g. carbon dioxide, ozone) and in which, therefore, the mean molecular weight of dry air is effectively constant. It forms a contrast with the overlying HETEROSPHERE.

horizon.   In meteorology, this term signifies the line where the earth's surface and the sky apparently join. Neglecting REFRACTION, the distance of the horizon from an observer at height h_0 is $(2h_0a)^{1/2}$ where a is the earth's radius. For a height of 100 ft (30 m) the corresponding distance is 20 km; the actual distance is about 4 km greater on account of refraction. If h_c is the altitude of a cloud top *just visible* to a sea-level observer at a distance d, then $h_c = d^2/(2a)$. For example, a cloud top at a distance of 200 km would be at a height of 10 000 ft (3050 m), or, conversely, a cloud top of height 10 000 ft on the horizon would be at a distance of 190 km. Distances corresponding to other heights may be obtained from Figure 32.

horse latitudes.   The belts of variable, light winds and fine weather associated with the subtropical anticyclones at about 30–40° latitudes. The belts move slightly north and south after the sun. The name arose from the old practice of throwing overboard horses, which were being transported to America or the West Indies, when the ship's passage was unduly prolonged.

hot-wire anemometer.   See ANEMOMETER.

hour angle.   The hour angle of a heavenly body at any instant is the angle (usually expressed as a time on the scale 24 hours $= 360°$) between the observer's MERIDIAN and the meridian through the body. The angle is measured westwards from the observer's meridian.

Hovmüller diagram.   A diagram with one axis representing time and the other longitude on which are shown isopleths of an atmospheric variation, such as pressure

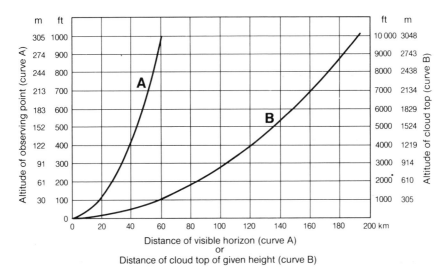

FIGURE 32. Relations between (curve A) the height of an observing point and the distance of the visible horizon, and (curve B) the height of a cloud (or other distant object) and the distance at which it is just visible on the horizon.

or thickness, usually averaged over a band of latitude. It usefully demonstrates the progression over weeks or months of large-scale atmospheric features.

humidity. The condition of the atmosphere in respect of its WATER VAPOUR content. The word 'humidity' used alone generally signifies RELATIVE HUMIDITY, but various other measures are employed such as HUMIDITY MIXING RATIO, VAPOUR CONCENTRATION, VAPOUR PRESSURE, SPECIFIC HUMIDITY, DEW-POINT, etc.

humidity mixing ratio. The humidity mixing ratio (r) — or, more generally, simply the 'mixing ratio' — of moist air is the ratio of the mass (m_v) of water vapour to the mass (m_a) of dry air with which the water vapour is associated, i.e.

$$r = m_v/m_a.$$

If e is the vapour pressure, p the total pressure and ϵ the ratio of the densities of water vapour and dry air ($\epsilon = 0.62197$), then

$$r = \epsilon e/(p - e).$$

Since e is small compared with p, r is a small quantity of about 0.01 kg kg^{-1}. For convenience, therefore, r is usually expressed in grams per kilogram.

humidity slide-rule, tables. The slide-rule or tables, based on the PSYCHROMETRIC FORMULA (with an assumed value of the product Ap in this formula appropriate to the conditions of exposure of the instruments), from which the DEW-POINT, VAPOUR PRESSURE, and RELATIVE HUMIDITY of an air sample may be obtained from readings of the dry- and wet-bulb hygrometer. See PSYCHROMETER.

humilis (hum). A CLOUD SPECIES. (Latin for low.)
    'CUMULUS clouds of only a slight vertical extent; they generally appear flattened' [2, p. 19]. See also CLOUD CLASSIFICATION.

hurricane. A name (of Spanish or Portuguese origin) applied to the intense TROPICAL CYCLONES which occur in the West Indies and Gulf of Mexico regions and off the Queensland coast. They are of essentially the same type as the western Pacific TYPHOON and Bay of Bengal CYCLONE.

In the BEAUFORT SCALE of wind force the name 'hurricane force' is given to a surface wind of force 12, corresponding to a mean speed of 64 knots or more. Mean speeds of this magnitude are very rarely attained in the British Isles, but the speed is frequently exceeded in gusts.

hurricane wave. The raising of the level of the sea by some 3–6 m in a restricted region near the centre of an intense TROPICAL CYCLONE.

hydrodynamic stability. An alternative for DYNAMIC STABILITY.

hydrogen. A gas, of molecular weight 2.02, with VOLUME FRACTION and MASS FRACTION, relative to dry air, of 5.0×10^{-7} and 4.0×10^{-8}, respectively.

Hydrogen is very abundant in the universe but occurs in only minute concentration throughout most of the ATMOSPHERE because its extreme lightness causes it to diffuse upwards to high levels.

hydrography. The study of water, more especially in the open seas and oceans, both from the physical standpoint and from that affecting the safety of navigation. In so far as tidal heights in estuaries and rivers are affected by river water, the science overlaps with that of HYDROLOGY.

hydrolapse. The rate of decrease with height of the amount of water vapour per unit mass or volume of air. It is not a term with a generally agreed standard meaning and, if used, needs precise definition in terms of HUMIDITY MIXING RATIO, VAPOUR CONCENTRATION, or other related quantity.

hydrological balance. The hydrological balance, or 'hydrological budget', is the relationship between the EVAPORATION (E), PRECIPITATION (P), RUNOFF (R), and the change in water storage (ΔS), for a specified land area and period of time, and is expressed by the hydrological balance equation

$$P = E + R + \Delta S$$

in which ΔS may be positive or negative.

hydrological budget. An alternative for HYDROLOGICAL BALANCE.

hydrological cycle. The full cycle of events through which water passes in the earth–atmosphere system, comprising EVAPORATION from land and water surfaces, CONDENSATION to form clouds, PRECIPITATION on to the earth's surface, movement and accumulation in the soil and bodies of water, and finally re-evaporation. 'Short-circuiting' of the complete cycle occurs in the form of evaporation of products of condensation and precipitation within the atmosphere.

hydrology. The study of the incidence and properties of water on and within the ground, including that held in rivers and lakes. It comprises studies of RAINFALL, EVAPORATION, RUNOFF, GROUND WATER, SOIL MOISTURE, the HYDROLOGICAL BALANCE, snow and ice accumulation, and the chemistry of natural waters.

The practical applications of hydrology are extremely wide and the number of local and government agencies and other bodies concerned with the subject in the United Kingdom is correspondingly large. The most comprehensive regular publication is *Hydrological data United Kingdom* [33].

hydrometeor. A generic term for products of CONDENSATION or SUBLIMATION of atmospheric water vapour. Hydrometeors include: ensembles of falling particles which may either reach the earth's surface (rain, snow, etc.) or evaporate during their fall (virga); ensembles of particles suspended in the air (cloud, fog, etc.); particles lifted from the earth's surface (drifting or blowing snow, spray); particles deposited on the earth or on exposed objects (dew, hoar-frost, etc.).

hydrometer. An instrument for measuring the density of liquids. In marine meteorology, hydrometers are used for determining the density of sea water.

hydrosphere. That part of the EARTH which is covered by water substance.

hydrostatic approximation. In dynamical meteorology, the assumption that the vertical component of the EQUATIONS OF MOTION is given by the HYDROSTATIC EQUATION. It has the effect of filtering out those solutions of the equations which correspond to vertically propagating sound waves, these being of no meteorological interest.

hydrostatic equation. In an atmosphere at rest with respect to the earth, the variation of pressure with geometric height (z) is given by

$$\frac{\partial p}{\partial z} = -\rho g.$$

In terms of GEOPOTENTIAL height (Z) the equation is

$$\frac{\partial p}{\partial Z} = -9.80665\rho,$$

the various quantities being expressed in SI UNITS.
  The hydrostatic equation is very closely approximated when the air has horizontal motion relative to the earth and is therefore used as the basis for computations of height from vertical soundings.

hydrostatic equilibrium. The state of balance between the force of GRAVITY and the vertical component of the PRESSURE GRADIENT FORCE in the atmosphere. It is a state in which there is no vertical acceleration and the HYDROSTATIC EQUATION holds. See also HYDROSTATIC APPROXIMATION.

hydroxyl. Hydroxyl radicals, of chemical formula OH, consist of one oxygen and one hydrogen atom. Hydroxyl has been identified in the spectrum of the AIRGLOW. It is considered to be formed by the DISSOCIATION of water vapour in the high atmosphere to form hydrogen atoms and hydroxyl.

hyetograph. A proprietary pattern of self-recording RAIN-GAUGE in which the recording pen is actuated by a series of stops attached to a vertical float rod.

hygrogram. The record made by a HYGROGRAPH.

hygrograph. A recording HYGROMETER. The type most familiar is the 'hair hygrograph' which uses the fact that human hair increases in length with increasing RELATIVE HUMIDITY. The changes in hair length are not linear, being proportionately less at high relative humidity than at low, but the recorded changes are generally made linear by the design of the mechanism. Hair hygrographs are not very precise, being subject in particular to gradual changes, but have the advantage of operating both above and below the freezing-point.

hygrometer. An instrument for measuring the humidity of the air. Among the many types of hygrometer are:

(i) The 'dry- and wet-bulb hygrometer' which is used in routine surface observations (see PSYCHROMETER).
(ii) Those hygrometers which use the property of expansion and contraction of certain materials with changing RELATIVE HUMIDITY as, for example, the 'hair HYGROGRAPH' in surface observations and GOLD-BEATER'S SKIN in upper-air observations.
(iii) 'DEW-POINT hygrometers' of the type designed by Daniell, Regnault and others in which artificial cooling of a polished surface, whose temperature is measured, is continued until dew is seen to condense on it (see also FROST-POINT HYGROMETER).
(iv) Chemical hygrometers in which the quantity of moisture in a given mass of air is determined by direct weighing.
(v) Resistive hygrometers in which the effect of humidity on the electrical resistance of films of hygroscopic substances is determined.

hygrometric formula. An alternative for 'psychrometric formula'. See PSYCHROMETER.

hygroscope. An instrument for showing whether the air is dry or damp, usually by the change in appearance or dimensions of some substance. Hygroscopes are frequently sold in the form of weather predictors, e.g. 'weather houses' in which the appearance of the 'old man' or the 'old woman' is determined by the twisting and untwisting of a piece of catgut in response to changes of humidity.

hygroscopic. A hygroscopic substance is one which tends to absorb moisture by accelerating the condensation of water vapour.

hygrothermograph. A combined HYGROGRAPH and THERMOGRAPH, i.e. an instrument in which variations of atmospheric humidity and temperature are continuously recorded by separate traces on a single sheet. The instrument is sometimes termed a 'thermohygrograph'.

hypsometer. Literally, an instrument for measuring height. In meteorology, however, the term is applied exclusively to an instrument in which the boiling-point of water is measured and the corresponding atmospheric pressure deduced from the boiling-point–pressure relationship. If, as is normal, the instrument is used on a mountain and the pressure is known at some known (low) level, the pressures deduced at higher levels may be converted to heights by means of the ALTIMETER equation. An accuracy of 0.01 K in temperature measurement is required to obtain height to within 3 m.

hypsometric formula. An alternative for 'altimeter equation'. See ALTIMETER.

hythergraph. A CLIMAGRAM in which the selected meteorological elements are temperature and humidity or temperature and rainfall.

I

ICAO. Abbreviation for International Civil Aviation Organization.

ICAO standard atmosphere. The STANDARD ATMOSPHERE adopted by the International Civil Aviation Organization (ICAO). Its chief specifications are: mean-sea-level (MSL) pressure 1013.25 mb; MSL temperature 288.15 K (15 °C); temperature lapse rate -6.5 K km^{-1} up to 11 km where the temperature is 216.65 K (-56.5 °C) and with an isothermal lower stratosphere above to 20 km. The extension, agreed in 1964 [5], of this atmosphere to 32 km specifies a lapse rate of -1.0 K km^{-1} which gives a temperature, at 32 km, of 228.65 K (-44.5 °C). Differences between the ICAO standard atmosphere and the ICAN (International Commission for Air Navigation) atmosphere which was previously adopted by many countries are generally negligibly small.

ice. Water substance in solid form. It occurs in the atmosphere and/or on the earth's surface in many forms such as ICE CRYSTALS, SNOW, HAIL, HOAR-FROST, RIME, GLAZE, GLACIER, etc.

The density of ice is generally about 0.917 g cm^{-3} (917 kg m^{-3}). At 0 °C the latent heat of fusion (pure ice to water) is 334×10^3 J kg^{-1} and the latent heat of sublimation (pure ice to vapour) is 2834×10^3 J kg^{-1}; these large values imply that the formation and disappearance of large masses of ice are important items in the heat budget of the earth–atmosphere system. The coefficient of linear expansion of pure ice at -10 to 0 °C lies between 0.000050 and 0.000054; its specific heat at 0 °C is 0.5. The physical properties of sea ice — in particular the values of latent and specific heats — may be very different from those of pure ice and vary greatly with temperature and SALINITY.

When a natural water surface cools, the water cooled at the surface is continually replaced by warmer water from below until the whole mass has fallen to 4 °C (39 °F), at which temperature water has its greatest density. The surface water then cools undisturbed until the freezing-point is reached and ice begins to form.

Three kinds of ice are produced in rivers:

(i)   sheet ice, which forms on the surface of the water, first of all near the banks and extending gradually towards the centre,
(ii) FRAZIL ICE, and
(iii) ground ice. Frazil ice and ground ice both form in the rapidly moving stream in the centre of the river in very cold weather.

Ground ice (or anchor ice) forms at the bottom, adhering to rocks and other substances in the river bed. It often rises to the surface, dragging with it masses of rock, and may destroy river structures.

When the sea freezes, the crystals formed contain no salt, but cannot easily be separated from the brine which is mixed up with them. When, however, the ice forms hummocks under the action of pressure, the brine drains out and leaves pure ice.

The *International meteorological vocabulary* [34, pp. 249–251] contains the Abridged International Ice Nomenclature which defines the terms used for sea ice.

ice accretion. The formation and building up of a layer of ICE on terrestrial objects (usually GLAZE (CLEAR ICE)) or on an aircraft in flight. See ICE FORMATION ON AIRCRAFT.

ice-age.   A geological period during which great GLACIERS and ICE SHEETS extended from the polar regions to as far equatorwards as about latitude 50°. Individual ice-ages wax (glacial phases) and wane (interglacial phases) in strength. Present-day climate is considered to be that appropriate to an interglacial phase of the Quaternary Ice-Age which began in the northern hemisphere about 2 to 3 million years ago; since then, the hemisphere has suffered predominantly glacial conditions for about 90 per cent of the time. Glacial phases lasting about 100 thousand years are separated by much warmer interglacial phases lasting about 10 to 12 thousand years. Antarctica has, however, been glaciated continuously for at least 15 million years.

At the peak of a glacial phase, land ice is considered to have covered an area about twice as large as at present. The term 'Little Ice-Age' is applied to the period from about 1550 to 1850 which was conspicuous for its low mean temperature and equatorward extension of glaciers.

Other great ice-age epochs occurred in the Permo-Carboniferous (250 million years ago), the late Ordovician (450 million years ago), the early Cambrian (650 million years ago) and there were two in the early Proterozoic (900 and 1250 million years ago). The Permo-Carboniferous Ice-Age is thought to have developed mainly within the present tropics, in South America, Africa, India and Australia. See also CLIMATIC CHANGE AND VARIABILITY.

iceberg.   A mass of ice, broken from a GLACIER ('glacier berg') or an ice shelf ('barrier berg' or 'tabular berg'), which floats in the sea; the WMO nomenclature specifies a minimum height of 5 m above sea level. Glacier bergs are greenish in colour, irregular in shape and are typical of the Arctic; barrier or tabular bergs are whitish in colour, more regular in shape and typical of the Antarctic.

Erosion of icebergs proceeds most rapidly at and below the water line. Favourable winds sometimes carry icebergs to latitudes 40–50° or even below.

Iceberg ice is as pure as distilled water, containing only four parts of solid per million. Icebergs also contain a considerable quantity of air, ranging in various specimens examined from 7 to 30 per cent.

iceblink.   A typical whitish glare on low clouds above an accumulation of distant ice. See also WATER SKY.

ice-bulb temperature.   That temperature (T_i) at which pure ice must be evaporated into a given sample of air, adiabatically and at constant pressure, in order to saturate the air at temperature T_i under steady-state conditions. The temperature recorded by the ice-covered 'wet bulb' of a PSYCHROMETER may not exactly accord with this definition. See also THERMODYNAMIC TEMPERATURES.

ice-crystal cloud.   A cloud which is composed (almost) exclusively of ice crystals, as opposed to water droplets. The CLOUD GENERA of CIRRUS, CIRROCUMULUS and CIRROSTRATUS are normally ice-crystal clouds.

ice crystals.   Ice crystals form in the atmosphere on ICE NUCLEI at temperatures appreciably below freezing-point. They exist in simple form, notably needles and hexagonal columns and plates, and in more complex star-shaped and branched forms. They may multiply by SPLINTERING and grow, by diffusion of water vapour on to them, into forms and at rates which depend on a variety of conditions, for example the internal and surface structure of the crystals and the prevailing conditions of temperature and supersaturation. The crystals collide and coalesce with other crystals or supercooled water droplets to form SNOWFLAKES or HAIL.

ice-crystal theory.   An alternative term for the BERGERON (or BERGERON–FINDEISEN) THEORY.

ice day. An ice day is defined as a period of 24 hours, beginning normally at 0900 GMT, on which the maximum temperature is less than 0 °C (32 °C).

ice formation on aircraft. Ice may form on aircraft either on the ground or in flight. ICE ACCRETION in flight may constitute a danger by affecting the aerodynamic characteristics, engine performance, or in other ways.

There are four types of airframe icing:

(i) RIME. A light, white, opaque deposit which forms, generally at temperatures well below 0 °C, in clouds of low water content consisting of small, supercooled water droplets.

(ii) GLAZE (CLEAR ICE). A coating of clear ice which forms in clouds of high water content consisting mainly of large, supercooled water droplets, or which forms when rain (usually supercooled) falls on an aircraft whose temperature is close to 0 °C or lower.

(iii) Cloudy ice or mixed ice. In flight, rime and glaze are often mixed because a cloud contains a large range of drop sizes and because ice crystals, liquid droplets and snowflakes may all be present in a cloud. These conditions affect the formation of ice and result in a rough, cloudy deposit called 'cloudy ice' or 'mixed ice'.

(iv) HOAR-FROST. A white crystalline coating of ice which forms in clear air, by sublimation of water vapour, when an aircraft surface is at a temperature below 0 °C and is colder than the frost-point of the air with which it is in contact. This may occur on a parked aircraft or when an aircraft moves rapidly (normally by descent) from a level at which the temperature is below 0 °C into a layer of warmer and relatively moist air.

When a supercooled water drop impinges on an aircraft, part of the drop freezes immediately at the leading edges, part may be lost before freezing, while the remainder freezes subsequently. For a small drop the final freezing occurs quickly, giving opaque ice on and near the leading edge; a light deposit of this kind may, however, be lost by the action of wind. For a large drop the final freezing is less rapid and the drop has the opportunity to spread over the wing, leading to the more dangerous type of clear ice.

Observations indicate that icing risks are greatest at temperatures not too far below freezing-point, from say 0 to −12 °C. No lower temperature limit can, however, be set for possible ice formation which has been reported, for example, at a temperature of −60 °C. Icing rate tends to increase with liquid water content of clouds and with cloud-particle size (greater for cumuliform clouds than for layer clouds). Clear ice is associated with the passage of fronts, chiefly warm fronts in winter, and is comparatively rare in the British Isles.

Engine and air-intake systems are liable to icing ('carburettor icing') in a variety of conditions which may include ambient temperature substantially above 0 °C and clear air.

Icelandic low. The 'Icelandic low' has a value of about 994 mb on the January chart of mean surface pressure. This mean depression is centred between Greenland and Iceland and represents the aggregate of many deep depressions. In summer, depressions in this region are less intense. Fluctuations of several millibars in depth, and small fluctuations in position, have been observed on decadal and longer time-scales, especially in winter.

ice nucleus. A generic term which includes both 'freezing nucleus' and 'sublimation nucleus'. See NUCLEUS.

ice pellets. Precipitation, generally from layered cloud, of transparent pellets of ice which are spherical or irregular, rarely conical, and which have a diameter of less than 5 mm.

ice prisms. A fall of unbranched ice crystals, in the form of needles, columns or plates, often so tiny that they seem to be suspended in the air. These crystals may fall from a cloud or from a cloudless sky. (Rarely observed in the British Isles.)

Ice Saints. St. Mamertus on 11 May, St. Pancras on 12 May and St. Gervais on 13 May known on the Continent as the 'cold Saints' days'. It is said in France that these three days do not pass without a frost. A Buchan cold spell from 9 to 14 May covers this period — see BUCHAN SPELLS.

ice sheet. A large area of land ice with a dome-shaped, almost smooth surface. The largest ice sheets now existing are those of the Antarctic and Greenland.

icing. See ICE ACCRETION and ICE FORMATION ON AIRCRAFT.

IGC. Abbreviation for INTERNATIONAL GEOPHYSICAL CO-OPERATION.

IGY. Abbreviation for INTERNATIONAL GEOPHYSICAL YEAR.

Illuminometer. The common name for a DAYLIGHT recorder.

incus (inc). A supplementary cloud feature. (Latin for anvil.)
'The upper portion of a CUMULONIMBUS spread out in the shape of an anvil with a smooth, fibrous or striated appearance' [2, p. 22]. See also CLOUD CLASSIFICATION.

independence. A term used in its natural sense in statistical work, but often with the implication that the COVARIANCE or CORRELATION between corresponding values of the elements concerned does not differ significantly from zero.

index. The pointer or other feature in an instrument whose position with regard to the scale determines the reading. The term is also sometimes applied to the fixed mark which constitutes the zero; thus the 'index error' of a barometer is the error which is due to faulty positioning of the scale.

index correction. The quantity which (with proper sign) has to be added to an instrumental reading to correct for INDEX error. It is of the same magnitude as the index error, but of the opposite sign.

index cycle. A term often applied to alternating periods of predominantly ZONAL FLOW and MERIDIONAL FLOW. The corresponding variations of the ZONAL INDEX are, however, only roughly cyclic in nature. See CYCLE.

index error. See INDEX, INDEX CORRECTION.

Indian summer. A warm, calm spell of weather occurring in autumn, especially in October and November. The earliest record of the use of the term is at the end of the 18th century, in America, and it was introduced into the British Isles at the beginning of the 19th century. There is no statistical evidence to show that such a warm spell tends to recur each year.

indirect circulation. See DIRECT CIRCULATION.

inert gases. The elements HELIUM, NEON, ARGON, KRYPTON, XENON and RADON, termed 'inert gases' (sometimes also 'noble gases' or 'rare gases') because of their chemical inactivity. Apart from argon (VOLUME FRACTION 9.3×10^{-3}) these gases occur in the air in only minute quantities.

inertial flow. Flow in the absence of external forces. See CIRCLE OF INERTIA.

inertial stability. A type of DYNAMIC STABILITY, associated with the earth's rotation, in which an air particle, embedded in a wind flow along a circle of latitude, tends to return to this latitude on being subjected to a small displacement from it. Inertial instability arises when such a displacement results in an acceleration of the particle away from its original latitude.

The condition for stability is that the CORIOLIS PARAMETER should exceed the northward increase of the geostrophic west-wind component, i.e. $f > \partial u_g / \partial y$. Similarly, instability arises when $f < \partial u_g / \partial y$.

Similar conditions apply to air currents which are not east–west and may be extended to displacements along sloping isentropic surfaces which are most likely to give rise to instability. The criterion for stability in such currents is $f > (\partial v / \partial n)_\theta$ where v is the velocity and n the distance measured normal to the flow along the isentropic surface.

Strictly, these criteria apply only in the simple case of straight flow with no variation of velocity along the flow. In the more general case of curved flow with velocity variations the criteria become much more complicated.

inertia wave. A type of stable atmospheric wave which arises because of the inertia of a mass moving over the earth's surface and the associated CORIOLIS FORCE arising from the earth's rotation.

infra-red radiation. ELECTROMAGNETIC RADIATION in the approximate wavelength range from about 0.7 to 1000 μm; 52 per cent of the total solar RADIATION intensity is contained within this range of wavelengths, the amount at wavelengths greater than 4 μm being very small.

In meteorology, the term is often used loosely as an alternative for LONG-WAVE RADIATION.

initialization. The process of obtaining suitable initial values of all the independent variables used in carrying out a numerical forecast from the original observations (assumed corrected for instrumental and transmission errors). Forecast models based on the PRIMITIVE EQUATIONS have as their solutions high-speed GRAVITY WAVES as well as slower-moving disturbances which are of meteorological significance. The feeding into a primitive equation model of grid-point values interpolated from an OBJECTIVE ANALYSIS, produced without consideration of the controlling equations of that model, would generally excite spurious gravity waves of large amplitude which would render the forecast useless. This difficulty may be overcome in various ways of which two of the most commonly used are the following:

(i) *Non-linear normal-mode initialization*

The dynamical forecast equations, when their forcing terms are removed, have as solutions (for disturbances of small amplitude) an infinite set of what are known as 'normal modes'. (These, though much more complicated, are analogous to the free oscillations of a taut elastic string or of the surface of a liquid in a rectangular tank.) The low-order modes represent large-scale flow patterns in quasi-geostrophic balance, and the high-order modes represent fast-moving gravity waves. The results of the objective analysis may be analysed into a set of normal modes, each with a different amplitude, of which

157

the high-order components represent 'noise' which must be eliminated. Because the actual forecast equations are non-linear, and the normal-mode analysis is derived from linear approximations to them, it is not possible to produce a properly initialized field simply by throwing away all the high-order components, because non-linear interactions would tend to regenerate them. Instead, a procedure like a very short forecast is gone through in which the 'rates of change' of the unwanted high-order gravity waves are equal to zero, and this yields a suitably balanced set of values from which the actual forecast can be run. Normal-mode initialization is used as routine by the European Centre for Medium-range Weather Forecasts.

(ii) *Continuous data assimilation*

In this method the objective analysis and initialization processes are effectively combined into one. The forecast equations themselves will dissipate spurious gravity waves, given sufficient time, so that a forecast run even from a non-initialized analysis would, after a week or so, look quite realistic.

In continuous data assimilation a forecast model, possibly modified to enhance damping of gravity waves, is nudged each time step towards consistency with the observations. As long as the excitation of spurious gravity waves by the nudging process is less than the forecast model's dissipation of them, then the model state will remain close to the ideal initialized state.

This is the method used for the Meteorological Office operational forecast model. At the end of each time step the deviations of the model state from the observations valid near that time are calculated. A correction field to the model state, such as to reduce these deviations, is calculated, and a fraction (λ) of this correction field is added to the forecast field. Excitation of gravity-wave noise by this nudging process is kept to a minimum by using a small value for λ, and by incorporating some geostrophic and hydrostatic coupling in the correction fields.

insolation. A term which is used in various senses:

(i) The intensity at a specified time, or the amount in a specified period, of direct solar RADIATION incident on unit area of a horizontal surface on or above the earth's surface.

(ii) The intensity at a specified time, or the amount in a specified period, of total (direct and diffuse) solar radiation incident on unit area of a specified surface of arbitrary slope and aspect.

The insolation depends on (a) SOLAR CONSTANT, (b) calendar date, (c) latitude (involving variations of length of day and degree of obliquity of sun's rays), (d) slope and aspect of surface, and (e) degree of transparency of atmosphere.

The purely astronomical factors (including variations of length of day but omitting all effects of atmospheric attenuation which depend in part on latitude) are accounted for in the tables and charts contained in the *Smithsonian meteorological tables* [15, pp. 418/9]. These show the daily solar radiation totals on a horizontal surface at various latitudes and dates, on the assumption of a perfectly transparent atmosphere; seasonal and annual totals are also given. Their main features are a primary daily maximum at the summer pole, a secondary daily maximum at about 45° latitude in the summer hemisphere, a minimum (zero) at and near the winter pole, and an annual total at the equator about 2.4 times that at either pole.

These theoretical results have only partial relevance to what is found in practice. If, for example, account is taken of attenuation of the radiation by the atmosphere, the maximum at the summer pole disappears because of the oblique nature of the radiation received in high latitudes. If, further, account is taken of systematic

158

latitudinal variations of atmospheric attenuation of radiation by clouds, the secondary maximum at latitude 45° moves to the semi-arid climatic zone at about latitude 35°.

instability.   See STABILITY.

instability index.   A numerical measure of the likelihood of thunderstorm development based on the static stability of the atmosphere — see STABILITY.

Various such indices have been proposed, most of them based on empirical studies relating the distribution of temperature, usually also of humidity, at selected levels within the troposphere, to thunderstorm incidence. See BOYDEN INDEX and SHOWALTER INDEX.

instability line.   In synoptic meteorology, a line or belt, generally some hundreds of kilometres in length, along which slight or moderate instability (convective) phenomena exist but which is not marked by a surface front; such a line may, however, be marked by an UPPER-LEVEL TROUGH. When the instability phenomena are of a violent nature the term SQUALL LINE is generally used.

instant occlusion (or **pseudo-occlusion**).   A lambda-shaped cloud pattern visible in a satellite photograph when a cloud band in a low-level CONVEYOR BELT associated with a polar trough interacts with an upper-level cloud band associated with the polar front JET STREAM. The appearance of the 'instant occlusion' in the photograph is closely similar to that of the classical OCCLUSION associated with a well developed frontal depression; without knowledge of the way in which the cloud patterns had developed, therefore, 'instant' and 'classical' occlusions might well be confused.

intensification.   In synoptic meteorology, 'intensification' of a pressure system signifies an increase with time of the pressure gradient around the centre of the system. The converse term is 'weakening'.

inter-diurnal variation.   Day-to-day change (of a specified element).

interglacial phase.   A period of an ICE AGE during which (as now) permanent ice at low levels is confined to latitudes polewards of about 75°. See GLACIAL PHASE.

internal energy.   For PERFECT GASES internal energy (E) is a function only of temperature. The change of internal energy per unit mass for a temperature increment dT is

$$dE = c_u dT.$$

For an atmospheric column of unit cross-section, extending from the surface to height h, the internal energy is given by

$$E = c_u \int_0^h \rho T dz$$

where z is the height. See also POTENTIAL ENERGY.

International Geophysical Co-operation (IGC).   Period from 1 January to 31 December 1959, during which the major part of the observational programme of the INTERNATIONAL GEOPHYSICAL YEAR was, by international agreement, continued.

International Geophysical Year (IGY).   An international programme of observation in nearly all branches of GEOPHYSICS, on a world-wide scale, from 1 July 1957 to 31 December 1958, near sunspot maximum.

International Polar Year. See POLAR YEAR.

International Quiet Sun Years (IQSY). An international programme of observation in nearly all branches of GEOPHYSICS, on a world-wide scale, from 1 January 1964 to 31 December 1965, near sunspot minimum.

inter-quartile range. See QUARTILE.

intertropical convergence zone. The intertropical convergence zone (ITCZ) is a relatively narrow, low-latitude zone in which AIR MASSES originating in the two hemispheres converge.

Over the Atlantic and Pacific Oceans, where it is closely related to the DOLDRUMS, the ITCZ is the boundary between the north-easterly and south-easterly TRADE WINDS. Over the continents it is replaced by the boundary between other wind systems with components directed towards the equator, for example in Africa between the HARMATTAN and the south-west MONSOON. The ITCZ moves northwards in the northern summer and southwards in the southern summer, its mean position being somewhat north of the equator. Over the oceans the range of movement is small, but over the continents it may be very large.

The horizontal convergence associated with the ITCZ implies generally upward motion in the lower troposphere and cloudy, showery weather. The zone is subject to day-to-day oscillations of position and variations of activity. Shallow low-latitude depressions, which may on occasion greatly intensify, are often associated with the zone.

intortus (in). One of the CLOUD VARIETIES. (Latin for twisted.)
'CIRRUS, the filaments of which are very irregularly curved and often seemingly entangled in a capricious manner' [2, p. 20]. See also CLOUD CLASSIFICATION.

intransitive system. See TRANSITIVE AND INTRANSITIVE SYSTEMS.

inversion. An inversion (of temperature) is said to occur at a point, or through a layer, at which or through which temperature increases with increasing height; such a feature is an inversion of the condition of a positive LAPSE of temperature which normally exists in the atmosphere.

An inversion is marked on a normal temperature–height diagram (height the vertical ordinate) by a line which slopes upwards to the right. It implies great static STABILITY and absence of turbulence at the level concerned. See also INVERSION LAYER.

inversion layer. An atmospheric layer in which there is an INVERSION of temperature. Vertical motion of air through such a layer is inhibited by considerations of static STABILITY. If the air under the inversion is relatively moist, the layer itself often lies above a layer of stratified cloud. An inversion layer near the earth's surface ('surface inversion') occurs on a RADIATION NIGHT; inversion layers at higher levels are often associated with the tropopause and anticyclones, and sometimes with fronts.

ion. The name (selected by Faraday) for an electrically charged atom or molecule; the presence of ions in a gas or electrolyte renders it an electrically conducting medium.

Ions are formed in the low atmosphere mainly by radiations from radioactive material in the air and by cosmic radiation; on a local scale, the rupture of water drops or the friction of wind-driven snow or sand may be important in producing atmospheric ions. Cosmic rays are an increasingly important ionizing source with increase of height above sea level. Solar ultraviolet radiation and X-rays are mainly responsible for the intense ionization of the atmosphere above about 50 km.

160

In the process of ion formation, an electron is ejected from a neutral particle (molecule or atom) which thus acquires a positive charge, while the electron attaches itself to another neutral particle which acquires a negative charge; an 'ion pair' is thus formed. These so-called 'small ions' are highly mobile; they may disappear by recombination or, near the earth's surface, by becoming attached to electrically neutral (e.g. pollution) particles, forming 'large ions' of much lower mobility, with a resulting decrease in air conductivity.

ionization. The process of ION formation.

ionization potential. The minimum energy (ELECTRON-VOLTS) required to remove an electron from an atom or molecule. It is also termed 'ionization energy'.

ionogram. In radio-echo sounding of the IONOSPHERE an automatic record of corresponding values of wave frequency and VIRTUAL HEIGHT.

ionosphere. That portion of the earth's ATMOSPHERE, extending upwards from about 60 km to an indefinite height, which is characterized by a concentration of IONS and free ELECTRONS high enough to cause reflection of radio waves.

The IONIZATION, which is caused at these levels mainly by solar ultraviolet and X-ray radiation, reaches peak values, separated by shallow troughs, in rather well defined regions of the atmosphere, giving rise to the ionospheric E-, F_1- and F_2-layers situated at about 110, 160 and 300 km, respectively. There is sometimes no peak value but only a 'ledge' in which the height gradient of electron concentration is small. Radio reflections are sometimes also obtained from heights of 65 to 80 km — the so-called D-layer which is not, however, characterized by a maximum ionization concentration.

The E- and F_1-layers are, to a first approximation, 'regular' in that their peak values of ion and electron densities and corresponding heights have systematic latitudinal, diurnal, seasonal and sunspot-cycle variations in reasonable accord with CHAPMAN LAYER theory (electron density highest near the equator, at midday, in summer, and at sunspot maximum, respectively, when the intensity of the ionizing radiation is greatest). The corresponding features of the F_2-layer, on the other hand, show many anomalies due to solar and lunar tidal effects and the influence of the earth's magnetic field. The large, short-period changes of electron height distribution and concentration in this layer are closely correlated with geomagnetic storms and are termed 'ionospheric storms'.

The properties of the ionosphere are regularly investigated by a world-wide network of stations using an automatic radio-echo technique ('ionospheric echo sounding') in which short bursts of radio waves of different frequencies are emitted in rapid succession at vertical incidence and the resulting reflected waves registered on an 'ionogram'. Since the electron concentration required for reflection of radio waves increases with the frequency of the waves the technique yields a CRITICAL FREQUENCY of penetration of the regions of electron peak concentration and the corresponding VIRTUAL HEIGHT. Rockets and satellites are also used to sound the ionosphere and have yielded information, in particular, of the electron concentrations in the troughs separating the various layers and at levels above the F_2-layer.

Observations show that the ionosphere is in a state of constant and complex motion which is in part systematic (including tidal), and in part random, in nature. In the lower ionosphere the charged particles are generally moved bodily with the uncharged particles; at higher levels the motion of the charged particles is mainly independent of that of the uncharged particles and is much affected by the forces exerted by the earth's magnetic field.

IQSY. Abbreviation for INTERNATIONAL QUIET SUN YEARS.

iridescence. A word formed from the name of Iris, the rainbow goddess, to indicate rainbow-like colours; an alternative is 'irisation'.

Iridescence in the form of tinted patches of red and green, sometimes of blue and yellow, is occasionally observed on high clouds, generally within about 30° of the sun. The boundaries of the tints are not circles with the sun as centre but tend to follow the outlines of the cloud. Iridescent clouds are considered to be parts of CORONAE, the coloration being caused by DIFFRACTION of sunlight by very small cloud particles.

irrotational motion. Motion in which there is no VORTICITY. It is defined by the equation curl $\mathbf{V} = 0$ where $\mathbf{V}$ is the three-dimensional velocity vector which defines the motion. In meteorology, where the vertical component (ζ) of vorticity is dominant, it is defined by $\zeta = 0$.

isallo-. A prefix used to denote lines drawn on a map or chart to display the tendency (time rate of change) of any element, each line being drawn through places at which the element has the same tendency, e.g. ISALLOBAR, isallotherm.

isallobar. A line of constant BAROMETRIC TENDENCY. Such lines are sometimes drawn on synoptic charts, mainly as an aid to forecasting the movement of features of the pressure distribution.

isallobaric wind. That part of the AGEOSTROPHIC WIND which is associated with local time change of the pressure gradient.

One of the conditions for the application of the GEOSTROPHIC WIND equation is that the pressure distribution should be steady. In the treatment by D. Brunt and C.K.M. Douglas of the effect on the wind of a changing horizontal pressure gradient, it was shown that the main term additive to the geostrophic wind corresponds to a wind directed normally to the ISALLOBARS towards the isallobaric low and of magnitude $(1/\rho f^2)\mathrm{grad}_H(\partial p/\partial t)$ where $\mathrm{grad}_H(\partial p/\partial t)$ is the horizontal gradient of the isallobars. This, now generally known as the 'isallobaric wind', rarely exceeds 5 m s^{-1}, but is at times important in its effects on horizontal convergence and rainfall.

isanomaly. A line of constant anomaly (i.e. difference from normal) of a meteorological variable.

isentropic. Without change of ENTROPY, generally equivalent in meaning to ADIABATIC. Isentropic surfaces in the atmosphere are surfaces of constant POTENTIAL TEMPERATURE.

isentropic analysis. An analysis of the physical and dynamical processes taking place in the free atmosphere on the basis of the location and configuration of the various ISENTROPIC surfaces and distribution of atmospheric properties (e.g. temperature, humidity mixing ratio) and motion on them. 'Isentropic charts' are the main tool of this work.

iso-. A prefix meaning 'equal', extensively used in meteorology to denote lines drawn on a map or chart to display the geographical distribution of any element, each line being drawn through places at which the element has the same value, e.g. ISOBAR, ISOTHERM. The words 'isogram' or 'isopleth' are sometimes used as generic names of this type.

isobar. A line of constant (atmospheric) pressure.

'Surface isobars' are based on pressure readings corrected to a common level (see REDUCTION to sea level). Such isobars, drawn at intervals of one millibar or more (depending on scale of chart), produce characteristic patterns such as are illustrated in

162

the entries on ANTICYCLONE, COL, DEPRESSION, RIDGE, SECONDARY DEPRESSION, and TROUGH.

isobaric analysis.   Analysis along a surface of constant pressure; the analysis usually includes variations of GEOPOTENTIAL, temperature, humidity and wind.

For study of conditions in the upper air there are theoretical and practical advantages in employing isobaric analysis rather than analysis along surfaces of constant geopotential. The GEOSTROPHIC WIND speed is related to contour spacing in a way that is independent of air density and permits the use of the same wind scale at different isobaric levels.

isobaric surface.   Surface of constant (atmospheric) pressure.

isoceraunic line.   A line of constant percentage frequency of days in a year on which thunder is heard.

isochasm.   A line of constant frequency of visible AURORA.

isochrone.   A line drawn on a map in such a way as to join places at which a phenomenon is observed at the same time, e.g. lines indicating the places at which rain commences at a specified time.

isodynamic line.   A line of constant total magnetic intensity.

isogon.   A line of constant wind direction.

isogonic (or isogonal) line.   A line of constant magnetic declination. See DECLINATION, MAGNETIC.

isogram.   See ISO-.

isohaline.   A line of constant SALINITY.

isohel.   A line of constant sunshine duration.

isohyet.   A line of constant rainfall amount.

isohypse.   An alternative for CONTOUR.

isokinetic sampling.   Air sampling which takes place at a rate such that the flow velocity in the sampling tube is equal to that of the environment being sampled. To ensure that no error occurs in the sampling of particulate matter, isokinetic conditions should hold.

isomeric line.   The distribution of monthly rainfall over an area may be represented by a plot of percentage of average annual rainfall attained in the month at each station (isomeric values). A line of constant percentage on such a chart is an isomeric line or isomer.

isoneph.   A line of constant cloud amount.

isopleth.   See ISO-.

isopycnic surface.   A surface of constant (atmospheric) density. Such surfaces intersect ISOBARIC SURFACES in a BAROCLINIC, but not in a BAROTROPIC, atmosphere.

isosteric surface. A surface of constant (atmospheric) SPECIFIC VOLUME. Since specific volume is the inverse of density, an isosteric surface is also an ISOPYCNIC SURFACE.

isotach. A line of constant wind speed.

isotherm. A line of constant temperature.

isothermal. Of equal temperature. The term 'isothermal layer' was originally applied to the region now known as the STRATOSPHERE. The term is, in this restricted sense, now obsolete but is still used to denote any layer with zero LAPSE rate of temperature, revealed by vertical sounding of the atmosphere.

isovel. An alternative for ISOTACH.

J

Jacob's ladder. See CREPUSCULAR RAYS.

jet streak. A core of strong winds in the middle or lower TROPOSPHERE, decreasing sharply above and below and on either side, with a length of not more than a few hundred kilometres. Jet streaks may occur in association with intense and rapidly developing DEPRESSIONS and are temporary features in comparison with the main JET STREAM.

jet stream. A fast narrow current of air, generally near the TROPOPAUSE, characterized by strong vertical and lateral wind shears. A jet stream is usually some thousands of kilometres in length, hundreds of kilometres in width and some kilometres in depth. The vertical and horizontal shears are as a rule of the order 5 to 10 m s^{-1} (10 to 20 kn) per km and 5 m s^{-1} (10 kn) per 100 km, respectively, but much stronger shears can occur with more intense jet streams. For certain purposes, arbitrary minimum speeds varying from about 25 to 50 m s^{-1} (50 to 100 kn) have been used to define a jet stream.

The distance from a jet-stream core to the nearest point at which the wind velocity falls to one half of the peak value is termed the 'half-width' of the jet stream.

Two main types of jet stream are recognized, (i) subtropical (westerly) and (ii) polar front (westerly). The subtropical jet stream is relatively constant in position in a given season and dominates mean seasonal wind charts; in contrast, the polar-front jet stream is highly variable in position from day to day over a wide range of temperate latitudes and is therefore masked on such charts. In addition, a westerly 'polar-night jet stream' (high latitudes in winter) occurs at times within the stratosphere above 50 mb, and an easterly jet stream occurs in the stratosphere in the equatorial belt of the eastern hemisphere.

There also exist at much lower levels in the atmosphere other narrow, fast currents of air which are only some tens or hundreds of kilometres in length and are usually referred to as jets rather than jet streams; for these see LOW-LEVEL JET and WARM CONVEYOR BELT.

joule (J). The unit of work or energy in SI UNITS and in the M.K.S. SYSTEM of units. This unit was also recommended at the Ninth General Conference of Weights and Measures (1948) for use as the unit of HEAT.

The joule is the work done by a force of one NEWTON in moving its point of application one metre in the direction of the force. The dimensions are ML^2T^{-2}.

$$1 \text{ J} = 10^7 \text{ERG}.$$

K

Kármán's constant. See VON KÁRMÁN'S CONSTANT.

katabatic wind. On a 'radiation night' of clear skies and slack pressure gradient, LONG-WAVE RADIATION from the earth's surface causes a layer of cold air to form near the ground, with an associated INVERSION of temperature. If the ground is sloping, the air close to the ground is colder than air at the same level but at some horizontal distance. Downslope gravitational flow of the colder and denser air beneath the warmer and lighter air produces the 'katabatic wind'. It is known also as the 'drainage wind' or 'mountain breeze'. This downslope wind has important agricultural effects. On a larger scale it is experienced, for example, as the fiord winds of Norway; on a still larger and more violent scale it is the main component of the outflowing winds from the Greenland and Antarctic continents. See also DOCTOR.

katafront. A FRONT (warm or cold) along which the warm air is descending relative to the cold air. Downward motion of the warm air at most levels is generally implied, with frontal activity feeble or absent.

Kelvin effect. An effect, relating to the saturation vapour pressure over liquid surfaces, discovered by Lord Kelvin.

The saturation VAPOUR PRESSURE (p) at a curved liquid surface (e.g. spherical water droplet) exceeds that over a similar plane surface (p_∞) by an amount which increases with decreasing radius of curvature (r) of the surface. The relationship between them is given by

$$\log_e \frac{p}{p_\infty} = \frac{2M\sigma}{\rho R^* Tr}$$

where M is the molecular weight of the vapour, σ the surface tension of the liquid, ρ the liquid density, R^* the universal gas constant and T the temperature.

The effect of surface curvature is important for droplets of diameter less than about 1 μm; at 10 °C, for example, the ratio p/p_∞ is 1.023 for a water droplet of diameter 0.1 μm, and 1.002 for a diameter of 1 μm. The effect is of fundamental importance in the initiation of CONDENSATION in the atmosphere. See also NUCLEUS.

Kelvin–Helmholtz instability. Instability arising from a strong vertical shear of wind through a narrow atmospheric layer across which there is a sharp gradient of temperature and density, e.g. at an inversion. A wave-like perturbation may be set up which gains energy at the expense of the large-scale flow. Kelvin–Helmholtz instability is often made visible in the phenomenon of BILLOW CLOUDS.

Kern's arc. A faint and very rare HALO PHENOMENON which is centred on a point opposite to, and on the same circle as, a CIRCUMZENITHAL ARC.

Kew-pattern barometer. A portable marine BAROMETER designed by P. Adie in 1854 for the Kew Committee of the British Association. The scale is graduated to take account of changes of the level of the mercury in the steel cistern so that it is only necessary to read the top of the mercury column. The tube is constricted to minimize

166

PUMPING when the barometer is used at sea. Similar barometers, known as 'station' barometers, but without the constriction, were subsequently adopted for use on land and were regularly employed at British meteorological stations, but have now been generally replaced by PRECISION ANEROID barometers.

khamsin. A southerly wind blowing over Egypt in front of depressions passing eastwards along the Mediterranean or north Africa, while pressure is high to the east of the Nile. Because this wind blows from the interior of the continent it is hot and dry, and often carries much dust. It is most frequent from April to June.

Gales from the south or south-west in the Red Sea are also known as khamsin.

kilocalorie. The metric thermal unit, being the heat required to raise the temperature of 1 kilogram of water from 14.5 to 15.5 °C. See also CALORIE, JOULE.

kilogram. The unit of mass in the M.K.S. SYSTEM of units. See also GRAM, SI UNITS.

kilomole. The kilogram-molecular weight of a substance, i.e. the weight of it in KILOGRAMS which is numerically equal to its MOLECULAR WEIGHT.

kinematical analysis. Analysis of the field of atmospheric wind flow.

kinetic energy. The ENERGY possessed by a body by virtue of its motion. It is a scalar quantity of magnitude $\frac{1}{2}mv^2$, where m is the mass and v the velocity of the particle.

D. Brunt estimated that the total kinetic energy of the large-scale motion of the atmosphere is of the order 3×10^{20} joules (8×10^{13} kilowatt-hours). He also estimated that, in the absence of solar radiant energy, dissipation of the atmosphere's kinetic energy by turbulence, generated mainly at the earth's surface, would be almost complete after 6 days.

Kirchoff's law. See RADIATION.

kite balloon. A captive balloon, of such a form as to have aerodynamic lift.

knot. A speed of one nautical MILE per hour (approximately 0.5 m s^{-1}).

Kolmogoroff similarity hypotheses. These hypotheses are concerned with the transfer of energy in a turbulent fluid from large to small eddies and its eventual dissipation at the molecular level. Kolmogoroff considered the large anisotropic eddies to be the prime sources of energy which is then transferred down the size scale. At some point the eddies become isotropic and homogeneous, hence similar to one another, and from then on their energy is determined only by the rate of transfer from the larger eddies and the rate of dissipation by the smaller ones.

The hypotheses are:

(i)   At large REYNOLDS NUMBERS the local average properties of the small-scale components of any turbulent motion are determined entirely by kinematic viscosity and the average rate of dissipation per unit mass.

(ii) There is an upper 'inertial sub-range' to the small eddies in which the local average properties are determined only by the rate of dissipation per unit mass. It may be shown from these hypotheses that in the inertial sub-range the energy is partitioned among the eddies in proportion to $K^{-5/3}$ where K is the wave number.

The Kolmogoroff hypotheses do not apply to atmospheric weather systems whose horizontal extent is much greater than their depths.

kona storm. A storm over the Hawaiian Islands, associated with a depression passing north of the islands and characterized by strong or gale force southerly winds and heavy rain.

konimeter. An instrument for counting the dust particles in a known volume of air. See DUST COUNTER.

Köppen classification. A classification of climate according to the way in which the fauna and flora of regions are dependent on the local variations of temperature, wind, and rainfall. Each climate is described by a short group of letters, the most important groups designated by capitals with subdivisions by lower-case letters. The main groups are:

A   Tropical rain climates
B   Arid climates
C   (Warm) temperate rain climates
D   Boreal forest and snow climates
E   Cold snow climates (treeless)
BS   Steppe grassland climate
BW   Warm desert climates

The most important subdivisions are:

s   Summer dry season
w   Winter dry season
x   Early summer wet, late summer fine
f   Moist — no marked dry season
m   Monsoon climates, with short dry season in winter

For example, central and western Europe, including the British Isles, is classified as Cf, and Mediterranean countries as Cs.

The Köppen classification allows considerable elaboration, and a good description is given by Lamb [35]. It is probably open to fewer objections than any other simple system of classifying climates and is widely used.

kosava. A RAVINE WIND which occurs on the Danube, south-east of Belgrade.

Koschmieder's law. A law which states that the apparent brightness (B_s) of a black object at distance d, due entirely to scattered light, is related to the brightness of the horizon at the same azimuth (B_h) by the equation

$$B_s = B_h (1 - e^{-\beta d})$$

where β is the SCATTERING coefficient, assumed constant throughout the part of the atmosphere concerned.

See also VISIBILITY, AIRLIGHT.

krypton. One of the INERT GASES, with VOLUME FRACTION and MASS FRACTION, relative to dry air, of 1.0×10^{-6} and 3.0×10^{-6}, respectively. Its molecular weight is 83.80.

K-theory. The K-theory, also termed the EXCHANGE COEFFICIENT theory, of turbulent mixing is based on the assumption that the vertical flux of a conservative air-mass property (E) which is effected by atmospheric eddies may be expressed as the product of an exchange coefficient (K) and the vertical gradient of the property concerned, i.e.

$$\text{flux} = K \, \partial E / \partial z.$$

The value of the theory is limited by the extent to which K is dependent on height and atmospheric stability, on the atmospheric property concerned (heat, water vapour, momentum, etc.), and, most of all, by the source distribution of E. A typical value of K is 1 m^2 s^{-1}. See also TURBULENCE, DIFFUSIVITY, EDDY CONDUCTIVITY.

Kuro Shio. Meaning 'black salt-water current' or 'black tide', this is a warm-water current of a characteristic dark blue colour, the main branch of which flows north-eastwards along the south coast of Japan before merging in the general drift of the North Pacific. This current is analogous to the GULF STREAM of the Atlantic in that it carries warmth to higher latitudes. A branch of the Kuro Shio travels into the Sea of Japan.

kurtosis. A statistic describing the peakedness of a symmetrical FREQUENCY DISTRIBUTION, defined in terms of the MOMENTS as $\beta_2 = \mu_4 / \mu_2{}^2$. For the NORMAL (FREQUENCY) DISTRIBUTION $\beta_2 = 3$.

The measure ($\beta_2 - 3$) is sometimes called the 'excess of kurtosis'. A distribution for which $\beta_2 > 3$ is termed 'leptokurtic' and has a high peak, deficient shoulders and long tails, compared with the normal distribution; while one for which $\beta_2 < 3$ is 'platykurtic' and has comparatively large shoulders and deficiencies in peak and tails.

L

labile. A term sometimes used to denote a condition of static instability or of neutral equilibrium in the atmosphere, i.e. a LAPSE rate equal to or exceeding the ADIABATIC. In German, the word is used to denote instability only.

lacunosus (la). One of the CLOUD VARIETIES. (Latin for having holes.)
'Cloud patches, sheets or layers, usually rather thin, marked by more or less regularly distributed round holes, many of them with fringed edges. Cloud elements and clear spaces are often arranged in a manner suggesting a net or a honeycomb.
This term applies mainly to CIRROCUMULUS and ALTOCUMULUS; it may also apply, though very rarely, to STRATOCUMULUS' [2, p. 21]. See also CLOUD CLASSIFICATION.

lag. A delay between a change in conditions and its indication by an instrument. While instrumental construction is usually designed to minimize lag, it is in some cases deliberately introduced, e.g. in mercurial BAROMETERS (by narrowing of tube) to avoid PUMPING, and in SOIL THERMOMETERS (by immersion of bulb in micro-crystalline wax) to ensure negligible change of indicated temperature on withdrawal of the instrument from the earth for reading.
In statistics, the lag is the displacement in time between terms being compared in the same or different series.

lag correlation. An alternative for AUTOCORRELATION. See also CORRELOGRAM.

Lagrangian change. The Lagrangian change of an element (e.g. pressure) is the time rate of change of the value of the element possessed by an individual moving fluid parcel ('change following the motion' or 'substantial change' or 'individual change') and is usually written, for example, dp/dt.
Lagrangian change is related to EULERIAN CHANGE (e.g. $\partial p/\partial t$) by the relationship

$$\frac{d}{dt} = \frac{\partial}{\partial t} + u\frac{\partial}{\partial x} + v\frac{\partial}{\partial y} + w\frac{\partial}{\partial z} \ .$$

Lagrangian similarity. Dimensional treatment in which the rate of vertical spread of particles by TURBULENCE is related to the characteristic scales of velocity and length referred to under SIMILARITY THEORY OF TURBULENCE.

Lamb catalogue of daily weather types. In 1950 Lamb [36, 37] introduced a classification of daily weather types over the British Isles which was an elaboration of an earlier scheme invented by Levick [38]. With some later modifications the classification consists of eight directional types (e.g. north-easterly, southerly) plus a non-directional type, each type being further divided into three categories namely anticyclonic, cyclonic, and unspecified. The choice of directional type depends on the broad flow over the whole area rather than the wind at any one point. Lamb and his fellow workers have classified each day from 1861 to date.
The catalogue is convenient for computer analysis, and additional simple climatological indices may be derived for the categorization of whole weeks or months, e.g. the 'PSMC' indices devised by Murray and Lewis [39]. There are applications to the study of climatic variation and long-range forecasting.

Lambert's law. See ABSORPTION.

Lamb wave. A horizontally propagating acoustic atmospheric wave with zero vertical velocity. Its chief characteristics may be found by the PERTURBATION METHOD which shows that the amplitudes of both pressure and momentum oscillations are at a maximum at the ground. Lamb waves may therefore be eliminated, or filtered, from numerical solutions of the equations of motion by requiring that $\omega \equiv dp/dt = 0$ at the lower boundary; this is a natural and simple boundary condition when use is made of PRESSURE COORDINATES. They can also be eliminated by use of the ANELASTIC APPROXIMATION.

laminar boundary layer. A shallow layer of air, adjacent to a fixed boundary, in which the air velocity increases from the boundary (where it is zero) to the free airstream where the flow is laminar. See LAMINAR FLOW.

laminar flow. A type of fluid flow in which small elements of the fluid retain their identity without mixing with their surroundings, i.e. the flow is predominently 'smooth' although it may be unsteady.

land- and sea-breezes. Local winds caused by the unequal diurnal heating and cooling of adjacent land and water surfaces; under the influence of solar radiation by day and radiation to the sky at night a gradient of pressure near the coast is produced. During the day the land is warmer than the sea and a breeze, the sea-breeze, blows onshore; at night and in the early morning the land is cooler than the sea and the land-breeze blows offshore. The breezes are most developed when the general pressure gradient is slight and the skies are clear. In such circumstances the sea-breeze usually sets in during the forenoon and continues until early evening reaching its maximum strength during the afternoon; the land-breeze may set in about midnight or not until the early morning. The land- and sea-breezes are much influenced by topography and vary considerably from one part of the coast to another. In the British Isles the pure sea-breeze rarely exceeds Beaufort force 3, but in lower latitudes it may reach force 4 or 5. The land-breeze is usually less developed than the sea-breeze. The sea-breeze does not usually extend more than 30 to 40 km either side of the coastline, and often its extent is considerably less. Its farthest penetration inland is sometimes marked by a line of horizontal convergence of wind and vigorous convection, termed a 'sea-breeze front'. The current is shallow at its commencement, but in favourable circumstances the depth may exceed 0.5 km at the time of its maximum development.

langley. An obsolescent unit of energy per unit area

$$1 \text{ langley} = 1 \text{ cal}_{15} \text{ cm}^{-2}.$$

Laplacian. A mathematical operator, designated ∇^2 and defined in three-dimensional CARTESIAN COORDINATES by

$$\nabla^2 = \frac{\partial^2}{\partial x^2} + \frac{\partial^2}{\partial y^2} + \frac{\partial^2}{\partial z^2}.$$

The two-dimensional Laplacian is defined by

$$\nabla^2 = \frac{\partial^2}{\partial x^2} + \frac{\partial^2}{\partial y^2}.$$

For a scalar field (ϕ) the quantity $\nabla^2 \phi$ at a point is a measure of the difference between the value of the scalar ϕ at that point and the average value of ϕ in the neighbourhood of the point (in either three or two dimensions). For example, the

171

factor $\triangledown^2 z$ in the expression for the GEOSTROPHIC VORTICITY at a point 0 is, to a close approximation, given by

$$\triangledown_0^2 z = \frac{z_1 + z_2 + z_3 + z_4 - 4z_0}{b^2}$$

where z_0 is the height at point 0 and z_1, z_2, z_3 and z_4 are the heights at points each a (small) distance (b) from the point, the lines $z_1 z_0 z_3$ and $z_2 z_0 z_4$ being at right angles to each other.

lapse. 'Temperature lapse' (or often simply 'lapse') denotes decrease of temperature with increasing height in the atmosphere. 'Temperature lapse rate', or 'lapse rate', signifies decrease of temperature per unit increase of height, taken positive when temperature decreases with height.

The temperature lapse rate, as denoted for example by a 'lapse line' drawn on a temperature–height diagram, is of fundamental significance in atmospheric ADIABATIC vertical motion and static STABILITY. The lapse rate averages about 6 °C km^{-1} in the TROPOSPHERE, where, however, it is very variable in space and time; over a limited interval of height it may be negative, corresponding to an INVERSION. The lapse rate is very small in the lower STRATOSPHERE.

The term 'lapse' is also used, as in 'humidity lapse rate', to signify the rate of decrease with height of certain meteorological variables other than temperature.

latent heat. The quantity of heat absorbed or emitted, without change of temperature, during a change of state of unit mass of a material. The dimensions are $L^2 T^{-2}$.

In meteorology, the energy transformations associated with changes of state of water are very important; heat is absorbed in the changes from ice to water to water vapour and is released in changes which are in the opposite sense.

The latent heat of fusion (ice to water) at 0 °C is 334×10^3 J kg^{-1}. The latent heat of sublimation (ice to water vapour) at 0 °C is 2834×10^3 J kg^{-1}, varying with temperature by less than 0.2 per cent in the meteorological range. The latent heat of vaporization (water to water vapour) at 0 °C is 2501×10^3 J kg^{-1} and at 40 °C is 2406×10^3 J kg^{-1}. See also HEAT, ENTHALPY.

latent instability. See STABILITY.

latitude. The geographical latitude is defined as the angular elevation of the celestial POLE above the surface tangential to the spheroid which represents the earth. It is also equal to the angle between the normal to this surface and the plane of the equator. The geographical latitude (ϕ) at any point differs only slightly from the geocentric latitude (ϕ'), the latter being the angle between the radius of the earth passing through the point and the plane of the equator. The difference $\phi - \phi'$ amounts to as much as 11′ in middle latitudes, but zero at the equator and poles. Geocentric latitude is found to be more useful than geographical latitude in seismology as the calculation of distance from place to place is facilitated.

law of storms. A nautical expression denoting the mariners' rules for avoiding the so-called 'dangerous half' of a TROPICAL CYCLONE (that part of the cyclone which is to the right of the cyclone's path in the northern hemisphere, to the left in the southern hemisphere).

layer clouds. CLOUDS of no marked vertical development, also termed 'sheet clouds', as opposed to 'heap clouds'. The layer cloud types comprise CIRRUS, CIRROCUMULUS, CIRROSTRATUS, ALTOCUMULUS (except altocumulus castellanus), ALTOSTRATUS, NIMBOSTRATUS, STRATUS, and STRATOCUMULUS.

leader stroke. See LIGHTNING.

least squares. When a set of observations or frequencies can be represented by values calculated from a formula containing adjustable parameters, the criterion of least squares (i.e. the least sum of the squared discrepancies) is generally used to determine the 'best' values for the parameters, usually with satisfactory results. Other ways of measuring an overall discrepancy include the use of weighted least squares (if it is more important to fit some of the data than others) or the minimizing of the CHI-SQUARE STATISTIC.

lee depression. An alternative for OROGRAPHIC DEPRESSION.

leeward. On or towards the downwind side.

lee waves. A system of stationary air-waves ('standing waves') forming, under certain conditions, to the lee of a hill or mountain which presents a mechanical obstruction to the wind. The existence of such waves is sometimes shown by the presence of cloud near the wave crests ('lee-wave clouds'). The 'lee wavelength', i.e. the distance between adjacent wave crests, is usually between 4 and 40 km. The 'lee-wave amplitude', i.e. half the vertical distance from the wave trough to crest, is very small at very low and high levels and reaches a maximum at an intermediate level.

Meteorological conditions suitable for lee-wave formation are:

(i) a stable layer of air between less stable layers at the ground and at a higher level,
(ii) a wind of at least 15 kn across the top of the ridge, and
(iii) constancy of wind direction up to the top of the stable layer.

R.S. Scorer expressed the required conditions mathematically as an upward decrease of l^2 where

$$ l^2 = \left(\frac{g\beta}{V^2} - \frac{1}{V}\frac{\mathrm{d}^2 V}{\mathrm{d}z^2} \right) $$

where β is the static stability ($\theta^{-1}\partial\theta/\partial z$, where θ is the potential temperature and z the height), g the gravitational acceleration, and V the wind-speed component in the direction across the hill. The first term is the more important.
See STANDING WAVE, MOUNTAIN WAVE.

Lenard effect. The separation of electric charges that accompanies the shattering of water drops. In pure water the larger drops are found to carry a positive charge to earth while the very fine droplets carry a negative charge into the surrounding air. Experiments show, however, that the magnitude of the effect and even the charge signs may be altered by various factors, such as the presence of impurities.

The Lenard effect forms the main basis of G.C. Simpson's theory of charge generation and distribution in a thunder-cloud — see THUNDERSTORM.

lenticularis (len). A CLOUD SPECIES (Latin, *lens* lentil).

'Clouds having the shape of lenses or almonds, often very elongated and usually with well defined outlines; they occasionally show IRIDESCENCE. Such clouds appear most often in cloud formations of orographic origin, but may also occur in regions without marked orography.

This term applies mainly to CIRROCUMULUS, ALTOCUMULUS and STRATOCUMULUS' [2, p. 18]. See also CLOUD CLASSIFICATION.

leptokurtic. See KURTOSIS.

leste. A hot, dry, southerly wind occurring in Madeira and northern Africa in front of an advancing depression.

levanter. A humid easterly wind in the Strait of Gibraltar. It is most frequent from June to October, but may occur in any month. At Gibraltar it causes standing waves and complex eddies which may cause dangerous conditions to the lee of the Rock. It is usually associated with stable air under an inversion and BANNER and ROTOR cloud to leeward. Strong winds and / or day heating cause the cloud base to lift and the cloud to thin or disperse.

leveche. A hot, dry, southerly wind which blows on the south-east coast of Spain in front of an advancing depression. It frequently carries much dust and sand, and its approach is indicated by a strip of brownish cloud on the southern horizon.

level. A surface is 'level' if it is everywhere at right angles to the direction of the force of gravity which is indicated by a plumb-line; along such a surface the GEOPOTENTIAL is constant. In general, a level surface in the atmosphere does not quite coincide with one of equal geometric height above mean sea level.

libeccio. A strong, squally south-westerly wind in the central Mediterranean, most common in winter.

lidar. A device of the radar type which uses pulses of light produced by a laser, instead of radio waves; the name is derived from Light Detection And Ranging. Lidars may be installed on the ground or in aircraft, and have been used for detecting clouds and atmospheric pollution. The wavelengths employed range from ultraviolet through visible to infra-red.

life cycle. A term used in synoptic meteorology to include the processes of formation, deepening, occluding and filling-up which are typical of a frontal DEPRESSION. The cycle usually occupies some three days but is subject to a wide range of variation of length.

lifting condensation level. See CONDENSATION LEVEL.

lightning. A visible electric discharge ('lightning flash') associated with a THUNDERSTORM. The various types include the 'cloud discharge' occurring within the THUNDER-CLOUD, the 'air discharge' between a part of the cloud and the adjacent air, and the 'ground discharge' between cloud and ground. Centres of charge of opposite sign are neutralized by the occurrence of each flash, that in the air being the SPACE CHARGE.
  High-speed recording of lightning discharges and also studies of associated electric field charges have shown that the ground discharge comprises at least two 'strokes' or 'streamers' which are generally as follows: first, a 'leader stroke' from cloud to ground in which the luminosity is relatively faint, the time of travel variable but often between 10^{-1} and 10^{-2} seconds, the progress sometimes in distinct steps ('stepped leader') and the path sometimes tortuous and with many branches of luminosity extending out from the main channel; second, an immediate 'return stroke' from ground to cloud which is very rapid and vivid and which lights up the main channel and any branches. Observations have shown, however, that leader strokes directed upwards from ground to cloud may predominate from very high structures. Sometimes (in the British Isles in a minority of cases) the ground discharge comprises a series of double strokes separated by a time interval of the order of microseconds; any subsequent leader stroke

('dart leader') is much more rapid than the first, since it follows the relatively ionized path to ground made by the first leader.

The typical lightning flash discharges a quantity of about 20 coulombs and involves a potential difference of some 10^8 or 10^9 volts. The sign of the charge carried by discharges to earth is predominantly negative.

See also BALL LIGHTNING, FORKED LIGHTNING, PEARL-NECKLACE LIGHTNING, RIBBON LIGHTNING, ROCKET LIGHTNING, SHEET LIGHTNING, STREAK LIGHTNING.

light waves.   ELECTROMAGNETIC RADIATION contained within the VISIBLE SPECTRUM, i.e. between about 0.4 and 0.7 μm.

limb.   A term used generally in astronomy to describe the rim or edge of a heavenly body that presents a visible disc to an observer. In meteorology, it is often applied to the outer edge of the atmosphere.

limb scanning.   The scanning by a radiometer carried in an artificial satellite of radiation received from the LIMB of the earth's atmosphere, i.e. more or less at right angles to the local vertical.

line convection.   The band of almost vertical convection that gives rise to a narrow ana-cold-frontal rain band. It occurs at the position of the surface cold front and, though it may be continuous over long distances, it is more likely to be broken into line elements. It occurs in the absence of pronounced convective instability.

line squall.   A phenomenon accompanying the passage, at any particular place, of a SQUALL LINE or active COLD FRONT. The characteristic features of a line squall are:

  (i)   arch or line of low black cloud,
  (ii)   rapid rise of wind speed and veer of wind,
  (iii) rapid drop in temperature,
  (iv)   rapid rise in pressure, and
  (v)   severe thunderstorm, often with hail.

liquid-water concentration.   In a cloud, the weight per unit volume of water condensed in the form of droplets, usually measured in grams per cubic metre. Its experimental determination is a matter of considerable difficulty. See DISDROMETER.

lithometeor.   A little-used generic term for non-aqueous solid particles suspended in the air or lifted from the earth's surface, e.g. haze, smoke, dust, drifting sand.

lithosphere.   That part of the EARTH which is solid.

local time.   Time reckoned from the epoch noon which at any place is the time of transit across the MERIDIAN of either the true sun or mean sun (according as local apparent time or local mean time is being used). A local time variation is one based on the mean sun, while a sunshine recorder or sundial indicates local apparent time. See TIME.

logarithmic velocity profile.   The theoretical variation of mean wind speed near the earth's surface (within the 'surface BOUNDARY LAYER'), derived under various restrictive conditions. For aerodynamically rough flow — the meteorologically significant case — the wind profile is given by the equation

175

$$\frac{\bar{u}}{u_*} = \frac{1}{k} \ln\left(\frac{z}{z_0}\right)$$

where $\bar{u}$ is the mean wind velocity at height z, u_* the FRICTION VELOCITY, z_0 the ROUGHNESS LENGTH, and k is VON KÁRMÁN'S CONSTANT (about 0.4). The equation applies only for $z \geqq z_0$. Observations show that the equation applies only in conditions of neutral stability.

longitude. The longitude of any place is the angle between the geographical MERIDIAN through that place and a standard or 'prime meridian', which is taken to be the meridian of Greenwich.

long-range forecast. See FORECAST.

long wave. In meteorology an alternative for ROSSBY WAVE.

long-wave radiation. RADIATION (also termed 'terrestrial radiation' or, loosely, 'infra-red radiation'), which is emitted by the earth and atmosphere in the approximate temperature range 200–300 K. The radiation is confined within the wavelengths of about 3 and 100 μm and has maximum intensity at about 10 μm. Since the earth is nearly a perfect radiator, the radiation from its surface varies in close accord with the STEFAN–BOLTZMANN LAW.

looming. An optical phenomenon, associated with a greater-than-normal rate of decrease of air density with height near the surface, in which objects which are normally below the HORIZON become visible. See MIRAGE.

low. A term commonly used in synoptic meteorology to denote a region of low pressure, or DEPRESSION.

Lowitz, arcs of. On rare occasions arcs slightly concave towards the sun extend obliquely downwards and inwards from the parhelia of the 22° halo. They are formed by REFRACTION through ice crystals oscillating about the vertical. See also HALO PHENOMENA.

low-level jet. A JET STREAM at a level well below the high troposphere. Examples are the jet over east Africa at a height of about 1.5 km, the NOCTURNAL JET and the jet associated with the WARM CONVEYOR BELT.

lumen (lm). The derived unit of luminous flux in the SI UNITS system. The lumen is the flux emitted within unit solid angle (1 STERADIAN) by a source of unit luminous intensity (1 CANDELA) radiating equally in all directions.

luminance. The luminance of a surface is expressed by its luminous intensity per unit projected area, the plane of its projection being perpendicular to the direction of view. The unit is the CANDELA per square metre (sometimes known as 1 nit).

luminosity. For a given wavelength, the ratio of the luminous flux to the radiant energy flux. It varies throughout the VISIBLE SPECTRUM from zero at either end of the spectrum to a maximum at wavelength 0.555 μm. The ratio is also termed 'luminous efficiency', or 'visibility ratio'.

lunar. Relating to *luna*, the moon; thus a lunar rainbow is a rainbow formed by the rays of the moon.

lustrum. A period of five consecutive years, which is sometimes used for grouping meteorological statistics which extend over a long period of years.

lux (lx). The derived unit of illumination in SI UNITS, being 1 LUMEN per square metre. It is related to the obsolescent unit, the foot-candle, by

1 foot-candle = 10.76 lux
1 kilolux   = 92.9 foot-candles.

lysimeter. An instrument for measuring the rate of PERCOLATION of rain through soil.

M

Mach angle.   See SHOCK WAVE.

Mach lines.   See SHOCK WAVE.

Mach number.   A pure number, significant in the movement of bodies through the air, defined as the ratio of the air speed of a body to the speed of sound at the corresponding temperature. Neglecting a very small variation with water-vapour concentration, the speed of sound varies directly with the square root of the THERMODYNAMIC TEMPERATURE of the air; in the ICAO STANDARD ATMOSPHERE it is, for example, 341 m s^{-1} at mean sea level and 295 m s^{-1} at 11 km.

mackerel sky.   A sky covered with CIRROCUMULUS or, occasionally, high ALTOCUMULUS clouds, arranged in a regular pattern of 'waves' and small gaps and resembling the scales of a mackerel; the cloud variety is VERTEBRATUS.

macroclimate.   The general CLIMATE of a substantial part of the earth's surface, as for example all or most of a continent.

macroclimatology.   The study of MACROCLIMATES.

macrometeorology.   The study of large-scale processes in the atmosphere occurring over substantial regions of the earth's surface, up to and including the GENERAL CIRCULATION of the atmosphere itself.

macroviscosity.   A quantity denoted N and defined by O.G. Sutton as $N=u_*z_0$ where u_* is the FRICTION VELOCITY and z_0 the ROUGHNESS LENGTH. Aerodynamically rough flow — the meteorologically important case — signifies a value of N greater than about 10^{-2} m^2 s^{-1}, i.e. a value some 10^3 times greater than the kinematic VISCOSITY (v). The parameter N plays a part in fully rough flow analogous to that of v in smooth flow.

maestro.   A north-westerly wind in the Adriatic. It is most frequent on the western shore and in summer. North-westerly winds in other parts of the Mediterranean, notably in the Ionian Sea, and on the coasts of Sardinia and Corsica are also known as maestro.

magnetic storm.   See GEOMAGNETISM.

magnetohydrodynamics.   The study of the motion of an electrically conducting fluid in the presence of a magnetic field. Geophysical applications in which such study is important include those concerned with motions in the earth's core and in the earth's high atmosphere.

magnetosphere.   Term used for that composite region of the earth's high atmosphere, including the EXOSPHERE and much of the underlying IONOSPHERE, in which the earth's magnetic field exerts strong control over the motion of ionized particles.

mamma (mam). A supplementary cloud feature, previously termed 'mammatus' (Latin for udder).

'Hanging protuberances, like udders, on the under surface of a cloud.

This supplementary feature occurs mostly with CIRRUS, CIRROCUMULUS, ALTOCUMULUS, ALTOSTRATUS, STRATOCUMULUS and CUMULONIMBUS' [2, p. 22]. See also CLOUD CLASSIFICATION.

manometer. An instrument for measuring differences of pressure. Ordinarily the weight of a column of liquid is balanced against the pressure to be measured. The mercury BAROMETER is, therefore, a form of manometer.

map projection. See PROJECTION.

mares' tails. The popular name for tufted CIRRUS clouds.

marine climate. An alternative for MARITIME CLIMATE.

maritime climate. A type of CLIMATE which is dominated by the near presence of the sea and is characterized by small diurnal and annual ranges of air temperature. Such a climate prevails over islands and windward parts of continents, e.g. the British Isles, more especially the extreme west.

Marsden square. Mainly used for identifying the geographical position of meteorological data over the oceans. A system which divides a Mercator chart of the world into squares of 10° latitude by 10° longitude. Each square is numbered, and is again subdivided into 100 one-degree squares which are numbered from 00 to 99 so that, given the position of the square, the first figure of the sub-square denotes the latitude and the second figure the longitude.

mass concentration. See SPECIFIC HUMIDITY.

maximum. In meteorology, the highest value reached for a specified element in a given period. See EXTREMES.

maximum entropy method. A method of STATISTICAL INFERENCE, used to obtain a description of a sequence of data in the form of a TIME SERIES as a spectrum of harmonic frequencies. The 'entropy' referred to is not the familiar entropy of physical thermodynamics but a mathematical function with similar abstract properties used in statistical mechanics and information theory. It describes the 'average information per symbol' produced by a discrete information source which yields a sequence of symbols from a finite set $y_1, y_2, ..., y_n$. If the symbols occur independently, with probabilities $P(y_i)$, the 'entropy' (H), or the average amount of information per symbol, is defined by

$$H = - \sum_1^n P(y_i) \log P(y_i) \text{ bits per symbol.}$$

(Clearly, the occurrence of a very rare symbol conveys much more information than the occurrence of a very common one.)

The maximum entropy method (MEM) uses a form of the entropy function to provide a constraint on the estimate of the POWER SPECTRUM of a time series. Any estimate of the power spectrum of a discrete finite series must make *some* assumption about the projection of the series to infinity on either side of the available data; the MEM uses the assumption that H is maximized, which is equivalent to *minimizing* 'information' artificially injected by arbitrary assumptions about the projection of the series beyond the observed data.

179

The spectra produced by the MEM can be very sharp and it can appear to have a greater resolving power than other methods; unfortunately, if the detailed statistical model used is not adequate, and the observed autoregressions have been carried too far for the number of terms available, then this apparent resolving power is spurious. Because of this sampling instability, and various other reasons, the technique has fallen somewhat into disfavour since the 1970s. For further details see Bloomfield [40] and Lacoss [41].

The ideas of the MEM have also been applied to fields other than those of geophysical time series, e.g. to the analysis of observations of OZONE made using the UMKEHR EFFECT [42].

maximum thermometer. Usually, a mercury-in-glass THERMOMETER which has a constriction in the bore, close to the bulb. If the thermometer is exposed horizontally the mercury is able, on expansion, to flow from the bulb past the constriction but not, on subsequent contraction, in the opposite direction. The end of the mercury column farther from the bulb indicates the maximum temperature attained since the last 'setting' of the thermometer; the setting is effected by shaking the mercury past the constriction.

mean. If $x_1, x_2, ..., x_n$ are a set of measurements of the same kind, the most important mean values are defined as follows:

$$\text{Arithmetic mean} \quad \bar{x} = \frac{1}{n} \sum_{j=1}^{n} x_j$$

$$\text{Geometric mean} \quad \bar{x}_g = (x_1 \times x_2 \times , ..., x_n)^{1/n}$$

$$\text{Harmonic mean} \quad \frac{1}{x_m} = \frac{1}{n} \sum_{j=1}^{n} \frac{1}{x_j}.$$

The term 'mean' used without qualification generally implies the arithmetic mean. However, an expression such as 'the mean temperature of the atmosphere' is ambiguous unless it is accompanied by a statement of the positions at which the temperature has been measured or estimated. In a vertical air column, the 'mean temperature with respect to pressure' implies that the averaging process has been applied to a set of temperatures which are at equal intervals of pressure. The 'mean temperature with respect to height' is an average of another set of temperatures, namely those at equal height intervals.

mean deviation. In a series of n values, the mean deviation (e) is the mean of all the deviations (x) from the arithmetic mean, taken without regard to sign, i.e. $e = \Sigma|x|/n$. In a series with approximately NORMAL (FREQUENCY) DISTRIBUTION, $e \approx 0.8\sigma$, where σ is the STANDARD DEVIATION.

mean free path. Mean distance travelled by the molecules or atoms of a gas between consecutive collisions with other molecules or atoms. Mean free path is a function of gas pressure. At low atmospheric levels it is of the order 10^{-9} m, increasing with height to about 10^{-2} m at 85 km, to 10^2 m at 180 km, and to still greater values at higher levels.

mean sea level. See SEA LEVEL.

measurement. See OBSERVATION.

median. When a series of n observations is arranged in order of magnitude, the central value (n odd), or the mean of the two central values (n even), is termed the 'median'. The median may differ appreciably from the MEAN, in which case the frequency distribution is said to be 'skew'.

median volume diameter. See CLOUDS, PARTICLE DISTRIBUTION IN.

mediocris (med). A CLOUD SPECIES. (Latin for medium.)
'CUMULUS clouds of moderate vertical extent, the tops of which show fairly small protuberances' [2, p. 20]. See also CLOUD CLASSIFICATION.

Mediterranean-type climate. A distinctive type of subtropical CLIMATE, included in the KÖPPEN CLASSIFICATION, which is characterized by dry, hot, sunny summers and mild, moderately rainy winters. The type is found in the land regions bordering the Mediterranean, central and coastal southern California, central Chile, extreme south of South Africa, and parts of southern Australia.

medium-range forecast. See FORECAST.

megathermal climate. A CLIMATE of high temperature; more specifically, in the KÖPPEN CLASSIFICATION, one in which no month has a mean temperature below 18 °C. Such conditions are found in the more humid of the tropical or subtropical regions.

Meltemi. See ETESIAN WINDS.

melting band. A conspicuous horizontal band often observed in vertical cross-section radar displays of precipitation.
On reaching the melting (0 °C) level, falling snowflakes begin to melt. Since an ice particle with a wet skin reflects almost as well as if it were an entirely liquid particle, the strength of the echo returned from the region below 0 °C is greater than that above. Since, further,

   (i)   the melting particles are spheroidal in shape and have a larger surface area than raindrops, and
   (ii) the rate of fall of the particles increases and their volume concentration decreases when completely melted,

the radar reflectivity of the melting particles exceeds that of the raindrops below. The various effects combine to produce a bright band which has a maximum intensity a few hundred metres below the 0 °C level. See also RADAR METEOROLOGY, BRIGHT BAND.

melting-point. That temperature, characteristic of a given substance at a given pressure, at which the change of state from solid to liquid occurs. For pure ice at standard atmospheric pressure the melting-point is 0 °C.

Member. A Member of the WORLD METEOROLOGICAL ORGANIZATION, as defined in its Convention; primarily a State, Territory or group of Territories having or maintaining a Meteorological Service.

member. A person elected or designated and serving on the Executive Committee or other committee or on a Technical Commission or on a Working Group of the WORLD METEOROLOGICAL ORGANIZATION.

meniscus. The curved upper surface of liquid in a tube. The meniscus is concave for water, convex for mercury. The curvature effect arises from SURFACE TENSION. Scales and measures are graduated on the assumption that readings are taken at the centre of the meniscus in either case, i.e. lowest point of water meniscus, highest point of mercury meniscus.

Mercator's projection. See PROJECTION.

mercury. Mercury is a metallic element of great value in the construction of meteorological instruments. In the mercury BAROMETER its great density enables the length of the instrument to be made moderate, while the low pressure of its vapour at ordinary temperatures makes possible a nearly perfect vacuum in the space above the top of the barometric column. In the mercury THERMOMETER there is no risk of condensation in the upper end of the stem, as there is with the spirit thermometer.

Specific gravity	=	13.5951 at 0 °C
Specific heat	=	0.0335 at 0 °C
Vapour pressure	=	0.00021 mb at 0 °C
	=	0.00343 mb at 30 °C
Freezing-point	=	234.2 K
Coefficient of expansion	=	0.000182 per °C

meridian. The (geographical) meridian at any point of the earth's surface is the semi-GREAT CIRCLE which passes through the point and terminates at the geographical POLES. The 'prime meridian' is that which passes through Greenwich.

The meridian of an observer is that semi-great circle of the CELESTIAL SPHERE which passes through the observer's ZENITH and terminates at the celestial poles.

The magnetic meridian at any point of the earth's surface is the direction of the compass at that point. The 'geomagnetic meridian' or 'dipole meridian' at any point is the meridian which passes through the point and terminates at the geomagnetic poles.

meridional cell. See MERIDIONAL CIRCULATION.

meridional circulation. Generally, a closed circulation in a vertical plane oriented along a geographic MERIDIAN. It is also termed 'meridional cell' — see, for example, HADLEY CELL.

meridional extension. Marked elongation of upper RIDGES and TROUGHS in a north–south direction.

meridional flow. Airflow in the direction of the geographic MERIDIAN, i.e. south–north or north–south flow.

meso-. A prefix meaning middle, intermediate (Greek *mesos* middle).

mesoclimate. The CLIMATE of a moderately restricted region of the earth's surface; an urban district, as opposed to a neighbouring rural district, is near the lower end of the scale of areas encompassed by this term.

mesoclimatology. The study of MESOCLIMATES.

mesometeorology. The study of the atmosphere pursued on a geographical scale between those employed in MICROMETEOROLOGY and SYNOPTIC METEOROLOGY. A several-fold increase in the number of synoptic stations over a restricted region of southern England, say, would meet the observational needs of mesometeorology, the object of which is a study of those substantial local variations of meteorological phenomena which are missed on the normal synoptic network. However, since the 1970s the main operational tools in practice have been radar and satellite.

mesopause. The top of the MESOSPHERE, at a height of about 85 km, marked by a temperature minimum and hence temperature inversion.

mesoscale. The scale appropriate to atmospheric systems between individual cumulus clouds on the one hand and major depressions and anticyclones on the other, that is from about ten to a few hundred kilometres. Orlanski has defined a logarithmic scale for atmospheric processes in which the mesoscale is divided as follows:

	Range of size (km)
meso-α	200–2000
meso-β	20–200
meso-γ	2–20

mesosphere. That part of the ATMOSPHERE, between the STRATOPAUSE at about 50 km and the MESOPAUSE at about 85 km, in which temperature generally falls with increasing height. An alternative definition, in which the mesosphere was considered to include also the layer from about 20 to 50 km in which temperature generally increases with height, is not now favoured.

mesothermal climate. A CLIMATE of moderate temperature; more specifically, in the KÖPPEN CLASSIFICATION, a climate in which the mean temperature of the coldest month lies between −3 and +18 °C (see MICROTHERMAL CLIMATE and MEGATHERMAL CLIMATE). Such conditions are found mainly between latitudes 30 and 45° but extend up to about 60° on the windward side of continents.

meteor. As defined in the *International cloud atlas* [2, p. 3] 'a phenomenon, other than a cloud, observed in the atmosphere or on the surface of the earth, which consists of a precipitation, a suspension or a deposit of aqueous or non-aqueous liquid or solid particles, or a phenomenon of the nature of an optical or electrical manifestation'. Meteors so defined are classified into three groups, namely HYDROMETEOR, LITHOMETEOR and ELECTROMETEOR; the last two of these terms, in particular, are very little used.

In its more commonly used sense, a meteor, or 'shooting star', is a fragment of solid material (iron or stone) of undetermined origin, which enters the upper regions of the atmosphere and is there made visible by incandescence caused by compression of air in front of the meteor, the meteor itself evaporating in the air. (In another usage, the term 'meteor' is applied only to the visible trail, the particle itself being termed a 'meteoroid'.) In clear-sky conditions an individual observer normally sees a few meteors per hour; in large 'meteor showers', however, many more are to be seen. Most meteors are visible for only one or two seconds but a very large meteor may have a luminous trail that persists for half an hour or longer. The size of meteors ranges from about a centimetre downwards, most being very much smaller. Those too small to be detected visually or by the reflection of radio waves from the ionized trail which they leave in the high atmosphere are termed 'micrometeors' comprising, at very small particle size, 'meteoric dust'. The occasional bodies large enough to reach the ground are termed 'meteorites'. The large and very bright meteors are known as 'fireballs'; those whose destruction in the atmosphere is associated with an air explosion (audible up to a distance of about 50 miles) are termed 'bolides'.

Meteors are grouped into:

(i) 'meteor showers' which compose parts of great streams of particles orbiting the sun and are mostly associated with comets, and
(ii) 'sporadic meteors', of random occurrence, which account for the bulk of total meteor activity.

Evidence advanced by E.G. Bowen to the effect that meteor-shower occurrence is a significant factor in the time distribution of world-wide rainfall, through the formation of condensation nuclei, is not generally accepted as convincing.

From synchronized meteor observations at a known distance apart (now obtained photographically through wide-angle telescopes) the velocity and brightness and the heights of appearance and disappearance of individual meteors may be measured. Such observations show that most visible meteors appear at about 110 km and disappear at about 80 km, with a secondary maximum frequency of disappearance at about 45 km and minimum frequency at 55 km. Since the rate of burn-out of a meteor is a function of air density, the density and hence the temperature of the air may be calculated from such data. The results indicate a temperature maximum at about 60 km at least as high as that near the ground and a rapid increase of temperature above about 85 km.

The IONIZATON of the air which is produced along the path of a meteor is significant in the sporadic E-LAYER. The radio-echo technique has been used with the ionized trails produced by meteors in order to determine density, temperature and winds at the atmospheric levels concerned.

meteorite. A METEOR, or 'meteoroid', large enough to survive passage through the earth's atmosphere and so reach the earth's surface.

meteorograph. An instrument which gives an automatic record of two or more of the ordinary meteorological elements. The term has been used more especially of an instrument, now obsolete, attached to a KITE BALLOON to measure the pressure, temperature and humidity of the upper atmosphere.

meteorological office. In aeronautical terminology, an office designated to provide meteorological information for international air navigation. Subdivision is made in this terminology into:

(i) Regional meteorological office: headquarters of a meteorological region which directs, controls and inspects the various stations of the region; it is qualified to issue directives, technical instructions, regional forecasts and warnings.

(ii) Main meteorological office: a meteorological office competent to prepare forecasts, supply meteorological information and briefing to aeronautical personnel, and supply meteorological information required by an associated dependent or supplementary office.

(iii) Dependent meteorological office: a meteorological office competent to prepare forecasts under the guidance of a main meteorological office, supply meteorological information and briefing to aeronautical personnel, and supply meteorological information required by an associated supplementary office.

(iv) Supplementary meteorological office: a meteorological office competent to supply aeronautical personnel with meteorological information received from a main or a dependent office and with meteorological reports otherwise available.

(v) Meteorological watch office: a meteorological office competent to maintain watch over meteorological conditions within a defined area or along designated routes or portions thereof for the purpose of supplying meteorological information, in particular, meteorological warnings. A watch office may be an independent office or may be part of a main or dependent meteorological office.

meteorological optical range (MOR).   As recommended by the WMO in 1957, a quantity which is identical for practical purposes with VISIBILITY and is defined as the length of path in the atmosphere required to reduce the luminous flux in a collimated beam of light to 0.05 of its original value. (The 'light' is defined as that emanating from an incandescent lamp at a colour temperature of 2700 K.)

meteorological reconnaissance flight.   An aircraft flight made for the specific purpose of obtaining information in a region inadequately served by surface observations (generally over the sea). These flights usually include vertical ascents at selected points on the route.

meteorological rocket.   A small solid-fuel rocket designed to carry a payload of a few kilograms to a height of 60 to 90 km. This payload is often a ROCKETSONDE, CHAFF or grenades (see GRENADES SOUNDING). These rockets are usually unguided and the settings of the launcher have to be determined from BALLISTIC WIND measurements.

meteorology.   The science of the atmosphere (Greek, *meteoros* lofty or elevated, and *logos* discourse). Meteorology embraces both WEATHER and CLIMATE and is concerned with the physical, dynamical and chemical state of the earth's atmosphere (and those of the planets), and with the interactions between the earth's atmosphere and the underlying surface. The term was first used by Aristotle.

methane.   Gas, of chemical formula CH_4, which occurs in minute concentration in the atmosphere (VOLUME FRACTION of about 2.0×10^{-6}).
   Methane is released to the atmosphere mainly by the decay of biological products and is destroyed mainly by atmospheric OZONE. Its mean lifetime in the atmosphere is considered to be not greater than 200 years. It contributes to the GREENHOUSE EFFECT.

metre.   The unit of length in SI UNITS. It was intended to be equal to one forty-millionth of the Paris MERIDIAN, but errors entered into the calculation and it must now be considered as an arbitrary length. For many years it was defined as the distance, at the melting point of ice, between two lines engraved upon a platinum–iridium bar kept in Paris. In October 1960 it was decided at the International Bureau of Weights and Measures to abandon the ruled metre bar as a standard and to redefine the metre as being 1 650 763.73 wavelengths, in a vacuum, of the radiation corresponding to the transition between two specified energy levels ($2 P_{10}$ and $5 D_5$) of the krypton atom of mass 86. This is a measurement whose accuracy can be maintained to one part in 100 million.
   An Order in Council in 1898 defined the inch as 25.400 millimetres, from which it may be deduced that one metre is 39.370078 inches.

Mexican plume.   A plume of potentially warm air that flows from Mexico towards the USA forming a capping inversion. This inversion traps moisture below it by restraining small-scale convection which would otherwise spread the moisture up through the TROPOSPHERE. The moisture is continuously augmented by TRADE WINDS from the Gulf of Mexico leading to very high values of wet-bulb potential temperature in the lower layers. The interaction of this very moist low-level flow with a trough approaching from the west, combined with the eventual erosion of the capping inversion, often leads to the development of violent thunderstorms over Texas and Oklahoma [43].

micro-.   A prefix meaning 'small', e.g. microbarograph (Greek, *mikros* small). In units it signifies one-millionth and is designated μ, as in MICROMETRE, μm.

microbarograph. An instrument designed for recording small and rapid variations of atmospheric pressure. In the Shaw–Dines microbarograph, differences in pressure between a large and well lagged reservoir of air and the external atmosphere are reflected in movements of a bell-shaped float in a bath of mercury and recorded on a clockwork-driven chart by a pen which is directly linked with the float. The reservoir is provided with a slow leak to atmosphere in the form of fine capillary tubing. In all but rapid changes of atmospheric pressure the leak equalizes the reservoir and atmospheric pressures, and the pen remains at the centre of the chart.

In the more modern and very sensitive Jones–Forbes microbarograph, flexible metal bellows constitute an aneroid capsule and are made to drive a capacitance micrometer. The instrument is compensated for temperature changes and is protected against mechanical distortion and vibration. Its frequency response is made adjustable by the provision of variable leaks between the capsule and the atmosphere.

microburst. A small-scale DOWNBURST with horizontal dimensions of about 5–10 km.

microclimate. The physical state of the atmosphere close to a very small area of the earth's surface, often in relation to living matter such as crops or insects. In contrast to CLIMATE, microclimate generally pertains to a short period of time.

microclimatology. The study of MICROCLIMATES.

micrometeorology. The study of the fine structure of the physical processes occurring in the atmosphere. This branch of meteorology is much concerned with atmospheric conditions close to limited regions of the earth's surface but embraces also the detailed structure of physical processes, more especially TURBULENCE, at higher levels of the atmosphere.

micrometre. A measure of length, equal to 10^{-6} m, denoted μm; previously termed 'micron'.

micron. See MICROMETRE.

micropulsations. In GEOMAGNETISM, rapid variations of the magnetic elements of a (nearly) periodic nature. The period ranges from a fraction of a second to some hundreds of seconds, classification being made in terms of the period range.

Natural pulsations have a wide variety of characteristics and have been shown to be closely related to conditions in the IONOSPHERE and MAGNETOSPHERE. Pulsations have also been observed to result from a high-altitude nuclear explosion.

microseisms. Term usually restricted, by convention, to a type of quasi-periodic motion of the ground which is unrelated to earthquakes, explosions, or such artifical agencies as industry and traffic.

The periods of microseisms range from a fraction of a second to several minutes. Atmospheric phenomena — notably the effect of wind on the oceans and resultant effect on the ocean bottom — play an important part in their production which is, however, not well understood. Microseisms have long been used as an early indication of tropical-storm development over the sea. Tripartite stations, spaced in a triangle at distances a few kilometres apart, have more recently been developed as a useful aid in the tracking of tropical storms across sea areas.

microthermal climate. A CLIMATE of low temperature; more specifically, in the KÖPPEN CLASSIFICATION, a climate of long, cold winters and short summers, the mean temperature of the coldest month being lower than $-3\,°$C and that of the warmest month being higher than 10 °C. Such conditions are found in the interior and eastward parts of continents between about latitudes 40 and 65°.

microwaves. RADIO WAVES of short wavelength (generally in the range from a fraction to some tens of centimetres) employed, for example, in RADAR.

middle atmosphere. The STRATOSPHERE and MESOSPHERE taken together.

Mie scattering. SCATTERING of ELECTROMAGNETIC RADIATION by spherical particles. The theory of such scattering, developed by G. Mie in 1908, does not depend on the assumption that the scattering particle has a small radius compared with the wavelength of the radiation, and thus has wider application than the theory appropriate to RAYLEIGH SCATTERING. Extension of the theory to particles of various types and shapes, together with formulae and numerical data required to apply the theory, are contained in the book *Light scattering by small particles* [44].

mil. A unit, used in the preparation of standard BALLISTIC meteorological messages, defined by 6400 mils = 360°.

Milankovitch hypothesis. That variations in the earth's climate on the scales of 10^4 to 10^5 years are in large part due to variations in the earth's orbit about the sun. The orbital variations have three main components: changes in the eccentricity (period 95 000 years), changes in the tilt of the earth's axis (41 000 years), and the variation of longitude of PERIHELION relative to the EQUINOX (23 000 years). These variations produce changes both in the total amount of radiation per annum received by the earth and in its distribution according to hemisphere and season. Palaeoclimatic records, including the OXYGEN ISOTOPE METHOD of analysis of cores from the deep ocean bed, give some support to the hypothesis, and incorporation of the radiation variations in computer models also shows that significant effects might occur. Accounts are given by Imbrie and Imbrie [45] and by Lamb [35, 46].

mile. The 'statute mile' is defined as 1760 yd or 5280 ft. The 'geographical mile' is defined as the length of one minute of arc of longitude at the equator, i.e. 1/21 600 of the earth's equatorial circumference and equals 6087.2 ft. The 'nautical mile' was originally taken, for navigational purposes, as the length of one minute of arc of latitude. Owing to the spheroidal shape of the EARTH this length varies with latitude, being given by $(6076.8 - 31.1 \cos 2\phi)$ ft where ϕ is latitude. The precise length of 6080 ft was adopted by the British Admiralty as the nautical mile. The US nautical mile is 6080.21 ft, while the 'international nautical mile' is exactly 1852 m (6076.12 ft).

millibar. The thousandth part of a BAR. The millibar (mb), equivalent to 1 hectopascal (hPa), replaced the inch or millimetre of mercury as the unit of pressure in the Meteorological Office on 1 May 1914. See also PRESSURE.

$$1 \text{ millibar } = 10^3 \text{ DYNES cm}^{-2} = 10^2 \text{ NEWTONS m}^{-2}.$$

Mills period. A minimum period of leaf wetness in apple trees in spring which, expressed as a function of mean air temperature during the period, is favourable for the development of apple scab infection.

Some examples of critical period length at various temperatures are: 30 hours at 6 °C (43 °F); 14 hours at 10 °C (50 °F); 10 hours at 15 °C (59 °F).

minimum. In meteorology, the lowest value reached for a specified element in a given period. See EXTREMES.

minimum deviation.   In REFRACTION, the minimum change of direction suffered by RADIATION in passing through a refractive medium, as for example light rays through ice crystals. Since changes of direction of the refracted rays are least near the position of minimum deviation, rays emerging after total refraction at and near this position contribute most to the brightness seen by an observer. Minimum deviation is fundamental in the production of the simple refraction HALO PHENOMENON and in the refraction plus REFLECTION phenomenon of the RAINBOW.

The 'angle of minimum deviation' (D) in a plane normal to the faces of a prism intersecting at angle A is given by the formula:

$$n = \frac{\sin\{(A + D)/2\}}{\sin (A/2)}$$

where n is the REFRACTIVE INDEX of the medium. For the passage of light through hexagonal ice crystals ($A = 60°$, $n = 1.31$), D is nearly $22°$, corresponding to the radius of the small halo.

minimum temperature.   The lowest temperature attained, usually in the THERMOMETER SCREEN (screen minimum temperature) or on the ground (GRASS MINIMUM TEMPERATURE or CONCRETE MINIMUM TEMPERATURE), during a given period.

Because of their significance in the formation of frost and fog, much attention has been given to the problem of forecasting night minimum temperatures of either type. Various empirical formulae have been derived, using mainly the observed dry-bulb and wet-bulb temperatures and dew-point, and the observed values (or those estimated for the cooling period) of surface wind speed and cloud amount. Owing to variations of topographical influences and length of cooling period, such formulae may be applicable only to a specified location and time of year.

minimum thermometer.   Usually, an alcohol-in-glass THERMOMETER which carries a small index within the bore, below the surface of the liquid. Initial 'setting' of the instrument is effected by raising the bulb of the thermometer higher than the stem; the index then falls until its lower end meets the end of the spirit column, being prevented from further fall by surface tension. Contraction of the spirit on cooling causes the index to be dragged back closer to the bulb by the end of the spirit column, while expansion of the spirit on heating does not affect the position of the index. The position of that end of the index farther from the bulb thus indicates the minimum temperature attained since the last setting of the thermometer. The thermometer is supported in the screen so that the stem has a slight slope downwards towards the bulb. With this arrangement the movement of the index towards the bulb is slightly assisted by gravity.

mintra.   That temperature above which, at a given pressure, theory indicates that no CONDENSATION TRAILS will form. The variation of mintra temperature with pressure is shown by a line on the Meteorological Office TEPHIGRAM chart.

mirage.   An atmospheric optical phenomenon produced by REFRACTION of light in the layers of air close to the earth's surface due to large temperature gradients in the vertical and associated changes of REFRACTIVE INDEX.

Two main classes of mirage occur: (i) 'inferior' and (ii) 'superior' in which the virtual image is below and above the object, respectively. The inferior mirage is seen over a flat, strongly heated surface (e.g. desert or road) and gives the illusion of an expanse of water; it is caused by the strong upward refraction of light from the clear sky towards the observer. The superior mirage is seen above a flat surface of much lower temperature than the air above it; light from an object is in this case bent downwards towards the observer, as in LOOMING (see Figure 33). In such physical conditions

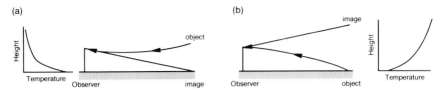

FIGURE 33. Paths of light caused by refraction with simplified height-temperature profiles for (a) inferior mirage, and (b) superior mirage.

multiple reflections may give rise to various images, some displaced laterally with respect to the object, as in FATA MORGANA.

mist. A state of atmospheric obscurity produced by suspended microscopic water droplets or wet hygroscopic particles. The term is used for synoptic purposes when there is such obscurity and the associated visibility is equal to or exceeds 1 km; the corresponding relative humidity is greater than about 95 per cent. The particles contained in mist have diameters mainly of the order of a few tens of micrometres. See also HAZE, FOG.

mistral. A north-westerly or northerly wind which blows offshore along the north coast of the Mediterranean from the Delta del Ebro to Genoa. In the region of its chief development its characteristics are its frequency, its strength and its dry coldness. It is most intense on the coasts of Languedoc and Provence, especially in and off the Rhône delta. On the coast, speeds are about 40 kn but in the Rhône valley a speed of over 75 kn has been reached.

mixed cloud. A cloud which is composed of both ice crystals and water droplets. The CLOUD GENERA, ALTOSTRATUS, NIMBOSTRATUS and CUMULONIMBUS are normally mixed clouds.

mixing condensation level. See CONDENSATION LEVEL.

mixing length. That distance (l) moved by a discrete EDDY in a turbulent fluid, carrying with it the momentum and other properties appropriate to the mean motion at the original point of the fluid, before mixing again with the general flow. This quantity, introduced in eddy diffusion by L. Prandtl on analogy with the 'mean free path' appropriate to the process of molecular diffusion, appears in expressions for the eddy velocity (u'), REYNOLDS STRESS (τ), and eddy VISCOSITY (K_M), as follows:

$$u' = l \frac{\partial \bar{u}}{\partial z},$$

where u' represents an instantaneous departure from the average velocity $\bar{u}$;

$$\tau = - \overline{\rho u' w'}$$

$$= \rho l^2 \left(\frac{\partial \bar{u}}{\partial z} \right) \left| \frac{\partial \bar{u}}{\partial z} \right|,$$

where the bar represents a time average;

$$K_M = l^2 \frac{\partial \bar{u}}{\partial z}.$$

mixing ratio. The ratio of the mass of a particular gaseous constituent (e.g. ozone) of the atmosphere to the mass of air with which the constituent is associated.

189

The term is most often used in respect of the admixture of water vapour with dry air in the atmosphere — see HUMIDITY MIXING RATIO.

mizzle. See SCOTCH MIST.

m.k.s. system. A system of units based on the metre, the kilogram and the second as FUNDAMENTAL UNITS.

Formerly much less widely used in meteorology than the C.G.S. SYSTEM, the m.k.s. units have been adopted, together with the ampere, the kelvin and the candela, as the basis of the International System of Units. See SI UNITS.

mock moon. An alternative for PARASELENE.

mock sun. An alternative for PARHELION.

mock sun ring. An alternative for PARHELIC CIRCLE.

mode. In a series of values, the value of most frequent occurrence. An approximate rule, when there is only one mode, is:

$$\text{mode} = \text{median} - 3 \, (\text{mean} - \text{median}).$$

model atmosphere. A simplified representation of the atmosphere devised for studying atmospheric behaviour and often for the purposes of producing NUMERICAL WEATHER PREDICTIIONS. The simplifications allow attention to be focused on those atmospheric properties which are thought to be of the greatest importance and usually permit these properties to be represented in the mathematical system of equations.

For many purposes the earth's atmosphere can be regarded as behaving as a stratified fluid and it can therefore be treated as consisting of a number of layers whose properties and interrelations can be expressed mathematically.

The simplest and most restrictive model is the BAROTROPIC model in which a single pressure surface, usually 500 mb, is considered. More sophisticated BAROCLINIC models involve consideration of many isobaric (or similar) surfaces. This enables vertical variations in winds and temperature to be represented more realistically. By considering the distribution of moisture, the effects of condensation and evaporation which are known to be important can also be incorporated.

modified refractive index. For convenience in radio meteorological work, in which the curvature of radio waves with respect to the curvature of the earth has to be considered, a 'modified refractive index' (MRI) is employed. The MRI is defined by the equation

$$M' = (n - 1 + h/a) \times 10^6 \text{ M units}$$

where M' denotes the MRI, n is the refractive index of the air, h the height above the earth's surface, and a the earth's radius. For explanation of M units see REFRACTIVE INDEX.

modon. A type of 'dipole' EDDY on a β-plane with equal and opposite amounts of cyclonic and anticyclonic angular momentum which is separated by a VORTICITY discontinuity (or free streamline) from the surrounding fluid. It has a minimum root-mean-square vorticity of $\beta R / \sqrt{2}$ where R is the radius of the free streamline. The flow is non-divergent. See Figure 34.

The modon as described is a theoretical construct; some workers have attempted to identify types of atmospheric blocking pattern as modons.

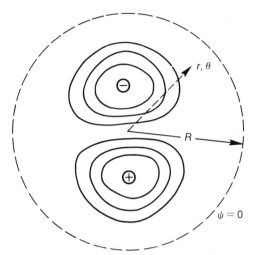

FIGURE 34. Schematic diagram of streamlines in a modon consisting of a cyclone ($-$), an anticyclone ($+$) and a free streamline ($\psi = 0$). The centre of the modon is at y $= 0$ on the β-plane, and θ measures the azimuthal distance of any point from the easterly direction.

moist air. A term which in physical meteorology usually signifies simply a mixture of dry air and water vapour. In synoptic meteorology and climatology the term is applied to air of high RELATIVE HUMIDITY.

moisture content. See SPECIFIC HUMIDITY.

mole (or mol). The gram-molecular weight of a substance, i.e. the weight of it in grams which is numerically equal to its MOLECULAR WEIGHT.

molecular weight. The weight of a molecule of an element, defined on a scale in which the molecular weight of the isotope carbon-12 is 12.0000.

The mean molecular weight (M) of moist air, as applied, for example, in the GAS EQUATION, is

$$M = \frac{m_1 + m_2}{m_1/M_1 + m_2/M_2}$$

where m_1 and m_2 are relative weights of dry air and water vapour, respectively, M_1 is the mean molecular weight of dry air up to about 25 km $= 28.9644$ and M_2 is the molecular weight of water vapour $= 18.0153$.

mole fraction. The mole fraction (N_i) of the ith component of a mixture of gases is defined by

$$N_i = \frac{m_i/M_i}{\Sigma(m_i/M_i)}$$

where m_i is the mass of the ith component in a given volume or mass of the mixture and M_i is its MOLECULAR WEIGHT, the summation indicated being made over all components. It is numerically equal to VOLUME FRACTION. See also AIR.

moment of inertia. The moment of inertia (I) of a body about an axis is the sum of the products of the mass (m) of each element of the body and the square of its perpendicular distance (r) from the axis, taken for all elements of the body, i.e. $I = \Sigma mr^2$.

moments.   In statistics, if $x_1, x_2, ..., x_n$ are a set of measurements of the same kind then the first moment (μ_1), second moment (μ_2), third moment (μ_3), etc. about the mean ($\bar{x}$) of their distribution are defined by

$$\mu_1 = \frac{1}{n} \sum_{j=1}^{n} (x_j - \bar{x}),$$

$$\mu_2 = \frac{1}{n} \sum_{j=1}^{n} (x_j - \bar{x})^2,$$

$$\mu_3 = \frac{1}{n} \sum_{j=1}^{n} (x_j - \bar{x})^3, \text{ etc.}$$

momentum.   The (linear) momentum of a particle is the product of its mass and its velocity. It has dimensions MLT^{-1}.

Monin–Obukhov length.   The length (L) defined by

$$L = - c_p \rho T u_*^{3} / kgQ$$

where Q is the vertical flux of heat ($= \rho c_p w' T'$), u_* the FRICTION VELOCITY and k von Kármán's constant. L is positive and small in stable conditions with light winds at night. It is small and negative (about -10 m) on strongly convective days, about -100 m on windy days with some solar heating, and tends to infinity in purely mechanical turbulence.

Monin–Obukhov similarity theory.   Used for studying the vertical fluxes of heat and momentum in the lower boundary layer (i.e. up to heights of a few tens of metres) when both are appreciable. Consideration of the physical quantities likely to be important, and of their DIMENSIONS, leads to the hypothesis that the height Z enters the equations for the vertical gradients of wind and potential temperature only in the non-dimensional form Z/L where L is the MONIN–OBUKHOV LENGTH. The theory cannot predict the precise functional forms of the equations, which have to be derived empirically from experimental measurements, but is provides a valuable methodology for treating a large mass of data in a unified and systematic manner. Qualitatively, if Z/L is strongly negative, heat convection is dominant; if small and negative, mechanical turbulence is dominant; if strongly positive, there is mechanical turbulence severely reduced by temperature stratification, i.e. at night with downward heat flux.

monochromatic radiation.   RADIATION of a single wavelength. 'Monochromatic flux', 'monochromatic intensity', etc. signify the flux, intensity, etc. per unit wavelength interval.

monsoon.   (Arabic, *mausim* season) The term originally referred to the winds of the Arabian Sea which blow for about 6 months from the north-east and for 6 months from the south-west, but is now used also of other markedly seasonal winds. The essential cause is the differential heating of large land and sea areas, altering with season.

Monsoon conditions are best developed in the subtropics, as in east and south-east Asia (north-east and south-west monsoons in winter and summer, respectively). The rainy season associated with the south-west monsoon is the outstanding feature of the climate of these regions and the term 'the monsoon' is popularly used there to denote the rains, without reference to the winds.

Monsoon conditions occur also, but to a lesser degree, in northern Australia, parts of western, southern and eastern Africa, and parts of North America and Chile. The term is also employed, for example, in north-west Germany. See also INTERTROPICAL CONVERGENCE ZONE.

Monte Carlo methods. A term used to describe various techniques involving the use of RANDOM NUMBERS, or random sequences, for objectives which are difficult or impossible to achieve by wholly systematic methods. For example, the distribution of the CHI-SQUARE STATISTIC can be determined by a Monte Carlo method when the expected frequencies are too small to validate a direct mathematical determination, while a multi-dimensional integral can be evaluated approximately by a Monte Carlo process when a systematic evaluation is impracticable.

moon. The earth's only (natural) satellite. Its radius is 1738 km, mass about 7.38×10^{22} kg, mean density $3^1/_3$ times that of water, mean distance from the earth 384 400 km. Its SIDEREAL PERIOD of revolution averages about $27^1/_3$ days, but varies as much as 3 hours from this value on account of the eccentricity of its orbit and its 'perturbations'. The SYNODIC PERIOD of revolution has a mean value of about 29.53 days, but varies by about 13 hours on account of the eccentricities of the orbits of the moon and earth.

The brightness of the moon is caused by its reflection of direct sunlight falling on it; while the faint illumination of the dark segment is caused by reflected sunlight which has been first reflected to the moon by the earth's surface and atmosphere ('earth-light') — see ALBEDO. The moon's albedo is about 10 per cent; this value and the considered absence of appreciable atmosphere imply a mean temperature of the moon's surface, in radiative equilibrium with absorbed solar radiation, of 267 K. The moon causes a measurable, though very small, tidal movement of the earth's atmosphere but has no other substantiated meteorological effect.

moon, phases of. The appearance of the moon, by custom restricted to the particular phases of 'new moon' when nothing is visible, 'first quarter' when a semi-circle is visible with the illuminated bow on the west, 'full moon' when a full circle is visible and 'last quarter' when a semi-circle is visible with the bow on the east. The changes of phase are caused by changes in the relative positions of earth, moon and sun. The moon rotates on its axis once in each orbital revolution and so the same face of the moon is always turned towards the earth.

morning glory. A type of SQUALL, or succession of squalls, occurring early in the morning, mainly in the spring, at places on the southern coast of the Gulf of Carpentaria in northern Australia. The squalls are accompanied by spectacular 'roll clouds' whose form is often accentuated by the height of the rising sun.

mother-cloud. A cloud from which another cloud develops. See CLOUD CLASSIFICATION.

mother-of-pearl clouds. An alternative term for NACREOUS CLOUDS. The name was given to them by H. Mohn because of their brilliant IRIDESCENCE, similar to that shown by mother-of-pearl.

motion, equations of. The equations of motion of the atmosphere are obtained by applying Newton's second law of motion, which equates the total force acting per unit mass, to the acceleration produced. For meteorological purposes the equation is required in terms of 'relative motion', i.e. in terms of accelerations and velocities measured on the rotating earth.

In vector notation the equation is

$$\dot{\mathbf{V}} + 2\,\Omega \times \mathbf{V} = -\frac{1}{\rho}\,\nabla p + \mathbf{g} + \mathbf{F}$$

where $\dot{\mathbf{V}}$ is 'relative acceleration' and $\mathbf{F}$ the frictional force.

In orthogonal curvilinear coordinates the equations of relative motion are:

$$\frac{du}{dt} + 2\,\Omega\,(w\cos\phi - v\sin\phi) - \frac{uv\tan\phi}{a} + \frac{uw}{a} = \frac{1}{\rho}\frac{\partial p}{\partial x} + F_x$$

$$\frac{dv}{dt} + 2\,\Omega\,u\sin\phi + \frac{u^2\tan\phi}{a} + \frac{vw}{a} = -\frac{1}{\rho}\frac{\partial p}{\partial y} + F_y$$

$$\frac{dw}{dt} - 2\,\Omega\,u\cos\phi - \frac{u^2+v^2}{a} = \frac{1}{\rho}\frac{\partial p}{\partial z} - g + F_z$$

where F_x, F_y, F_z are the components of friction in the positive x, y, z directions respectively, due to viscosity or, in mean flow on the meteorological scale, to Reynolds stresses, and $z \ll a$.

mountain breeze. An alternative for KATABATIC WIND.

mountain wave. An air wave which is stationary, or almost stationary, with respect to the earth's surface. Such a wave sometimes has cloud in the wave crest. It is formed over and/or to leeward of a hill or mountain which obstructs the airflow. See STANDING WAVE, LEE WAVES.

mountain wind. See VALLEY WIND.

moving averages. In a series of numbers a_1, a_2, a_3, etc., the '3-term moving averages' (also termed 'running means' and 'overlapping means') are $(a_1 + a_2 + a_3)/3$, $(a_2 + a_3 + a_4)/3$, etc. Similarly, n-term moving averages may be formed where n is any integer less than the total of numbers in the series; the greater n is, the greater the SMOOTHING of the original series. A moving average differs from the corresponding moving sum only by a constant factor.

Where the original numbers form a TIME SERIES, moving averages may be used either

(i) to eliminate from the original series a variation of known periodicity (e.g. by the formation of 12-term running means from a series of successive mean monthly temperatures), or

(ii) to smooth out short-period fluctuations and so reveal to better effect any long-period fluctuation present in the original series. Theory and experiment show, however, that spurious PERIODICITY may be introduced into data which have been smoothed by the formation of moving averages.

See also FILTERING for a discussion of weighted moving averages.

M-regions, solar. The hypothetical restricted solar regions which are thought to emit, usually during a few successive months, SOLAR CORPUSCULAR STREAMS. The existence of such M- (magnetically active) regions is inferred from a marked tendency for small and moderate magnetic storms to recur at intervals of about 27 days, which is the solar rotation period. M-regions have not yet been identified with any visible solar feature.

mutatus. See CLOUD CLASSIFICATION.

194

N

nacreous clouds. A rare type of stratospheric cloud, also termed 'mother-of-pearl clouds'. The reported occurrences of this cloud have been mainly in Norway and Scotland in winter on occasions of strong and deep west/north-west flow (deep depression over northern Scandinavia). The clouds are somewhat lenticular in form, very delicate in structure, and show brilliant IRIDESCENCE at angular distances up to about 40° from the sun's position. The colouring is most brilliant shortly after sunset (or before sunrise) and endures for a considerable time after sunset. The mean of C. Störmer's height measurements of the cloud is 24 km. The clouds show little or no movement; this fact, together with the circumstances of their occurrence, strongly suggests that they are in the nature of MOUNTAIN WAVE clouds. The nature of the particles is not known but the associated optical effects suggest diffraction by spherical particles of diameter less than 2.5 μm.

natural coordinates. A system of coordinates in which the motion is referred to a right-handed set of axes $\mathbf{t}$, $\mathbf{n}$, $\mathbf{k}$, where $\mathbf{t}$ is a unit vector in the direction of the flow, $\mathbf{n}$ a unit vector at right angles and to the left of $\mathbf{t}$, and $\mathbf{k}$ a unit vector directed vertically upwards.

In this system $\partial/\partial s$ represents differentiation along the STREAMLINES and $\partial/\partial n$ represents differentiation along the ORTHOGONAL LINES. Also

$$\text{grad} \equiv \triangledown = \mathbf{t}\frac{\partial}{\partial s} + \mathbf{n}\frac{\partial}{\partial n}$$

$$\text{div}\mathbf{V} = \frac{\partial \mathbf{V}}{\partial s} + \mathbf{V}\frac{\partial \psi}{\partial n}$$

$$\text{curl}\mathbf{V} = \mathbf{V}\frac{\partial \psi}{\partial s} - \frac{\partial \mathbf{V}}{\partial n}$$

where ψ is the angle measured positive anticlockwise from east to the velocity vector; $\partial\psi/\partial s$ is the streamline curvature usually denoted by K_s, and $\partial\psi/\partial n$ is the ORTHOGONAL CURVATURE usually denoted by K_n.

Navier–Stokes equations. The equations of motion appropriate to a viscous fluid. The component of acceleration, with reference to axes fixed in space, which acts in the x direction of the Cartesian system of coordinates is related to the corresponding forces acting per unit mass by the equation

$$\frac{du}{dt} = -\frac{1}{\rho}\frac{\partial p}{\partial x} + v\left(\frac{\partial^2 u}{\partial x^2} + \frac{\partial^2 u}{\partial y^2} + \frac{\partial^2 u}{\partial z^2}\right) + X$$

where v is the kinematic VISCOSITY and X the component of the external forces per unit mass in the x direction. Analogous expressions prevail in the y and z directions. See also MOTION, EQUATIONS OF.

nebule. A measure of atmospheric opacity, defined by the statement that a screen of 100 nebules transmits the fraction 1/1000 of the incident light. The definition implies that a screen of opacity 1 nebule has a TRANSMITTANCY of 0.933 and that 1 nebule per kilometre is equivalent to an EXTINCTION COEFFICIENT of 0.069 per kilometre. The number of nebules per kilometre of air varies from about 1 in very good visibility to about 10 000 in thick fog.

Reference to the unit DECIBEL shows that an opacity of 100 nebules corresponds to a difference of flux density of $10 \log_{10} (1000/1) = 30$ decibels.

nebulosus (neb). A CLOUD SPECIES. (Latin for fog covered.)

'A cloud like a nebulous veil or layer, showing no distinct details.

This term applies mainly to CIRROSTRATUS and STRATUS'[2, p. 18]. See also CLOUD CLASSIFICATION.

neon. One of the INERT GASES, with VOLUME FRACTION and MASS FRACTION, relative to dry air, of 1.8×10^{-5} and 1.2×10^{-5}, respectively. Its molecular weight is 20.183.

nephanalysis. The interpretation of SATELLITE cloud pictures in terms of cloud types and amounts. The analysis is transmitted either as a FACSIMILE picture or in the form of a five-figure code.

nephoscope. An instrument for determining the direction of motion of a cloud and its angular velocity about a point on the ground directly below it; the product of the angular velocity (radians per hour) and the cloud height, measured or estimated, in miles gives a measure of the cloud speed in miles per hour.

net pyrradiometer. A RADIOMETER designed to measure the net flux of upward and downward long- and short-wave radiation. It is sometimes termed a 'radiation balance meter'.

net radiation. An alternative for RADIATION BALANCE.

neutercane. A term that has been used in the USA to describe a subtropical CYCLONE with winds of storm or hurricane strength, with a radius much less than 100 nautical miles, and with its core intermediate in thermal structure between the typical cold-core subtropical and warm-core tropical cyclone.

neutral atmosphere. See ADIABATIC ATMOSPHERE.

neutral point. See POLARIZATION.

newton (N). The SI UNIT of force. It is the force that produces an acceleration of 1 metre per second squared when applied to a mass of 1 kilogram. The dimensions are MLT^{-2}.

nightglow. The night-time AIRGLOW emission, being the feeble light of the night sky emitted continuously by the upper atmosphere. It has also been termed 'night skylight' and 'permanent aurora'.

The light of the nightglow is measured by a photometer or a spectroscope. It is estimated to account, on average, for nearly half the intensity of light which is present on a clear, moonless night.

The 'auroral green line' (5577 Å of atomic oxygen), the SODIUM D line (5893 Å), and HYDROXYL bands are among the prominent identified emissions in the complex nightglow spectrum. Photometer measurements reveal the presence of systematic solar-cycle and shorter-period time variations of intensity similar to those of the electron density in the IONOSPHERE. Measurements indicate that emissions are generally most intense at 80–100 km but occur also at 200–300 km.

Nightglow is caused by the radiation emitted in the night-time chemical reactions of the ionized and dissociated products which are formed during the day by solar ultraviolet radiation in the atmospheric gases of the ionosphere.

night-sky lights. The light received near the ground on a clear, moonless night is estimated to be composed, on average, of direct light from stars and nebulae (about 40 per cent), ZODIACAL LIGHT (15 per cent), NIGHTGLOW (40 per cent), and light from all these sources scattered by the earth's atmosphere (5 per cent).

night-sky recorder. A camera with a long-focus lens which is mounted with its axis directed towards one of the celestial poles. The camera shutter is arranged, by a clockwork mechanism, to be open only when the sun is 10° or more below the horizon. As the earth rotates, images of the stars trace arcs of concentric circles on the film. Quantitative assessment may be made of the incidence of clouds (or fog or thick haze) which cause interruptions to the record.

Use of this instrument, which is also termed a 'starshine recorder', is very limited.

nimbostratus (Ns). One of the CLOUD GENERA (Latin, *nimbus* rainy cloud and *stratus* spread out).

'Grey cloud layer, often dark, the appearance of which is rendered diffuse by more or less continuously falling rain or snow, which in most cases reaches the ground. It is thick enough throughout to blot out the sun.

Low, ragged clouds frequently occur below the layer, with which they may or may not merge' [2, p. 37]. See also CLOUD CLASSIFICATION.

Nipher shield. A form of screen, based on a suggestion by F.E. Nipher in 1897, which is fitted to a rain-gauge (or snow-gauge), for the purpose of eliminating, as far as possible, wind eddies at the mouth of the gauge and also enabling a truer catch of rain (or snow) to be made. Such screens are not used in the Meteorological Office.

nitric oxide. A gas, of chemical formula NO, which occurs in minute quantities (VOLUME FRACTION of the order 10^{-9}) in the low atmosphere where it is of industrial origin. It is produced in the high atmosphere by DISSOCIATION and subsequent chemical reactions.

nitrogen. A chemically inactive gas, of molecular weight 28.016, with VOLUME FRACTION and MASS FRACTION relative to dry air, of 0.7809 and 0.7554, respectively.

Nitrogen exists only in the molecular form (formula N_2) in the lower atmosphere, but has been identified in atomic form in the high atmosphere where it suffers DISSOCIATION to a small extent.

nitrogen cycle. A complex circulation of nitrogen involving the soil, plants, animals and the atmosphere. In that part of the circulation which involves the atmosphere, nitrogen is released to the atmosphere from the soil (as NITROGEN and NITROUS OXIDE) by the action of nitrogen-fixing bacteria. R.M. Goody and C.D. Walshaw estimate the average magnitude of the flux in the earth–atmosphere nitrogen cycle to be not less than 10^7 molecules $m^{-2} s^{-1}$.

nitrogen dioxide. A gas, of chemical formula NO_2, which occurs in minute quantities (VOLUME FRACTION of the order 10^{-9}) in the low atmosphere where it is of industrial origin. It is produced in the high atmosphere by DISSOCIATION and subsequent chemical reactions.

nitrous oxide. A gas, of chemical formula N_2O, which occurs in the approximately uniform VOLUME FRACTION of about 4×10^{-7} throughout the TROPOSPHERE and lower

STRATOSPHERE. N_2O is considered to be supplied to the atmosphere from the soil as part of the NITROGEN CYCLE and to be destroyed by DISSOCIATION in the high atmosphere.

***n*-method.**   An alternative for SUPERPOSED-EPOCH METHOD.

noble gases.   An alternative for INERT GASES.

noctilucent clouds.   Tenuous but at times brilliant clouds in the very high atmosphere. These clouds have been observed during the midnight hours of the summer months at latitudes higher than about 50°. The clouds are to be seen in appropriate viewing conditions (direct illumination by sunlight against a dark sky and in the absence of lower clouds) more frequently than was previously thought. They resemble CIRROSTRATUS in appearance but have a bluish-white to yellow colour. In the British Isles they are usually seen towards the northern horizon, but extend on occasion to high elevations.

Measurements have shown the clouds to be at a height of 80–85 km and to have a movement from the north-east at speeds between about 100 and 300 kn. Pronounced wave formation is often visible and slower movement of the waves towards the north-east, contrary to that of the cloud material, has been reported. It is as yet uncertain whether the clouds are composed of dust or of ice particles. A particle radius of the order $10^{-1}\mu m$ is indicated by measurements of the strong POLARIZATION of the light from the clouds. No clear association between the appearance of the clouds and such occurrences as volcanic eruptions or meteor showers has yet been shown.

nocturnal jet.   A layer of strong wind occurring at night at a height of a few hundred metres. The cause is the decoupling of the flow in the FREE ATMOSPHERE from the friction layer when there is strong cooling near the ground and the formation of a marked INVERSION. The removal of frictional constraint often allows the wind, which is essentially on the local scale, to attain supergeostrophic speeds.

nocturnal radiation.   The excess RADIATION emitted by the earth's surface at night relative to that received by the earth's surface from the atmosphere (mainly from clouds and from atmospheric water vapour and carbon dioxide).

Since the earth radiates as a black body at its own temperature while the atmosphere is transparent over an important range of wavelengths at terrestrial temperatures (see ATMOSPHERIC WINDOW), the excess is nearly always positive and results in a fall in the temperature of the earth's surface at night. Nocturnal radiation is greatest when the air is cloudless and relatively dry.

The term is misleading because the same process occurs by day; the energy loss is then, however, generally small compared with the influx of solar radiation. It is better termed 'effective radiation'.

noise.   In the description of space or time fields a term used for random errors of observation and other unwanted and unco-ordinated effects which are irrelevant to the purpose of the measurements forming the field.

normal.   The name given to the average value over a period of years of any meteorological element such as pressure, temperature, rainfall or duration of sunshine.

Normal meteorological values are subject to 'uncertainty' owing to year-to-year variability of the observations; the computed STANDARD ERROR of the mean, decreasing with increasing length of period over which averages are taken, is a measure of this uncertainty. In the selection of a suitable length of period, a compromise must be struck in that the period must be long enough for the computed standard error of the mean to be small, but must not be so long that there is a risk that the period contains an appreciable SECULAR TREND of the observations. A period of about 30 years has in the past been thought to be a reasonable compromise and is generally used in deriving normals.

Climatological normals for Great Britain and Northern Ireland are contained in a number of Meteorological Office publications.

normal (frequency) distribution. The normal (also termed the Gaussian) distribution corresponds to the distribution of random errors about a population mean which is indicated by the NORMAL LAW OF ERRORS. The equation on which the distribution is based is

$$y = \frac{\exp(-x^2/2\sigma^2)}{\sigma(2\pi)^{\frac{1}{2}}}$$

where σ is the standard deviation. There is maximum frequency (y) of small errors (x), the frequency decreasing rapidly with increase of error size. The curve is therefore bell-shaped and is symmetrical about the y-axis along which the mean, mode and median all coincide.

The normal distribution is specified entirely by the values of the MEAN DEVIATION and STANDARD DEVIATION; the particular one in which the mean is zero and the standard deviation unity is termed the 'unit normal distribution'.

Probabilities of error (departure) occurrence greater than or less than a specified size (expressed as the ratio x/σ) are given by the 'probability integral', or 'error function' (erf), which is tabulated in books on statistics. For example, probabilities of a departure, of unspecified sign, greater than a given multiple of σ are: σ, 0.318; 2σ, 0.046; 3σ, 0.003; 4σ, 0.00006. The probabilities of a departure of a specified sign are half those quoted.

The normal distribution occupies a central position in statistical theory because of the number and variety of distributions which tend to normality and because several of the main statistical SIGNIFICANCE tests (e.g. STUDENT'S T-TEST, F-TEST and the test for the significance of a CORRELATION coefficient) are based on the assumed normality of the distributions of the data. The distributions of some of the meteorological elements, notably of temperature and pressure, often approximate to the normal distribution while those of other elements (e.g. daily rainfall) depart widely from it. The operations of averaging data, or converting them into departures from normal, tend to produce distributions which are nearer to normal; a reasonably close approach to normality is required before the above tests may properly be applied.

normalized series. In statistics, series obtained by subtracting the MEAN from each term of a series and dividing the result by the STANDARD DEVIATION. The use of normalized series simplifies the comparison of data expressed in disparate units.

normal law of errors. When an observation is subject to a large number of errors of measurement which are independent and individually small, then the central limit theorem states that repeated measurements will fall into the NORMAL (FREQUENCY) DISTRIBUTION. Such a measurement is sometimes said to conform to the normal law of errors.

Raw meteorological measurements rarely conform to the normal law of errors, but averages taken over many terms may do so well enough to allow the standard statistical tests to be applied. PERSISTENCE in the data or extreme SKEWNESS in the distribution of the individual data, can prevent the appearance of normality in means taken over samples of attainable size.

Normand's theorem. Of various propositions enumerated by C. Normand in relation to the thermodynamics of the atmosphere, that generally termed 'Normand's theorem' is to the effect that, on an AEROLOGICAL DIAGRAM, the dry adiabatic through the dry-bulb temperature, the saturated adiabatic through the wet-bulb temperature, and the saturation mixing ratio line through the dew-point temperature of an air sample, all meet in a point.

norte, norther.   A strong, cold, northerly wind which blows mainly in winter on the shores of the Gulf of Mexico. Here it is sometimes humid and accompanied by precipitation. The northers reach the Gulf of Tehuantepec as cold, dry winds, where they often set in suddenly and quickly raise a heavy sea.

North Atlantic Drift.   See GULF STREAM.

nor'wester.   A violent, convective type of storm, often accompanied by a LINE SQUALL, which occurs in Bengal and Assam in the months March to May. The storms are so named because of their pronounced tendency to move from the north-west.

nowcasting.   A term used to describe the immediate detailed description of current MESOSCALE weather patterns by methods in which REMOTE SENSING plays a dominant role, together with forecasts up to two or three hours ahead obtained by extrapolation of current trends. Of particular importance are observations from satellites and weather radars, normally co-ordinated by some form of central computer processing. Attempts have been made to use nowcasting data in mesoscale numerical forecast models.

n.s.r.t.   Abbreviation for near surface reference temperature, measured by a thermistor probe mounted in the ship's water intake just inboard of the sea valve.

n.t.p.   Abbreviation for normal temperature and pressure, signifying a temperature of 0 °C and a pressure of 760 mm of mercury (1013.2 mb), these being the selected standard conditions under which volumes of gases are compared. It is also termed s.t.p. (standard temperature and pressure).

nuclear winter.   A state of affairs first hypothesized in 1982 to occur following the extensive use of nuclear weapons involving the detonation of the equivalent of 5000 Megatons of TNT or more. The suggested course of events is as follows: huge industrial, urban and forest fires producing a widespread smoke path with large quantities of carbon aerosol of particle size ≤ 1 μm; some aerosol removed by precipitation (black rain); remaining smoke absorbs sunlight causing 'lofting' of the smoke; surface insolation falls drastically; surface cools; middle and upper atmosphere warm up suppressing low-level convection and further scavenging of smoke; changes induced in general circulation.
  A typical suggested result is that surface temperatures in the interior of continents would fall by about 20–30 °C in summer and remain low for some weeks. At other times of year, and at locations influenced by the oceans, the cooling would be less.
  The first papers on the subject contained simplifications of the science which probably led to some exaggeration of the effects. However, the very nature and range of variability of the possible initial conditions seem to preclude any final disproof of the hypothesis.

nucleation.   In meteorology, the initiation of either of the phase changes from water vapour to liquid water, or from liquid water to ice. The normal process in the atmosphere is one of 'heterogeneous nucleation' in which the phase change is initiated on minute foreign particles — see NUCLEUS. In the absence of nuclei, the phase change is one of HOMOGENEOUS NUCLEATION or 'spontaneous nucleation'.

200

nucleus. In meteorology, a minute solid particle suspended in the atmosphere.

Classification of nuclei is now generally made into 'Aitken nuclei' (radius $<0.2\ \mu$m), 'large nuclei' (radius from 0.2 to 1 μm) and 'giant nuclei' (radius $>1\ \mu$m). Nuclei are most numerous in the Aitken range, at about $0.05\ \mu$m. At the earth's surface, concentrations as low as about $1000\ \mathrm{cm}^{-3}$ are measured in country air and over the oceans but numbers may be up to nearly 1000 times greater than this in industrial areas. The particle mass is generally between 10^{-12} and 10^{-16} g. The concentration decreases with height, more rapidly in the case of the larger than of the smaller particles. The large and giant nuclei act as nuclei of CONDENSATION. The high saturation VAPOUR PRESSURE associated with drops of very small radius (see KELVIN EFFECT) prevents the Aitken nuclei from acting as condensation nuclei.

Nuclei are dispersed into the atmosphere by such processes as duststorms, volcanic eruptions, formation of sea-salt spray, and combustion. Chemical analysis shows that the nuclei contain the ions SO_4, NH_4, NO_3, Na and Cl.

A 'freezing nucleus' is one on which an ice crystal forms by the freezing of a water droplet. A 'sublimation nucleus' is one on which direct deposition of ice from water vapour occurs; the extent to which this sublimation process actually occurs in the atmosphere is doubtful. 'Ice nucleus' is a generic term which includes both freezing and sublimation nuclei. Such nuclei are much less common than condensation nuclei. Their number, or at least their effectiveness, increases as the temperature decreases below 0 °C. Measurement indicates that their size distribution is of the order 0.1 to several micrometres and that they are composed of volcanic dust and clay and other soil particles of crystalline structure similar to that of ice; some natural ice-nuclei are of organic origin. Ice crystals appear to form on such nuclei at saturation with respect to water and to grow by sublimation. Freezing of water droplets is generally considered to occur spontaneously (without the aid of nuclei) below a temperature of about -40 °C — so-called HOMOGENEOUS NUCLEATION.

numerical weather prediction. An OBJECTIVE FORECAST, in which the future state of the atmosphere is determined by the numerical solution of a set of equations describing the evolution of variables (e.g. pressure, wind, temperature and humidity) which together define the state of the atmosphere. The initial data are provided by an OBJECTIVE ANALYSIS. The governing equations comprise the EQUATIONS OF MOTION together with statistical representations of small-scale dynamical and physical processes which cannot be described explicitly and in detail (see PARAMETRIZATION), and forcing terms such as radiational heating and cooling.

Numerical weather prediction models have been used successfully to describe and predict the weather on all scales from local to global and for periods from a few hours to a week or more. Similar models are used to describe the GENERAL CIRCULATION of the atmosphere and can sometimes be used to provide guidance on the general character of the weather for a whole season.

Nusselt number. A non-dimensional number (Nu) which occurs in respect of the transfer of heat by free CONVECTION from a heated surface immersed in a fluid. It is given by

$$Nu = \frac{lq}{k\Delta T}$$

where l is a length characteristic of the system, q is the rate of heat flow per unit area of the surface, k is the thermal conductivity, and ΔT the characteristic temperature difference between the heated surface and the fluid.

O

objective analysis. In synoptic meteorology, in particular for the purposes of making a NUMERICAL WEATHER PREDICTION, an ANALYSIS of initial observational data by a predetermined numerical process. (This is in contrast with the traditional method in which a human analyst studies plots of the data on a synoptic chart and draws isobars and fronts by hand.) The result of such an objective analysis is normally expressed as a set, or sets, of numerical values at grid points. Modern methods of objective analysis are, however, open to intervention by a human analyst who can modify them by use of BOGUS OBSERVATIONS to make them consistent with non-numerical or other data not in the original set or, in certain circumstances, more synoptically 'reasonable'.

A method frequently employed in objective analysis is one variant or other of 'optimum interpolation'. This is a type of statistical regression analysis that extracts the maximum possible amount of information from all sources including current observations, climatological records, spatial correlations between the meteorological variables, etc. Variables may be analysed either separately — 'univariate' analysis — or simultaneously in groups of two or more — 'multivariate' analysis; a common instance of the latter is when geopotential heights and winds are analysed simultaneously, use being made of the geostrophic relationship.

The general idea may be illustrated by the univariate analysis of a single meteorological variable ψ. The analysis begins with a first guess at the result provided by the latest forecast available for the time of the new set of observations. Figure 35 shows a set of grid points at which forecast values of the variables are given, together

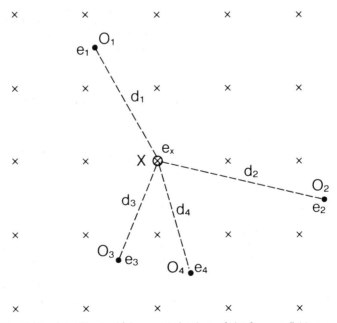

FIGURE 35. Grid points (X) at which corrected values of the forecast fields are required together with points (O_1, O_2, O_3, O_4) at which observations are available.

with a number of observing stations at which the new observations are available. Appropriate values ψ_1^F, ψ_2^F, etc. of the forecast variables are interpolated from the surrounding grid points to each observing station where the observed values are ψ_1^O, ψ_2^O, etc. The value $(\psi_1^O - \psi_1^F) = e_1$ is the calculated 'error' of the forecast at the observing stations. The appropriate 'observed' value of ψ at a particular grid point X can then be derived from the 'forecast' value at X plus an 'estimate' of the error (e_x) at X. This estimate e_x is obtained from the calculated errors e_1, e_2, etc. together with a knowledge of (i) the distance (and possibly bearing) of O_1, O_2, etc. from X, and (ii) the statistical nature (as shown by auto- and cross-correlations) of the spatial field of errors; the statistical nature of the error field is usually expressed by analytical formulae derived from a great mass of past data. The calculations amount to the derivation of a set of 'weights' for each grid point which are then applied to the neighbouring observations to provide the best possible estimate of ψ for that grid point.

Optimum interpolation is a flexible technique which may be applied with different degrees of elaboration and sophistication depending on the quantity of past data and computing power available.

objective forecast. A weather forecast which is entirely based on the application of a single rule or equation, or combination of rules or equations, to selected observed meteorological elements; personal judgement on the part of the forecaster is thus completely eliminated.

A NUMERICAL WEATHER PREDICTION is an example of a forecast which is (almost entirely) objective. Simpler examples include forecasts of night minimum temperature or radiation fog by the application of (largely) empirical rules based on observed values of such elements as temperature, dew-point, wind speed, and cloud amount.

oblique visibility. Oblique visibility, or 'slant visibility', is the greatest distance at which a given object can be seen and identified with the unaided eye along a line of sight inclined to the horizontal.

Oblique visibility in a downward direction, an important element in aircraft operation, is generally different from the VISIBILITY measured at the earth's surface owing to: (i) height variations of atmospheric EXTINCTION COEFFICIENT in the layer concerned, and (ii) the fact that objects are then viewed against a terrestrial background.

observation. In the meteorological context, a record of the measurement or assessment of one or more meteorological elements — e.g. temperature, pressure, cloud type and amount — at a particular time and place.

occlusion. A FRONT which develops during the later stages of the life cycle of a frontal DEPRESSION. The term arises from the associated occluding (shutting-off) of the warm air from the earth's surface.

As convergence takes place at the fronts and in the WARM SECTOR of the depression, the COLD FRONT closes in on the WARM FRONT. The warm sector is thus reduced to a TROUGH line called the line of occlusion and is subsequently lifted from the surface of the earth. The trough line is marked by a belt of cloud and precipitation and by a wind shift. In those cases where there is a substantial temperature difference between the cold air mass in advance of the warm front and that behind the cold front a 'warm occlusion' (less cold air behind) or 'cold occlusion' (less cold air in advance) forms; the effect is to extend the cloud and precipitation well in advance of the surface occlusion in the former case, and behind the occlusion in the latter case. Occlusions are common in north-western Europe, the warm type being the more common in winter, the cold type in summer.

ocean current. See CURRENT, OCEAN.

oceanity (or **oceanicity**).   In meteorology, a measure of the extent to which the climate at any place is subject to maritime, as opposed to land, influences. See CONTINENTALITY.

oceanography.   The study of the seas and oceans, including their physical, chemical, and dynamical properties (see CURRENT, OCEAN; TIDE).

The seas and oceans are by far the main source of atmospheric water vapour and are also a major reservoir of heat. Their interaction with the atmosphere is of great importance in controlling the distribution of climate over the earth as a whole, and also in affecting the day-to-day weather elements in neighbouring land areas.

ocean waves.   Away from coasts a wind-generated ocean-wave system normally covers a wide area of sea and changes its characteristics only slowly with distance. A general division of wave systems is made into 'sea' and SWELL. The system of waves raised by the local wind blowing at the time of observation is usually referred to as 'sea'. Those waves not raised by the local wind blowing at the time of observation, but due either to winds blowing at a distance or to winds that have ceased to blow, are known collectively as 'swell'. Sea and swell are separately reported only in clearly distinguished cases.

For each distinguishable system the reported characteristics are the direction from which the waves come (scale 01–36 as for wind direction), the period (seconds) and the height (metres). The reported height refers to 'characteristic' or 'significant' waves, being the mean height of the highest one-third of waves; the reported period also refers to significant waves.

ogive.   Another term for DISTRIBUTION FUNCTION. The word, referring to the inflected shape of the curve, is borrowed from architecture.

okta.   Unit, equal to area of one eighth of the sky, used in specifying cloud amount.

omega (ω) **equation.**   A differential equation for the variation with hydrostatic pressure of dp/dt — conventionally denoted by ω — which plays the part of vertical velocity when the equations of motion are written in PRESSURE COORDINATES. If we assume the hydrostatic relationship and quasi-geostrophic motion, the vorticity and thermodynamic equations yield the ω equation in the form

$$g\nabla^2(\sigma\omega) - f(\zeta + f)\frac{\partial^2 \omega}{\partial p^2} + \omega\nabla^2\frac{\partial^2 \phi}{\partial p^2} + \nabla\omega . \nabla\frac{\partial^2 \phi}{\partial p^2} + \nabla\frac{\partial \omega}{\partial p} . \nabla\frac{\partial \phi}{\partial p}$$

$$= f\frac{\partial}{\partial p}\{\mathbf{V}_g.\mathrm{grad}(\zeta_g + f)\} - \nabla^2\left(\mathbf{V}_g.\mathrm{grad}\frac{\partial \phi}{\partial p}\right)$$

where the atmospheric stability (σ) is given by

$$\sigma = (g\rho\theta)^{-1}\frac{\partial \theta}{\partial p}.$$

The ω equation is purely diagnostic and relates vertical velocities to existing pressure or contour patterns; it says nothing about rates of change.

opacus (op).   One of the CLOUD VARIETIES (Latin for shady).

'An extensive cloud patch, sheet or layer, the greater part of which is sufficiently opaque to mask completely the sun or moon.

This term applies to ALTOCUMULUS, ALTOSTRATUS, STRATOCUMULUS and STRATUS' [2, p. 22]. See also CLOUD CLASSIFICATION.

open system. An open (thermodynamic) system is one in which there is exchange of matter between the system and its environment, e.g. a precipitating cloud.

optical air mass. The length of the path of the sun's rays through the earth's atmosphere, measured in terms of the path length when the sun is in the ZENITH. For ZENITH DISTANCES (Z) up to about 80° the optical air mass (m) is approximately given by sec Z. Corrections are required, especially at zenith angles greater than 80°, for atmospheric refraction and for the earth's curvature.

Accurate values are given in the *Smithsonian meteorological tables* [15, p. 422].

optical mass. A term used, in calculations of emission or absorption of radiation, to signify the total mass of a given emitting or absorbing substance which lies in a vertical column of unit cross-sectional area between two specified levels (frequently, the earth's surface and the top of the atmosphere). It is also termed the 'optical thickness' or 'optical depth'.

optical phenomena. See ATMOSPHERIC OPTICS.

orographic cloud. Cloud which is formed by forced uplift of air over high ground. The reduction of pressure within the rising air mass produces ADIABATIC cooling and, if the air is sufficiently moist, CONDENSATION. Lenticular WAVE CLOUDS, including those formed well to leeward of the high ground, are common orographic clouds; STRATUS, CUMULUS and CIRRUS clouds are also sometimes orographic in origin.

orographic depression. A non-frontal DEPRESSION (or TROUGH of low pressure) formed by purely dynamical processes to leeward of a range of mountains which presents a barrier to the air flow. Well broken cloud usually characterizes the central region of such a depression because of the action of the FÖHN effect. It is also termed a 'lee depression'.

orographic rain. Rain which is caused, or enhanced, by the presence of high ground. The processes involved include:

(i) the forced uplift of moist air leading to the formation of OROGRAPHIC CLOUD and, if the uplift is strong enough, PRECIPITATION,
(ii) release of potential instability (see STABILITY), and
(iii) more complicated dynamical effects due to enhanced low-level CONVERGENCE which modify the general three-dimensional wind field.

The warm sector of a vigorous depression is the synoptic situation in which the orographic influence on rainfall is generally seen to best effect.

Even where rainfall is predominantly cyclonic or convectional in nature, the orographic influence is always present to some extent. Its dominant influence in mean RAINFALL distribution is readily seen on mean rainfall charts where, to a first approximation, high ground corresponds to high rainfall amount. In those (normal) cases in which a definite prevailing wind direction may be defined, larger rainfall amounts to windward than to leeward of high ground (over regions of average upward and downward air motion, respectively) are apparent. Empirical relationships, differing with locality, may be derived between mean rainfall amount and height of ground. Such relationships may, in the absence of detailed rainfall information, be usefully employed to obtain an estimate of spot or areal rainfall, provided either that the period is long enough to ensure, or it is otherwise confirmed, that departures of wind velocity from average values for the region were small during the period concerned.

orography. A term used in meteorology to signify the variation over a specified area of the height of ground above sea level.

orthogonal curvature. The curvature of ORTHOGONAL LINES. It is a measure of the extent to which STREAMLINES or CONTOURS are confluent or diffluent, and is taken as being positive for confluence. See NATURAL COORDINATES.

orthogonal functions. Functions are orthogonal over a field of points in one or more dimensions when the products of corresponding values at these points add up to zero when summed over the whole field. Examples are simple harmonic functions in one dimension and spherical harmonics in two.

The representation of fields of data in terms of the coefficients of a chosen set of orthogonal functions has great advantages provided that the functions chosen agree with the natural patterns present in the data; in this case the method enables the significant information to be separated from the NOISE. The most efficient separation is achieved by the use of 'empirical orthogonal functions' calculated from the data themselves by means of PRINCIPAL COMPONENT ANALYSIS.

orthogonal lines. Lines drawn so as to intersect the STREAMLINES or CONTOURS at right angles. See also ORTHOGONAL CURVATURE.

oscillations. Alternating departures in opposite senses from a mean value. A simple dynamical illustration is the motion of a pendulum.

In meteorology, the term is usually applied to processes which are not strictly periodic, e.g. oscillations of tropical stratospheric winds. Compare CYCLE.

oscillations, atmospheric. This term generally signifies the tidal movements undergone by the atmosphere. See ATMOSPHERIC TIDES.

overlapping means. An alternative for MOVING AVERAGES.

overseeding. In CLOUD SEEDING, the artificial production of an excessive number of ice crystals. The consequent multiple sharing out of available water among the crystals prevents any of them from growing big enough to fall through the cloud updraught and so inhibits precipitation.

overturning. A rapid exchange of air between different levels effected by BUOYANCY forces. Such an exchange occurs by vigorous CONVECTION in an atmosphere in which there is a SUPERADIABATIC LAPSE RATE.

oxygen. A chemically active gas which in the molecular form (O_2), of molecular weight 31.999, with VOLUME FRACTION and MASS FRACTION relative to dry air, of 0.2095 and 0.2314, respectively.

Oxygen also occurs in the atmosphere as atomic oxygen (O) and as OZONE (O_3). The dissociation of O_2 by ultraviolet radiation results in a rapidly increasing proportion of unattached atoms relative to molecules upwards of about 80 km, the atoms predominating above about 100 km. The dissociation process operates to a decreasing extent down to about 20 km, but in the denser air at these levels the atoms are quickly lost by attachment to oxygen molecules.

That part of the absorption spectrum of oxygen of chief meteorological interest is the strong Schumann–Runge region from about 0.13 to 0.17 μm, with a peak at 0.146 μm. Strong absorption by O_2 and O of wavelengths below 0.1 μm is important in the formation of the IONOSPHERE.

oxygen isotope method. A method of estimating palaeoclimatic temperatures which depends on the facts (i) that the ocean and atmosphere contain two stable isotopes of oxygen (^{16}O and ^{18}O) and (ii) that chemical and phase changes involving water act differentially with respect to these isotopes in a way that is temperature dependent. (A third isotope, ^{17}O, also exists but in amounts so minute it may be ignored.) Thus, from the surface of an ocean containing the two isotopes in a constant ratio, rather more ^{18}O evaporates when the surface is warm than when it is cold, so that the proportion of ^{18}O in atmospheric water vapour increases as the temperature of the sea surface rises. There is a corresponding differential effect when water vapour condenses to form rain. Again, tiny marine organisms absorb and deposit in their skeletons more ^{18}O when the water is cold than when it is warm. The ratio of ^{18}O to ^{16}O may be determined by mass spectrometry from cores bored out from the bed of the deep ocean or from the Greenland ice-cap and used to yield estimates of temperature, on both world-wide and local bases, over periods of up to 5×10^5 years. The technique is difficult, involving both careful experimental technique and elaborate statistical analysis. A particular complication during ice-ages is that the mean value of the isotope ratio in the oceans changes owing to the very large amount of fresh water locked up as ice. Because this ice is all derived from water evaporated from the oceans, it contains an undue amount of ^{16}O, and allowance has to be made for this in the interpretation of the records. A general account is given by Lamb [46].

ozone. The triatomic form (O_3) of OXYGEN, of molecular weight 47.998, which is present in the atmosphere in very small amounts ranging from about 0.2 to 0.6 cm equivalent thickness at normal temperature and pressure.

The presence of ozone is due mainly to photochemical processes in the high atmosphere. Downward diffusion brings the gas in very small concentration (VOLUME FRACTION generally less than 5×10^{-8}) to the lower atmosphere where it is reduced to oxygen by contact with various organic substances. Minor and local sources of ozone exist close to the earth's surface, produced by the oxidation by ultraviolet light of exhaust gases of motor vehicles. Local low-level formation in lightning discharges and in connection with radioactivity has also been suggested.

Ozone is formed and destroyed in the high atmosphere by a range of processes including the absorption of ultraviolet radiation by oxygen and ozone, respectively, and by subsequent particle-collision processes. The main absorption processes concerned are absorption by molecular oxygen of radiation in the Schumann–Runge region (wavelength about 0.13 to 0.17 μm) to form atomic oxygen, and by ozone in the Hartley region (about 0.20 to 0.30 μm) to form molecular and atomic oxygen. The main collision processes are:

(i) a triple collision between molecular and atomic oxygen and a third molecule to form ozone, and
(ii) collision between ozone molecules and oxygen atoms to form oxygen molecules.

The absorption processes are so intense that the associated temperature rise is largely concentrated near the top of the OZONE LAYER at about 50 km. The ozone which is formed at, or transferred to, levels below about 35 km is, in large measure, protected from destruction. The result is that some 90 per cent of atmospheric ozone is at levels below 35 km, with maximum concentration at about 25 km; and that total ozone or, more precisely, the ozone mixing ratio at various levels, is a useful tracer of horizontal and vertical air motion in the stratosphere.

The standard instrument for surface measurement of total ozone amount is the DOBSON SPECTROPHOTOMETER. The use of this instrument, in conjunction with the UMKEHR EFFECT method, yields a smoothed picture of the vertical distribution of ozone. Optical and chemical types of instrument, carried aloft by balloon, rocket or aircraft, have been used to obtain the ozone profile in finer detail and have shown, for example, a jump to higher ozone concentration on passage upwards through the tropopause.

The systematic space and time variations of total ozone also do not accord with conditions of photochemical equilibrium but largely reflect the corresponding large-scale vertical and horizontal transport mechanisms which are at work in the atmosphere. The main features are large amounts of ozone in high relative to low latitudes, especially in spring, and, in middle and high latitudes, a spring maximum and autumn minimum. Day-to-day changes which are correlated with surface and upper-air synoptic features also occur.

Ozone is an important gas in the radiation balance of the stratosphere. The main features of its absorption spectrum are: the intense Hartley region (0.20 to 0.30 μm, with a maximum at 0.25 μm) and the weak Huggins bands (0.30 to 0.35 μm) in the ultraviolet; the weak Chappuis bands (0.45 to 0.65 μm) in the visible; and bands centred at 4.7, 9.6 and 13.0 μm in the infra-red. See CHLOROFLUOROCARBONS.

ozone layer. That layer of the atmosphere, also termed the 'ozonosphere', in which the concentration of OZONE is greatest. The term is used in two ways to signify:

(i) the layer from about 10 to 50 km in which the ozone concentration is appreciable, and
(ii) the much narrower layer from about 20 to 25 km in which the concentration generally reaches a maximum.

ozonosphere. An alternative for OZONE LAYER.

P

pack ice. Term used in a wide sense to include any area of sea ice other than fast ice no matter what form it takes or how disposed.

Close pack ice: composed of floes mostly in contact.
Open pack ice: floes seldom in contact, with many leads and pools.

palaeoclimatology. The study of the nature of and reasons for the types of climate that are inferred, from a variety of evidence, to have prevailed over the earth in the course of geological time. See CLIMATIC CHANGE AND VARIABILITY.

palaeomagnetism. The study of the nature of and reasons for the changes (more especially directional changes) in the earth's magnetic field that are inferred, from studies of remanent rock magnetism, to have occurred in the course of geological time. Deductions of possible significance in the theory of CLIMATIC CHANGE AND VARIABILITY have been made from such studies. See GEOMAGNETISM.

pampero. A name given in Argentina and Uruguay to a severe storm of wind, sometimes accompanied by rain, thunder and lightning. It is a LINE SQUALL, with the typical arched cloud along its front. It heralds a cool south-westerly wind in the rear of a depression; there is a great drop of temperature as the storm passes.

pannus (pan). An accessory cloud. (Latin for shred.)
'Ragged shreds sometimes constituting a continuous layer, situated below another cloud and sometimes attached to it.
This accessory cloud occurs mostly with ALTOSTRATUS, NIMBOSTRATUS, CUMULUS and CUMULONIMBUS' [2, p. 24]. See also CLOUD CLASSIFICATION.

parallax. An apparent change in the position of an object caused by a change in the position of the observer. In connection with the reading of meteorological instruments, an error of parallax may arise whenever the indicator of the instrument (e.g. end of column of mercury or water, pointer, etc.) and the scale against which the indicator is to be read are at a distance from one another which is comparable with the length of the smallest readable scale division; for in such a case a movement of the observer's head may cause his line of vision to the indicator to intersect the scale at different points and so give rise to different readings. The error is eliminated by ensuring that the line of vision to the indicator is at right angles to the scale when the reading is made.

parameter. A quantity related to one or more variables in such a way that it remains constant for any specified set of values of the variable or variables.

parametrization. The mathematical device whereby the broad-scale effects of important physical processes occurring on scales less than the GRID LENGTH used for a NUMERICAL WEATHER PREDICTION are modelled in a statistical fashion. Such small-scale processes include the formation of CUMULUS and CUMULONIMBUS clouds and turbulent mixing in the BOUNDARY LAYER; the broad-scale effects include redistribution of momentum and water vapour and modification of the local

RADIATION BALANCE. Parametrization is carried out by developing semi-empirical equations which relate the broad-scale effects to values of the quantities (such as wind velocity, geopotential height and humidity) explicitly forecast at grid points by the PRIMITIVE EQUATIONS.

paranthelion (plural, **paranthelia**). A PARHELION ('mock sun') at the same elevation as the sun and in azimuth greater than 90° from the sun may be called a paranthelion. White paranthelia at 120° from the sun are fairly common. Paranthelia at about 140° from the sun have been recorded on rare occasions.

paraselene (plural, **paraselenae**). An image of the moon, also termed 'mock moon', produced in a way analogous to the PARHELION; because of the weak intensity of the moon's light relative to that of the sun the paraselene is more weakly coloured and less frequently observed than the parhelion.

parcel method. The estimation of vertical STABILITY in the atmosphere by a method based on the assumption that individual 'parcels' of air move upwards without disturbing their environment. See also ADIABATIC, SLICE METHOD.

parhelic circle. A bright but colourless circle passing through the sun parallel to the horizon. The phenomenon is explained by the REFLECTION of sunlight from hexagonal ice crystals whose axes are vertical. REFRACTION of light through such crystals produces PARHELIA with which, therefore, the parhelic circle is often associated.

parhelion (plural, **parhelia**). An image of the sun, coloured or white; it is also termed 'mock sun' or 'sun dog'.
The parhelia seen most frequently are at the same elevation as and on both sides of the sun and are coloured with red nearest the sun. When the sun is near the horizon the angular distance of the sun from the parhelia is equal to the radius of the ordinary halo, i.e. 22° (see HALO PHENOMENA). When the sun is higher the distance is greater so that if halo and parhelion are both seen the parhelion is outside the halo; for a solar elevation of 55° the angular difference is about 14°. Parhelia are occasionally seen at points on the PARHELIC CIRCLE other than near 22°, notably at 120° (PARANTHELION), less frequently at 46°, 90° and 140°.
Parhelia are caused by the REFRACTION of sunlight within hexagonal ice crystals whose axes are vertical. Oblique rays (sun above the horizon) do not lie in a plane perpendicular to the axes of such crystals and emerge at an angle greater than that corresponding to MINIMUM DEVIATION.

Parry arcs. Rare small arcs observed above and below the small halo (see HALO PHENOMENA) at angular distances varying with solar elevation. They are ascribed to REFRACTION of light through hexagonal ice crystals floating with principal axis and a pair of opposite sides horizontal.

partial pressure. In a mixture of gases, that part of the total gas pressure which is exerted by a specified constituent gas; it is the pressure that each would exert if it alone were present and occupied the same volume as the whole mixture.
According to Dalton's law, the total pressure of a mixture of gases is the sum of the partial pressures, as defined above.

pascal (Pa). The unit of pressure in SI UNITS. The pascal is thus 1 Newton per square metre and is equivalent to 10^{-2} mb.

pastagram. An AEROLOGICAL DIAGRAM, designed by J.C. Bellamy, in which the ordinate is a combined linear scale of height and corresponding pressure in the

STANDARD ATMOSPHERE, and the abscissa is the temperature anomaly $(T - T_p)/T_p$ where T is the actual temperature at pressure p and T_p the temperature at pressure p in the standard atmosphere.

path method. A term applied in synoptic meteorology to signify the method of extrapolation, for forecasting purposes, of the path of a pressure system or other set of isopleths (e.g. ISALLOBARS) drawn on a synoptic chart.

pearl-necklace lightning. A rare form of LIGHTNING, also termed 'chain lightning' or 'beaded lightning', in which variations of brightness along the discharge path give rise to a momentary appearance similar to pearls on a string.

pendulum day. The period of oscillation of a FOUCAULT PENDULUM, equal to the length of a sidereal DAY divided by the sine of the latitude.

pentad. A period of five consecutive days. Five-day means are often used in meteorological work, as five days form an exact subdivision (1/73) of the ordinary year, an advantage not possessed by the week.

percentile. A convenient term for denoting thresholds or boundary values in FREQUENCY DISTRIBUTIONS. Thus the 5 percentile is that value which marks off the lowest 5 per cent of the observations from the rest, the 50 percentile is the same as the MEDIAN and the 95 percentile exceeds all but 5 per cent of the values. When percentiles are estimated by ranking the items of a finite sample, the percentile generally falls between two of the observed values, and the midway value is often taken.

The terms TERCILE, QUARTILE and QUINTILE refer to the percentiles which divide the distribution into 3, 4 or 5 equal parts, respectively. See also QUANTILE.

percolation. The downward passage of surface water through the soil. Part of the rain which falls on the land surface re-evaporates, part runs off into streams and rivers to the sea, while part percolates through the soil. Usually the LYSIMETER consists of a cubic yard of natural earth inserted in a metal container and sunk in the hole formed by removing this earth. The rain-water which percolates through is drained off and measured regularly, access to the receiver being obtained by means of a trapdoor at the side of the gauge. Such or very similar types of gauge are sometimes referred to as 'drainage gauges' or 'seepage gauges'. See EVAPORATION.

perfect gas. A hypothetical gas which obeys the gas laws of Boyle and Charles perfectly. For practical purposes the gases which compose unsaturated air may be considered perfect gases.

perigee. That point of the orbit of an earth-orbiting satellite, natural or artificial, which is nearest the earth.

perihelion. That point of the orbit of a planet or comet which is nearest the sun. Perihelion for the earth occurs on about 1 January; the sun–earth distance is then 1.5 per cent less than the yearly mean distance.

period. A function which varies with time and which repeats itself exactly after a constant time interval (say, T) is said to be 'periodic', and T, the time of a complete oscillation, is termed the 'period' of the function; it is the reciprocal of the FREQUENCY.

periodicity. A time variation of a function comprising a single fixed PERIOD, or combination of fixed periods.

The standard methods of identifying periodicity in a variable quantity are HARMONIC ANALYSIS, PERIODOGRAM analysis, CORRELOGRAM analysis, FILTERING, and POWER SPECTRUM analysis. A (simple) periodicity requires for its complete determination the length of the period, the amplitude (i.e. half the total range) of the variation, and the time of occurrence of the maximum ('phase').

While periodicity uncomplicated by other factors is not found in any meteorological element, the process of averaging over a large number of periods tends to remove non-periodic factors and to allow certain periodic phenomena to emerge. Examples are the average diurnal variation of surface pressure and the average annual variation of surface temperature, both of which average variations are almost entirely explained by a combination of the first two harmonic components of periods 24 and 12 hours, and 12 and 6 months, respectively.

An exhaustive search for periodicity, other than diurnal or annual, in meteorological elements has been almost entirely unsuccessful. More probable than true periodicity in these phenomena is 'quasi-periodicity' of the type shown by annual SUNSPOT numbers (i.e. rather variable period and amplitude but apparently little or no change in phase), and by a RECURRENCE TENDENCY as shown, for example, by certain types of ionospheric and magnetic storms (little change in period but abrupt changes in phase). While some effects, more especially of the latter type, have been found, none is yet so striking as to be of much significance in the problem of long-range weather forecasting.

periodogram.   A diagram used in a method devised by Schuster for the investigation of possible hidden PERIODICITIES in a series of observations. The method is almost obsolete, having been replaced by POWER SPECTRUM analysis.

perlucidus (pe).   One of the CLOUD VARIETIES. (Latin for allowing light to pass through.)

'An extensive cloud patch, sheet or layer, with distinct but sometimes very small spaces between the elements. The spaces allow the sun, the moon, the blue of the sky or overlying clouds to be seen.

This term applies to ALTOCUMULUS and STRATOCUMULUS' [2, p. 22]. See also CLOUD CLASSIFICATION.

permafrost.   Soil which remains permanently frozen, summer heating being insufficient to raise above freezing-point the lower part of a frozen layer formed during the winter. The limit of permafrost is considered to accord very roughly with an annual mean air temperature of $-5\,°C$.

permanent gas.   A gas which is at a temperature above its 'critical temperature', i.e. at a temperature at which it cannot be liquefied by pressure alone. The gases in the air, other than water vapour, sulphur dioxide, and carbon dioxide, are permanent gases.

persistence.   In meteorology, a term used of a synoptic feature or meteorological condition that is unusually long-lasting.

In meteorological and other geophysical TIME SERIES, persistence (sometimes termed 'coherence' or 'conservation') signifies a greater-than-random tendency for relatively high (or low) values to occur in succession. The degree of persistence varies with the meteorological element and generally decreases with increase of time interval between successive members of the series.

The persistence which is inherent in most time series is fundamentally important in questions of statistical SIGNIFICANCE. Thus, for example, the estimate of average seasonal pressure at a specified place which is provided by a mean of, say, 20 successive daily pressure values is much less reliable than that provided by a mean of 20 such values selected at random in the season; in the former case the average may be biased by the dominance of a particular synoptic pattern during the period concerned.

In statistical investigations of time series, unnecessary labour may be saved by confining attention to statistically independent data, i.e. to data spaced at intervals greater than the 'persistence interval', which may be defined as that interval beyond which AUTOCORRELATION becomes negligibly small. When this is not done allowance must be made, in assessing significance, for the fact that the total number of observations employed may be far in excess of the number of statistically independent data. The 'equivalent number of repetitions' ($\epsilon(n)$) in a series of n values may be obtained from the formula

$$\epsilon(n) = 1 + 2/[n\{(n-1)r_1 + (n-2)r_2 + (n-3)r_3 + ... + r_{n-1}\}]$$

where r_1, r_2, r_3, etc. are the correlation coefficients between successive terms, terms two apart, terms three apart, etc. An approximate expression for $\epsilon(n)$ is

$$\epsilon(n) \approx 1 + 2/[n\{(n-1)r_1 + (n-2)r_1^2 + (n-3)r_1^3 + ... + r_1^{n-1}\}].$$

The 'effective number of independent observations' in the series of n values is given by $n/\epsilon(n)$.

persistence forecast. A type of FORECAST, often used as a basis of comparison in the assessment of the success attained in forecasts made by conventional methods, in which the assumption is made that meteorological conditions during the forecast period will remain unchanged from those existing at the beginning of the forecast period.

personal equation. An expression used to denote the error in an observer's readings of an instrument which is due to an unconscious tendency of the observer to read too high or too low. The tendency is usually nearly constant for any given observer reading a given instrument. PARALLAX is a common source of personal equation.

perturbation method. A method, widely applied in meteorology, by which a formal solution to the fundamental, non-linear equations of motion is obtained by the superposition of small disturbances on basic steady fluid flow. Such solutions have the form of waves, the stability and speed of which are found to depend on the wavelength and on the characteristics of the undisturbed flow. Though strictly applicable only to waves of very small amplitude, the solutions are often found to apply, with minor modifications, to disturbances of appreciable amplitude.

phase. The phase of a periodic function is its arrangement of maximum and minimum point or points with regard to a specified initial or starting point. It is measured by a 'phase angle' in which a complete revolution (360°) is equated to a complete PERIOD. Two periodic functions are said to display 'phase reversal' with respect to each other if a maximum value of one function corresponds to a minimum value of the other.

The term 'phase' is also used synonymously with 'state' of matter — solid, liquid, or gaseous.

phase velocity. The velocity of propagation of a pure sinusoidal wave of given frequency (or wavelength). In what is called a dispersive medium, the phase velocity varies with frequency. See GROUP VELOCITY.

phenology. The study of the sequence of seasonal changes in nature. All natural phenomena are included, seed-time, harvest, flowering, ripening, migration, and so on, but the observations are often limited to the time at which certain trees and flowering plants come into leaf and flower each year, and to the dates of the first and last appearances of birds and insects.

The phenology of plants flowering in the British Isles is contained in the *Phenological Reports* (1877–1948) of the Royal Meteorological Society. Long-term means are contained in a paper by E.P. Jeffree [47].

phenomenon. Word used in meteorology to denote either (i) an unusual intensity of some occurrence, e.g. 'ugly' sky, high rainfall, low temperature, high pressure, gale, or (ii) occurrences which are only occasionally noted, e.g. thunder, halo, fog, glaze.

photochemistry. The study of chemical reactions involved in the absorption or emission of RADIATION. In meteorology, such reactions are mainly confined to high atmospheric levels as, for example, in the absorption of radiation by oxygen to form ozone. Reactions also occur locally near the earth's surface in connection with certain products of atmospheric pollution.

photodissociation. The splitting of a molecule into atoms or into smaller molecules by the absorption of RADIATION.

photoelectron. An ELECTRON ejected from an atom or molecule which is exposed to RADIATION of a frequency higher than a critical value. The process of 'photoionization', important for example in the formation of the IONOSPHERE, is involved in such release of electrons.

photoionization. See PHOTOELECTRON.

photosphere. The bright disc of the SUN from which continuous emission of solar radiation takes place.

physical meteorology. Because of overlapping at many points with other 'branches' of meteorology this term cannot be defined precisely but is often used to signify all those directly physical aspects of meteorology which are not normally dealt with in DYNAMICAL METEOROLOGY.

piezotropy. In a change of the state of the pressure and density of an individual element of the atmosphere the condition that there exists a relationship

$$B = \frac{d\rho}{dp}$$

where $d\rho/dp$ is the change of density with pressure of the individual element and B the 'coefficient of piezotropy'.

B is a function of the thermodynamic variables and so depends on the initial state of a selected particle and varies for different particles. The condition of piezotropy is implicitly assumed in drawing an adiabatic 'path curve' of a selected particle on an AEROLOGICAL DIAGRAM. In this important special case of piezotropy, the coefficient B has the value $1/\gamma RT$.

pileus (pil). (Latin for cap).
'An accessory cloud of small horizontal extent, in the form of a cap or hood above the top or attached to the upper part of a cumuliform cloud which often penetrates it. Several pileus may fairly often be observed in superposition.

Pileus occurs principally with CUMULUS and CUMULONIMBUS' [2, p. 24]. See also CLOUD CLASSIFICATION, CAP CLOUD.

pilot balloon. This term is generally applied to the smaller meteorological balloons. Their main use is the determination of upper winds, although the smallest balloon, one weighing about 10 g, is most commonly used for the measurement of cloud-base height by timing the duration of flight below cloud base and assuming a constant rate of ascent. For this reason, 10 g balloons are sometimes known as 'ceiling balloons'. The theoretical rate of ascent (V) of a balloon filled with hydrogen is approximately

$$V = qL^{1/2}/(L + W)^{1/3}$$

where L is the FREE LIFT in grams and W the total weight in grams of the balloon and any attachments; q is a constant and has a value of 275 if V is expressed in feet per minute (1.40 if V is in metres per second).

Larger balloons, such as the 30 g balloon, are used for wind finding. In the 'single theodolite' method, the height of the balloon is estimated by timing as for cloud-base determination. The 'tail' method employs one (or two) piece(s) of paper attached to the balloon by a known length of string. The height of the balloon at any instant can then be deduced from the angle subtended by the tail (assumed vertical) at the theodolite; this angle is measured on a divided scale (graticule) fitted to the eyepiece of the theodolite. The 'double theodolite' method employs simultaneous observations of the balloon's position by two theodolites at known relative positions; greater accuracy is possible by this method since not only wind direction and speed but also height are determined.

The calculation of wind velocity from the theodolite readings is normally performed by a special slide-rule, although graphical methods can also be used.

When absence of cloud permits optical wind-finding at greater heights, balloons as large as 100 g may be used up to about 16 km.

Pitot tube. An instrument for determining the velocity of a stream of fluid by measuring the increase of pressure, above the 'static' or undisturbed pressure, in an open tube facing the stream. The velocity (v) is computed from the relationship $p = \frac{1}{2}\rho v^2$. Suitably mounted, a Pitot tube may be used as an ANEMOMETER.

pixel. The minimum area of the field of view capable of resolution as an individual entity by the optical and transmission systems of an artificial satellite or other REMOTE SENSING device.

Planck's law. See RADIATION.

planetary albedo. See ALBEDO.

planetary temperature. See RADIATION.

platykurtic. See KURTOSIS.

pluvial period. A geological period of large amounts of rainfall relative to earlier and later periods. Evidence of such periods, some of which are thought to have coincided in time with the onset phases of the glacial periods of ICE-AGES, is found in those land regions which lay immediately equatorwards of the ice-covered regions during the glacial periods. Pluvial periods appear to have also taken place in tropical regions, e.g. Africa, not necessarily in association with high-latitude glaciations.

poaching. The trampling of wet land, mainly by cattle, so that the ground becomes churned up and muddy with risk of damage to the soil structure.

Poisson distribution. If in a large number of trials an event occurs so rarely that the probability of its occurring during any one trial is small, then the frequency of occurrence of the event during a number of sets of trials follows a Poisson distribution. If over a large number of sets of trials the mean number of occurrences per set is m, then the fraction ($\Pi(n,m)$), of the number of sets in which the number of occurrences will be a chosen whole number (n) is given by

$$\Pi(n,m) = \frac{e^{-m}m^n}{n!}$$

where $m = np$ is the expected number of events in the period. It may be shown that m is both MEAN and VARIANCE of this FREQUENCY DISTRIBUTION.

The probabilities of many rare meteorological events vary with time and are often higher once an event has actually occurred. This feature limits or complicates the application of the Poisson distribution to meteorological events. See also BINOMIAL DISTRIBUTION, FREQUENCY DISTRIBUTION, NORMAL (FREQUENCY) DISTRIBUTION.

polar air. Air originating in high latitudes, normally subdivided in synoptic meteorology into 'polar maritime' (Pm) air and 'polar continental' (Pc) air, according to the nature of the surface over which the AIR MASS originates. Air which moves almost directly equatorwards from the ice-bound areas of the Arctic is now usually termed ARCTIC AIR.

During the motion of Pm air away from its source region, heat and moisture may be convected into the air mass from a warmer underlying sea surface. In the British Isles Pm air is associated generally with a west or north-west wind, steep lapse rate, low freezing-level, good visibility, instability showers (sometimes with thunder), and surface temperature below seasonal normal.

Polar maritime air which travels far to the south and then returns to the British Isles from the south-west is termed 'returning polar maritime air' (rPm); typical conditions associated with this air mass are much cloud but fewer showers than in Pm air, and temperatures close to the seasonal average in summer and above the seasonal average in winter.

Easterly winds are often associated in the British Isles with the less common Pc air mass. In winter, typical conditions are very cold but dry apart from light snow showers near the east coast; in summer, conditions are dry and rather warm except on the east coast where coastal fog ('haar') is often widespread. (The Pc air is warmed in winter, cooled in summer, by its passage over the North Sea.)

polar-air depression. A SECONDARY DEPRESSION of a non-frontal character which forms, more especially in winter, within an apparently homogeneous polar AIR MASS; near the British Isles the development is usually within a northerly or north-westerly airstream. The chief characteristics of this type of depression, which seldom becomes intense, are a movement in accordance with the direction of the general current in which the depression forms, and the development of a belt of precipitation near the depression centre and along a trough line which often forms on the side farther from the parent depression where also the pressure gradient (and surface winds) is increased.

polar climate. A type of CLIMATE which is prevalent in general, within the polar regions (polewards of 66° 33′N and S). In the KÖPPEN CLASSIFICATION, the polar climate is subdivided into TUNDRA climate (mean temperature of warmest month between 0 and 10 °C), and 'perpetual frost' climate (mean temperature of warmest month below 0 °C).

polar continental air. See POLAR AIR.

polar coordinates. A system of coordinates in which the position of a point is specified by its distance (r) from the origin ('pole') and by the angle (θ) made by the line joining point and origin with a reference line ('polar axis').

polar distance. An alternative for COLATITUDE.

polar front. A FRONT which divides 'polar' and 'tropical' AIR MASSES and on which most of the travelling disturbances of middle latitudes form. In the North Atlantic, for example, this front, which can often be traced as a continuous line over thousands of kilometres, extends in winter, on average, in a north-easterly direction from a position off the east coast of the USA (at about 30°N). In summer the front is less well marked and has little tendency for a preferred position.

polarization. A state of ELECTROMAGNETIC RADIATION in which the transverse vibrations which compose the wave motion occur wholly or in part ('partial polarization') in a specified manner — for example in a plane, circle or ellipse — that is, they do not occur in all the possible planes which contain the direction of propagation of the radiation. The plane of polarization is defined as that in which the wave motion is a minimum, or, in terms of electromagnetic theory, that in which the electric vector is a minimum.

Polarization of emitted radiation may be effected by a suitable aerial array. Since SCATTERING causes polarization of initially unpolarized radiation, the phenomenon is observed naturally in the atmosphere in the DIFFUSE RADIATION which reaches the earth's surface, such radiation being polarized in the plane which contains the sun, the observer and the observed point of the sky. In accordance with the theory of RAYLEIGH SCATTERING the polarization is strongest in the solar zenith (i.e. in the light scattered from a point 90° from the sun) and in the antisolar point. Because of multiple and non-molecular scattering, the polarization disappears at the 'neutral points' discovered by Arago, Babinet and Brewster — see ARAGO'S POINT.

polarization, electric. The separation of positive and negative charges within a particle in response to an electric field acting on the particle.

The occurrence of this effect within falling water and ice particles, due to the presence of the atmospheric electric field, was advanced by C.T.R. Wilson as leading to selective ion capture and hence the separation of charge in a thunderstorm.

polar low. An alternative for POLAR-AIR DEPRESSION.

polar maritime air. See POLAR AIR.

polar-night jet stream. See JET STREAM.

polar stratospheric clouds. These are large concentrations of ice crystals and other aerosols in the low stratosphere at high latitudes, especially over the Antarctic in winter. First reported by expeditions in the early 1950s, they have more recently been regularly identified by observations from meteorological SATELLITES as regions of low temperature ($< 195K$) and moderate extinction of $1\ \mu m$ solar radiation. Their exact composition is at present uncertain, but they are thought to play a significant role in stratospheric chemistry, particularly in the reactions affecting OZONE concentration.

polar vortex. The circumpolar westerly cyclonic circulation which affects middle and high latitudes of both hemispheres.

In summer the circulation affects the middle and upper troposphere; it decreases in the lower stratosphere and reverses at higher stratospheric levels, mainly because of maximum OZONE heating at high levels over the summer pole. In winter the westerly circulation in the higher troposphere is much stronger than in summer. Since, also, the upper atmosphere over the winter pole is a heat sink, the westerly circulation increases further with increasing height in the stratosphere in this season and comprises the 'polar stratospheric vortex' or 'polar-night jet'. Westerly winds of about 150 kn are attained, on average, in higher middle latitudes near the top of the stratosphere near midwinter; individual speeds in excess of 200 kn are not uncommon and values greater than 350 kn have been observed. (On the other hand, easterly winds of over 100 kn may occur in association with major stratospheric disturbances.) See also THERMAL WIND, JET STREAM.

polar wandering. Hypothetical movement of the earth's axis of rotation relative to the earth's surface, in the course of geological time. Such movement has been advanced as a possible cause of climatic changes on this time-scale. Among the evidence advanced in support of the hypothesis is that of remanent rock magnetism (see GEOMAGNETISM), on the assumption that the earth's magnetic axis has always been close to the axis of rotation. (Further evidence of the same kind is to the effect that CONTINENTAL DRIFT has also been involved.)

polar year. In the First International Polar Year (FPY), from 1 August 1882 to 1 September 1883, and the Second International Polar Year (SPY), from 1 August 1932 to 31 August 1933, near times of sunspot maximum and minimum, respectively, stations were manned by co-operating nations in the Arctic, and to a much smaller extent in the Antarctic regions. Observational programmes covered mainly meteorology, geomagnetism and aurora.

pole. The earth's geographical poles are the points of intersection of the earth's axis of rotation (polar axis) with the earth's surface.

pollution. See ATMOSPHERIC POLLUTION.

ponente. A westerly wind which blows in the Mediterranean.

population. A statistical term used to denote the (generally hypothetical) stock from which a sample of data available for analysis may be assumed to be drawn. Statistical arguments often seek to infer the properties of the population from those of the sample.

potato blight. A fungal disease in which the spores are spread to healthy plants by wind, RAIN SPLASH or other factors. Development is favoured by mild humid weather such as that defined by the SMITH PERIOD or BEAUMONT PERIOD.

potential energy. The ENERGY possessed by a body by virtue of its position. It is measured by the amount of work required to bring the body from a standard position, where its potential energy is zero, to its present position. A common example is that of 'gravitational potential energy', mean sea level being then the normal selected standard level.
　　The term 'total potential energy' was first employed by M. Margules and signifies the sum of gravitational potential energy, defined above, and INTERNAL ENERGY. See also AVAILABLE POTENTIAL ENERGY, KINETIC ENERGY.

potential evapotranspiration. See EVAPOTRANSPIRATION.

potential gradient. Atmospheric potential gradient is defined as the difference of electric potential between two points vertically disposed with respect to each other and separated by unit distance; it is expressed in volts per metre and is, by convention, reckoned positive if directed downwards.

Surface potential gradient — the potential difference between a conductor at a height of one metre and the level ground — is the most regularly measured element of ATMOSPHERIC ELECTRICITY.

potential instability. See STABILITY.

potential temperature. That temperature (θ), readily obtained from an AEROLOGICAL DIAGRAM, which a given sample of air would attain if transferred at the dry ADIABATIC lapse rate to the standard pressure, 1000 mb. If the pressure (mb) and absolute temperature of the air are p and T, respectively, then

$$\theta = T \left(\frac{1000}{p}\right)^{(\gamma-1)/\gamma} = T \left(\frac{1000}{p}\right)^{0.286}.$$

θ is related to ENTROPY (S) by the equation

$$S = c_p \log \theta + \text{constant.}$$

The 'partial potential temperature' (θ_d) of an air sample, as defined by C-G. Rossby and used in the ROSSBY DIAGRAM, is the potential temperature appropriate to the temperature and partial pressure (p_d) exerted by the dry air of the sample $(p_d = p - e,$ where $e =$ vapour pressure):

$$\theta_d = T \left(\frac{1000}{p_d}\right)^{0.286}.$$

potential transpiration. See TRANSPIRATION.

potential vorticity. If a column of air bounded by two isentropic surfaces is in ADIABATIC motion the ratio of the mean absolute VORTICITY of the column $(\zeta + f)$ to the pressure difference between the top and bottom of the column (Δp) is constant (potential vorticity theorem), i.e.

$$\frac{\zeta + f}{\Delta p} = \text{constant.}$$

The value of the absolute vorticity that corresponds to a standard value of Δp (say, 50 mb) and a standard latitude (such that $f = f_0$) is termed the potential vorticity and is conserved in adiabatic motion of an air column.

If, instead of a compressible column of air, we consider the motion of a column of incompressible fluid of variable depth (h) then

$$\frac{\zeta + f}{h} = \text{constant.}$$

These ideas were introduced by Rossby [48, 49].

The term 'potential vorticity' is often applied to the quantities $(\zeta + f)/\Delta p$ and $(\zeta + f)/h$ which do not have the dimensions of vorticity.

The idea of potential vorticity was generalized by Ertel to the case of a compressible, thermodynamically active inviscid fluid in adiabatic flow. If ζ is the relative vorticity of a fluid element, Ω the angular velocity of the rotating coordinate system, ρ the density, and S some conservative thermodynamic property (e.g. entropy or potential temperature), then the 'Ertel potential vorticity' (Π), where

$$\Pi = \rho^{-1}(2\Omega + \zeta).\nabla S,$$

is conserved along trajectories, i.e.

$$\frac{d\Pi}{dt} = 0.$$

(Π does not have the dimensions of vorticity (T^{-1}).) For a derivation of this theorem see Pedlosky [50].

Ertel's theorem is very general. If we assume: (i) the HYDROSTATIC APPROXIMATION, (ii) quasi-geostrophic flow, and (iii) that the RICHARDSON NUMBER (Ri)$\gg$1, then the above equation becomes

$$\frac{d}{dt}\left\{(\zeta+f)\frac{\partial\theta}{\partial p}\right\} = 0$$

where $(\zeta+f)$ is the vertical component of absolute vorticity and θ the potential temperature.

If we further assume: (iv) vertical advection is negligible, (v) $\partial\theta/\partial p$ and $(\zeta+f)$ may be replaced by $\partial\theta_s/\partial p$ and f_0 when they are not being differentiated (θ_s being a mean value over an isobaric surface), and (vi) that the flow is represented by a stream function ψ, then we find

$$\left(\frac{d}{dt}\right)_G q = 0$$

$$\text{where } q = \nabla_h^2\psi + f + \frac{\partial}{\partial p}\left(\frac{f_0^2}{\sigma}\frac{\partial\psi}{\partial p}\right)$$

$$\text{and } \sigma = -\frac{1}{\rho_s\theta_s}\frac{\partial\theta_s}{\partial p}.$$

q is known as the quasi-geostrophic potential vorticity, and is thus conserved in geostrophic adiabatic flow. A full derivation is given by Kuo [51].

power spectrum. A term mathematically equivalent to ENERGY SPECTRUM and applied to the specification of the character of a series of values of a quantity, measured at regular intervals of space or time, in terms of the partition of the 'power' of the variability of the series between oscillations of different frequencies. The power spectrum and the CORRELOGRAM of the series have a close mathematical relationship, one being the FOURIER TRANSFORM of the other. Power spectrum analysis is widely used to search for significant periodic or quasi-periodic components in long series of geophysical measurements.

PPI. Abbreviation for 'plan position indicator'. See RADAR METEOROLOGY.

praecipitatio (pra). (Latin for fall.)
'Precipitation (rain, drizzle, snow, ice pellets, hail, etc.) falling from a cloud and reaching the earth's surface.

This supplementary feature is mostly encountered with ALTOSTRATUS, NIMBOSTRATUS, STRATOCUMULUS, STRATUS, CUMULUS and CUMULONIMBUS'[2, p. 23]. See also CLOUD CLASSIFICATION.

Prandtl number. The non-dimensional ratio (σ) defined by the relationship $\sigma = v/a$ where v is the kinematic VISCOSITY and a the thermometric CONDUCTIVITY of a fluid; σ has the approximate value 0.7 in air near the earth's surface.

precipitable water. The precipitable water of a column of air is the depth of water (alternatively expressed as the total mass of water) that would be obtained if all the water vapour in the column, of unit area cross-section, were condensed on to a horizontal plane of unit area. 'Precipitable water' is a useful measure of the water vapour content of an air column. The term is not, however, to be regarded as implying that the amount of water can, in fact, be precipitated by an actual physical process.

The depth of precipitable water of an atmospheric column, of pressure p_1 (mb) at the bottom and p_2 (mb) at the top and of mean mixing ratio $\bar{r}$ (g kg^{-1}), is given, in tenths of millimetres, by the approximate formula

$$\text{precipitable water} = \frac{\bar{r}\,(p_1 - p_2)}{g}$$

where g is the gravitational acceleration (9.8 m s^{-2}).

The precipitable water for an entire atmospheric column is found by applying this formula to selected successive layers; a typical value for temperate-latitude summer conditions is 20–30 mm.

precipitation. Used in meteorology to denote any aqueous deposit, in liquid or solid form, derived from the atmosphere.

The main problem in the 'precipitation process' is the explanation of the manner of growth of drops from the size commonly associated with non-precipitating clouds (diameter about 15 μm) to that found in most forms of precipitation; growth by direct CONDENSATION on to a NUCLEUS is too slow to account for the transformation. Two mechanisms are considered to operate:

(i) Ice-crystal (Bergeron) process. In a cloud of predominantly supercooled water droplets ice crystals form, in increasing numbers with decrease of temperature below 0 °C, because of the presence of natural ice nuclei. The formation of the ice crystals occurs at saturation with respect to water. Since the saturation VAPOUR PRESSURE over ice is less than that over water at the same temperature (maximum deficit of 0.27 mb at −12 °C), the ice crystals increase rapidly in size because of diffusion of water vapour from neighbouring water droplets to the ice crystals. (Growth is subsequently accelerated by 'coalescence' of the relatively heavy ice crystals with other ice crystals or with water drops.)

(ii) Coalescence (accretion) process. If, within a cloud, there are some liquid drops appreciably larger than the great majority of drops, the slower rate of rise of such large drops in a cloud updraught leads to collisions and, in some cases, coalescence with the smaller liquid drops. Factors which promote this process are appreciable cloud depth and updraught speed which allow collision growth to a size sufficient to ensure that the drop will not evaporate at the top of the cloud but will fall back through the cloud, growing further by collision and reaching the ground as rain. The cause of the initial differences in drop size which are essential to the coalescence mechanism — whether, for example, the differences are due to the presence of rare giant nuclei or are associated with variations of updraught velocity and nuclei concentration near the condensation level — is as yet uncertain.

For many years it was considered that the ice-crystal process was much the more important of the two mechanisms. It is now clear that both play an important part. Attempts have been made, by appropriate CLOUD SEEDING, to promote precipitation by either mechanism.

precipitation static. See ST. ELMO'S FIRE.

precision aneroid. An ANEROID BAROMETER in which the movement of the face of the evacuated capsule is shown, not by means of a train of pivoted levers attached to a pointer, but by the deflection of one light, pivoted, counter-balanced electrical contact arm mounted in jewelled bearings. Electrical contact is established by manually rotating a micrometer screw. The reduction in backlash and friction to negligible proportion greatly increases the absolute accuracy of the instrument and versions have now generally replaced mercury barometers at SYNOPTIC STATIONS.

pressure. Pressure is the force per unit area exerted on a surface by the liquid or gas in contact with it. Pressure at any point in a fluid is exerted equally in all directions. The dimensions are $ML^{-1}T^{-2}$.

The pressure of the atmosphere at any point is the weight of the air which lies vertically above unit area centred at the point; on average, the weight above each square metre of the earth's surface is about 10^4 kg (10 tons). Pressure may be expressed in 'inches' or 'millimetres', being the equivalent height of a column of mercury of STANDARD DENSITY ($13\,595.1$ Kg m^{-3}) under conditions of STANDARD GRAVITY (9.80665 m s^{-2}) required to balance atmospheric pressure. In meteorology the MILLIBAR (mb) has almost entirely supplanted the inch or millimetre as the unit of pressure. In SI UNITS the unit is the PASCAL.

The pressure exerted by the wind is relatively small; a wind of force 6 on the BEAUFORT SCALE, for example, exerts only about one thousandth part of the pressure of the atmosphere.

Conversion formulae for pressure units are:

millibars $=$ inches $\times$ 33.86389
millibars $=$ millimetres $\times$ 1.333224
millibars $=$ 10^2 Pascals.

pressure altitude. The height of a given level in the ICAO STANDARD ATMOSPHERE above the level corresponding to a pressure of 1013.2 mb.

pressure coordinates. A system of coordinates in which the independent variables are x, y and p. x and y are horizontal rectangular coordinates while position in the vertical is defined by the hydrostatic pressure (p). This system has advantages over the closely similar CARTESIAN COORDINATE system where, as is normal in dynamical meteorology, isobaric analysis is preferred to constant-level analysis.

pressure gradient force. That force which acts on air by virtue of the variation of pressure in space. It is a three-dimensional vector, denoted $-\nabla p$ or $-$ grad p, and equals

$$-\left\{\frac{\partial p}{\partial x}\,\mathbf{i} + \frac{\partial p}{\partial y}\,\mathbf{j} + \frac{\partial p}{\partial z}\,\mathbf{k}\right\},$$

where $\mathbf{i}, \mathbf{j}, \mathbf{k}$, are unit vectors in the x, y, z directions. The force is normal to surfaces of constant pressure; the sign is negative because the force acts from high to low values of pressure.

The horizontal pressure gradient force is a horizontal vector which is perpendicular to horizontal isobars. It is denoted, for example, $-\nabla_H p$, and equals

$$-\left(\frac{\partial p}{\partial x}\,\mathbf{i} + \frac{\partial p}{\partial y}\,\mathbf{j}\right).$$

In meteorological dynamics, the pressure gradient force acting per unit mass of air is the significant force and is generally referred to simply as the 'pressure gradient force'. This force is $-(1/\rho)\nabla p$ or $(1/\rho)\nabla_H p$ in three or two dimensions, respectively, where ρ is the air density.

pressure jump. A sudden rise of pressure, of the order millibars per minute, which often accompanies the arrival of a LINE SQUALL or similar phenomenon.

pressure-pattern flying. The planning of a flight route in such a way as to complete the flight, for a given aircraft and load, in the shortest possible time. The normal synoptic aids in such planning are isobaric charts on which contours are drawn.

pressure-plate anemometer. See ANEMOMETER.

pressure-tube anemograph. See ANEMOMETER, ANEMOGRAPH.

prevailing wind. That direction of wind which, at a given place, occurs more frequently than any other during a specified period.
   Over all parts of the British Isles statistics show the prevailing wind to be, on an eight-point compass, south, south-west or west. There is, however, an appreciable annual variation, e.g. in some places the prevailing wind in spring and early summer has an easterly component.

primitive equations. The fundamental EQUATIONS OF MOTION of a fluid modified by the assumption of HYDROSTATIC EQUILIBRIUM. The term is used particularly in the context of numerical weather forecasting where the assumption of hydrostatic equilibrium is necessary to eliminate high-speed oscillations in the vertical from the solutions of the equations. See also QUASI-GEOSTROPHIC MOTION, NUMERICAL WEATHER PREDICTION.

principal component analysis. A powerful method for the statistical analysis of fields of data; it has the effect of representing each field as the linear sum of multiples of standard uncorrelated components, with the feature that the main patterns of variation are shown by the least possible number of components. The determination of the principal components, or 'empirical orthogonal functions', involves finding the eigenvectors of the covariance matrix derived from the data, and is only practicable by electronic computer. See also ORTHOGONAL FUNCTIONS.

probability. A numerical measure of the expectation that a particular event will occur. Its value ranges from 0 (impossibility) up to 1 (certainty).
   This fundamental statistical concept is developed from the probabilities that can be associated with apparatus such as dice and roulette wheels (with which a trial must have one of a number of results), these results being produced by processes which do not favour one result rather than another. For this limited class of probabilities, the laws for the addition and multiplication of probabilities can be built up, arguing from considerations of symmetry and the enumeration of cases. The calculus of probabilities defined in this way is found to be applicable to the much wider class of probabilities which cannot be estimated by the enumeration of cases but only in terms of relative frequencies. Thus if in a large number N of trials an event has occurred n times, and has failed to occur $(N - n)$ times, the probability of occurrence is n/N. Probabilities estimated in this way form the basis of most of STATISTICAL INFERENCE.
   The process of judging the SIGNIFICANCE of the result of an experiment consists in estimating the CHANCE EXPECTATION or 'probability' of the result, assuming the starting values of the quantities determining the result to have been chosen 'at random', that is in a way which does not favour one possible result rather than another. If the chance expectation is very small the result is regarded as 'significant'.
   Estimation of the chance expectation is usually a matter of mathematical inference from a certain FREQUENCY DISTRIBUTION, and is greatly simplified by the use of statistics which fall into the frequency distributions which have been studied and tabulated in advance. These include the CHI-SQUARE DISTRIBUTION, the F-distribution (see F-TEST), the NORMAL (FREQUENCY) DISTRIBUTION and the t-distribution (see STUDENT'S t-TEST). The experimenter, when deciding which statistic to use, is well advised to keep to statistics which conform to one of these distributions, or the

BINOMIAL DISTRIBUTION or the POISSON DISTRIBUTION for which the chance expectation can be derived from first principles.

probability distribution.   A somewhat more accurate term for what is often termed a FREQUENCY DISTRIBUTION, in which the probability density is plotted against each possible value of a variable, so that the PROBABILITY of the variable taking a value in a certain range is given by the integral of the probability distribution through the range.

probability integral.   See NORMAL (FREQUENCY) DISTRIBUTION.

probable error.   The range on either side of the MEAN which contains half the items in a NORMAL (FREQUENCY) DISTRIBUTION. It is found by multiplying the STANDARD DEVIATION by 0.6745. The probable error is now little used as a measure of dispersion since it has no advantage over the standard deviation from which it is derived.

probable maximum precipitation.   An estimate of the upper limit of precipitation, of specified duration, over a particular region. Such an estimate, used for example in dam design, may be based on physical reasoning or on past statistics; a very long RETURN PERIOD applies in either case.

profiler.   See WIND PROFILING.

prognostic chart.   A SYNOPTIC CHART depicting, for a specified future hour, the expected distribution of ISOBARS or CONTOURS and usually also of FRONTS. The term is often used of a forecaster-produced chart as distinct from a computer-produced forecast chart.

projection.   This term is used to denote any relationship establishing a correspondence between a domain of the earth's surface and a domain of a plane surface, the map, such that to each point of one corresponds one and only one point of the other. The projection is completely represented by constructing, on the plane surface, a graticule formed by two intersecting systems of lines, corresponding respectively to parallels of latitude and meridians of longitude on the earth. The position on the map of any features on the earth's surface is then determined by reference to this graticule.

The scale of the map is the ratio of the distance between two neighbouring points on the map to the corresponding distance on the earth. A perfect map in which the scale is uniform throughout is not possible. A class of projections termed 'orthomorphic' or 'conformal' has the property that, at any point, the scale in all directions is the same, though varying from point to point. This is equivalent to the property that the angle of intersection of any two lines on the earth (such as an ISOBAR and a MERIDIAN) is preserved unchanged on the map, or the shape of any small area is preserved. Orthomorphic projections are not much used generally but, owing to the above properties, they enter into meteorological practice as base maps for the representation of meteorological elements.

The following projections are suitable for weather charts:

(i)   The stereographic projection for the polar areas on a plane cutting the sphere at the standard parallel of latitude 60°.

(ii)   Lambert's conformal conic projection for the middle latitudes, the cone cutting the sphere at the standard parallels of latitude 10° and 40° or 30° and 60°.

(iii)   Mercator's projection for the equatorial areas, with true-scale standard parallel of latitude 22½°.

A chart required to give bearings for thunderstorm location uses the oblique gnomonic projection which has the valuable property that all GREAT CIRCLES on the globe are reproduced as straight lines.

proton. A constituent particle of all atomic nuclei, itself comprising the hydrogen nucleus, which carries unit positive charge (equal but opposite to that of the ELECTRON) and has a mass about 1850 times that of the electron.

pseudo-adiabatic (or **pseudo-adiabat**). Line on an AEROLOGICAL DIAGRAM representing the pseudo-adiabatic lapse rate.
　　See ADIABATIC.

pseudo-equivalent temperature. The pseudo-equivalent temperature (T_{se}) of a sample of air at any level is found from an AEROLOGICAL DIAGRAM by dry adiabatic expansion to the lifting CONDENSATION LEVEL of the sample, followed by ascent along the saturated adiabatic until all the water vapour is condensed, and finally dry adiabatic descent to the initial pressure level.
　　The 'pseudo-equivalent potential temperature' (θ_{se}) is found by progressing along the dry adiabatic line from T_{se} to the 1000 mb level.
　　T_{se} for an air sample exceeds the (isobaric) EQUIVALENT TEMPERATURE (T_e), and θ_{se} exceeds the equivalent potential temperature (θ_e), by an amount which is not negligible.
　　See also ADIABATIC.

pseudo-random numbers. A series of numbers, produced by a strictly defined numerical algorithm, which nevertheless have all the characteristics of truly RANDOM NUMBERS for the particular purpose in hand; for example, the AUTOCORRELATION characteristics of the series may be indistinguishable from those of a truly random one. Algorithms for pseudo-random numbers must be used with caution; for example, if successive groups of three pseudo-random numbers are used as three-dimensional coordinates, the points so generated may tend to lie in preferred planes. Pseudo-random number generators are used in computer simulations of MONTE CARLO METHODS.

pseudo wet-bulb temperature. A temperature (T_{sw}) obtained most readily from an AEROLOGICAL DIAGRAM by ascent of the sample at the dry adiabatic lapse rate until saturation is reached, followed by descent at the saturated adiabatic lapse rate until original pressure is reached.
　　The 'pseudo wet-bulb potential temperature' (θ_{sw}) is found by progressing along the saturated adiabatic from T_{sw} to the 1000 mb level.
　　T_{sw} for an air sample is less than the WET-BULB TEMPERATURE (T_w) and θ_{sw} is less than the WET-BULB POTENTIAL TEMPERATURE (θ_w), to a degree which is in practice negligible (usually less than 0.5 °).
　　See also ADIABATIC.

psychrograph. A recording PSYCHROMETER.

psychrometer. A type of HYGROMETER (also termed the 'dry- and wet-bulb hygrometer' or 'Mason's hygrometer') in which two similar thermometers are used; one, the 'dry-bulb', gives the air temperature (T), while the other, the 'wet-bulb', whose bulb is covered with muslin wetted with pure water, gives a reading (T_w). In unsaturated air T_w is lower than T by an amount (the 'wet-bulb' depression) which, at a specified temperature, depends mainly on the relative humidity of the air but also, to a small extent, on the degree of ventilation of the wet-bulb. The lower temperature is explained by the fact that the latent heat required to evaporate water from the muslin is supplied by the air which is in contact with the wet bulb.

At air pressure p, the vapour pressure (e) of an air sample is related to the saturation vapour pressure at the wet-bulb temperature (e'_w) and to the wet-bulb depression ($T - T_w$) by the semi-empirical 'psychrometric formula'

$$e = e'_w - Ap(T - T_w)$$

where A is a constant.

The *Hygrometric tables* [52] and humidity slide-rule issued by the Meteorological Office for the purpose of obtaining values of vapour pressure, dew-point and relative humidity from readings of T and T_w observed in a THERMOMETER SCREEN ('light air' conditions of ventilation) are based on a value of the product Ap of 0.799 for $T \geqq 0$ °C and of 0.720 for $T < 0$ °C for readings in degrees Celsius (0.444 for $T \geqq 32$ °F and 0.400 for $T < 32$ °F for readings in degrees Fahrenheit); in this product p is assigned the value 1000 mb. The saturation vapour pressure with respect to water is used as the standard at all temperatures.

In the ASSMANN PSYCHROMETER and WHIRLING PSYCHROMETER (or 'sling psychrometer') the rate of ventilation of the bulbs is controlled and values of A used in the reduction tables and slide-rule are different from those appropriate to thermometer screen readings.

psychrometric formula. See PSYCHROMETER.

pumping. Unsteadiness of the mercury in the barometer caused by fluctuations of the air pressure produced by a gusty wind, or due to the oscillation of a ship.

purga. See BURAN.

purple light. Shortly after the sun has set below the western horizon a brighter patch appears on the darkening sky about 25° directly above the position where the sun has disappeared. This patch appears brighter as the sky darkens and takes on a purple tone. The patch expands into a disc and when the sun is about 4° below the horizon it reaches its maximum brilliance, when it may be so bright that white buildings in the east which are lit up by glow with a purple colour which corresponds to the ALPINE GLOW seen on the peaks of snow-covered mountains. The disc of purple light sinks downwards at twice the rate at which the sun sinks while at the same time its radius expands and its light becomes less intense. It finally sets behind the bright segment of the TWILIGHT ARCH. Occasionally when the first purple light has passed below the horizon, the phenomenon repeats itself with less intensity. The second patch of light appears at a slightly lesser altitude than the first but otherwise follows the same course.

pyranometer. A term applied both to the type of instrument used in measuring the DIFFUSE RADIATION (direct sun excluded) on a horizontal surface and to that which measures the total radiation ('sun plus sky') received on a horizontal surface. The latter type of apparatus is also often incorrectly termed 'solarimeter', but this is the trade name of a particular commercially marketed pyranometer.

The instrument generally consists of a THERMOPILE which is under a protective hemispheric glass cover and is connected to a recorder.

pyrgeometer. An instrument for measuring the NOCTURNAL RADIATION. That designed by Ångström uses the fact that the radiation from a gilded strip of manganin is less than that from a blackened strip.

pyrheliometer.   An instrument for measuring the direct solar RADIATION (DIFFUSE RADIATION excluded) at normal incidence.

Four main types are used. In Abbot's silver-disc pyrheliometer, the rise of temperature in a silvered disc exposed normal to the sun's rays is measured directly and the intensity of radiation determined by reference to calibration data of the instrument. Another type uses a THERMOPILE covered by a flat glass plate. In Ångström's form, the rate of absorption of heat by a thin strip of blackened platinum normal to the sun's rays is found by measuring the electric current necessary to heat a similar strip to the same temperature. However, the cavity radiometer originally developed for measurements by satellites is becoming increasingly used. In this the radiation is directed through a small hole into a blackened cavity (an almost perfect absorber) in a metal cylinder. The cylinder can be warmed through the same temperature rise as is produced by the radiation by passing a known current through an electrical heating coil wrapped around it.

pyrocumulus.   A term sometimes given to the cumulus cloud that forms over an intense source of heat on the ground, such as a heath or forest fire. It does not form part of the official WMO cloud classification and is, in any case, an unhappy mixture of Greek and Latin.

Q

Q-code. A letter code used by aircraft in requests for information; it is also used in the supply of information to aircraft. Certain items in the code relate to meteorological information, e.g. QFE refers to station level pressure, QFF to mean-sea-level pressure, QNH to ALTIMETER SETTING. See *Handbook of weather messages* [53].

quantile. One of a set of critical values of a random variable which divide its DISTRIBUTION FUNCTION into ranges of equal total frequency.

quartile. One of three values of a random variable which divide its DISTRIBUTION FUNCTION into four equal parts. The second quartile is identical to the MEDIAN. The distance from the first to the third quartile is termed the inter-quartile range; half this distance is sometimes used as a crude measure of variability.

quasi-biennial oscillation (QBO). A well defined OSCILLATION of the zonal wind component in the equatorial stratosphere, the 'period' being about 27 months. It is also termed 'stratospheric oscillation'.
The range of the fluctuations, greatest at a level of about 25 km, decreases with distance from the equator. There is a phase lag from higher to lower levels. At a given level the phase does not appear to vary with longitude. Similar oscillations exist in stratospheric temperature and in total ozone amount.
The QBO has been explained as a consequence of the excitation of EQUATORIAL STRATOSPHERIC WAVES by the action of disturbances in the equatorial troposphere which convey momentum and energy upwards to the stratosphere through the tropopause. The energy and momentum are transmitted upwards at the GROUP VELOCITY of the waves while the phase moves down with the PHASE VELOCITY which has the opposite sign.

quasi-geostrophic motion. Air motion which is known, or more frequently assumed, to be closely approximate to the GEOSTROPHIC WIND. Such an assumption is generally valid in those cases where large-scale atmospheric motion is considered, but it becomes increasingly inaccurate as the scale of the motion decreases and cannot be made, for example, in considering FRONTS where the ageostrophic motions are very important. See also NUMERICAL WEATHER PREDICTION.
In describing the motion as 'quasi-geostrophic' rather than 'geostrophic', it is usually implied that derived quantities such as the horizontal DIVERGENCE which depend on small horizontal gradients of wind speed or direction cannot be taken from the gradients of the geostrophic wind, but must be determined in other ways.

quasi-stationary front. A FRONT whose position is (almost) unchanged on successive synoptic charts. There is a strong tendency in such cases for wave-like disturbances of the front to form.

quint. A term used in the Meteorological Office for the set of values included between two adjacent QUINTILES of a FREQUENCY DISTRIBUTION or those in the tails below the first and above the fourth quintile.

quintile. One of a set of four values of a random variable which divide its DISTRIBUTION FUNCTION into five equal parts.

Q-vector. The OMEGA EQUATION, as usually written, has on the right hand side the following two terms:

$$f\frac{\partial}{\partial p}\{(\mathbf{V}_g\,.\mathrm{grad}(\zeta_g+f)\} \;-\; \nabla^2\left(V_g\,.\mathrm{grad}\,\frac{\partial\phi}{\partial p}\right)$$

where $\mathbf{V}_g$ is the geostrophic wind and ζ_g the geostrophic vorticity. Hoskins *et al.* [54] have pointed out that the algebraic sum of these two terms is in general much smaller in absolute value than either taken separately, and that both terms change by equal and opposite amounts if referred to a second frame of reference moving with constant velocity relative to the first, whereas any term with real physical significance should be invariant in such circumstances. The two terms may, however, be combined and written as 2 div **Q**, where $\mathbf{Q} = (\mathrm{d}/\mathrm{d}t)_g\nabla_H\theta$, demonstrating that, under the HYDROSTATIC APPROXIMATION, vertical velocities are forced solely by the divergence of **Q**. Hoskins *et al.* show how patterns of **Q** may be applied to the analysis and understanding of synoptic situations. Hoskins and Pedder [55] show how Q-vectors are related to the Sutcliffe theory of development and also how this theory may be expressed by means of a vector $\mathbf{Q}_s$ analogous to, but not identical with, **Q**.

R

radar. A system of detection and location of 'targets' which are capable of reflecting high-frequency radio waves (microwaves), generally in the wavelength range from a fraction of a centimetre to some tens of centimetres.

The system consists of sending from a transmitter a narrow beam of radio waves and obtaining any reflected signal ('echo') in an adjacent receiver which generally uses the same aerial system as the transmitter. The distance of the target is obtained from the time interval between transmission and reception, and its direction from that of the aerial system employed. The information is presented visually on a cathode-ray tube. Detection of the echo requires a very sensitive receiver and portrayal requires great amplification since much of the emitted radio energy is lost by atmospheric absorption and scattering. See also RADAR METEOROLOGY.

radar meteorology. The main applications of RADAR in meteorology are in the measurement of RADAR WINDS and in the detection of cloud and precipitation elements. The latter is used as a direct forecasting aid and in detailed studies of the structure of precipitation regions.

Various types of radar display are used in the surface investigation of precipitating clouds. They include: a 'plan position indicator' (PPI), which employs an aerial system scanning at a constant elevation about a vertical axis and gives a picture of the distribution of precipitation regions in all directions round the observing point to a distance dependent on the characteristics of the set and the intensity of precipitation — for moderate precipitation this distance may be about 150 km; a 'range–height indicator' (RHI), in which the scanning is in the vertical plane; and an 'A-scope indicator', which shows the variation of amplitude of reflected signal with range of target.

Since the amount of back-scattered energy and the degree of atmospheric attenuation of the waves which travel to and from the target both increase with decrease of wavelength, a compromise choice of wavelength is required. For detection of precipitation a choice is made in the approximate range 3 to 20 cm, while detection of the larger cloud particles even at very short range requires a wavelength in the millimetre range.

For a given wavelength, the strength of the echo increases with the concentration, and very rapidly with the size (proportional to the sixth power of the diameter), of cloud or precipitation particles. The echo strength is, therefore, almost entirely due to the energy back-scattered by any precipitation particles, as opposed to the much smaller cloud particles, which may be present. Empirical relations have been obtained between precipitation rate and corresponding echo intensity.

The vertical and horizontal extent, development and movement of a precipitation region may be indicated by radar. Various complicating factors greatly reduce, however, the amount of quantitative work possible and make it very difficult to assess the precise nature of the reflecting particles. These factors include: particle size distribution relative to radio wavelength in use (determining whether RAYLEIGH SCATTERING or MIE SCATTERING is prevalent), particle shape (affecting 'radar cross-sections'), and particle phase (whether, for example, ice or water, or dry or wet hailstones). See also DOPPLER RADAR, BRIGHT BAND, MELTING BAND, RAYLEIGH SCATTERING.

Since the late 1970s, integrated networks of radars, which provide remote colour displays of composite data on visual display units at meteorological offices (Figure 36) so that the forecaster may diagnose and forecast the movement of rain areas over periods of 2–3 hours, have been established in various countries, including the United Kingdom. See NOWCASTING.

radar storm-detection. An obsolescent term for the detection of 'storms' — in practice, regions of precipitation — in the atmosphere, by means of ground-based or airborne RADAR. See RADAR METEOROLOGY.

radar wind. A wind in the upper atmosphere determined by means of radar reflections from a 'radar target' carried aloft by a free balloon, the observed elements being the range, elevation, and azimuth of the target.

radian. A unit of angular measure, being the angle subtended at the centre of a circle by an arc equal in length to the radius of the circle. Thus, π radians $= 180°$.

FIGURE 36. Picture from the United Kingdom weather radar network for 0100 GMT on 14 January 1989. Rainfall intensity (mm h^{-1}) is depicted as follows: white 0–⅛, dark blue ⅛–1, green 1–4, yellow 4–8, magenta 8–16, red 16–32 and cyan >32.

231

radiation. The transmission of energy by electromagnetic waves. See ELECTRO-MAGNETIC RADIATION. (The term 'radiation' is also used to signify the emission of particles by a source, as in 'cosmic radiation' and 'solar particle radiation'.)

Radiation flux across a surface is the energy which crosses unit area of the surface in unit time. The following relations hold for the various units of flux which have been used in meteorology (the SI UNIT is $W\,m^{-2}$ but WMO recommended practice is $mW\,cm^{-2}$):

$$1\ cal_{IT}\,cm^{-2}\,min^{-1} = 4.1868\ J\,cm^{-2}\,min^{-1}$$

$$= 69.8\ mW\,cm^{-2}.$$

A valuable concept in radiation is that of the 'black body'. A perfect black body is one which absorbs all the radiation falling on it and which emits, at any temperature, the maximum amount of radiant energy. The term arises from a correlation between darkness of colour and proportion of visible light absorbed. A body which appears white because it scatters the visible light falling on it may, however, act nearly as a black body to radiation of a different wavelength. Snow is an example, being effectively a black body for wavelengths greater than 1.5 μm.

The properties of a black-body radiator are expressed in a number of laws:

(i) Planck's law. The distribution of energy with temperature (T) and wavelength (λ) for a perfect radiator was represented by Planck as

$$E_\lambda = c_1\lambda^{-5}/(e^{c_2/\lambda T} - 1)$$

where E_λ is the energy emitted in unit time from unit area within unit range of wavelength centred on λ, and c_1 and c_2 are constants. The corresponding energy distribution curve is illustrated in Figure 37 for four radiation temperatures; it shows, for example, that the solar radiation spectrum barely

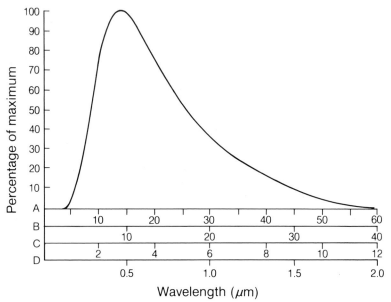

FIGURE 37. Distribution of energy in black-body spectrum. Scale A for 200 K (stratosphere), scale B for 300 K (ground), scale C for 1000 K (red-hot) and scale D for 6000 K (sun).

overlaps that of ground radiation. The area bounded by the curve, the wavelength axis and any pair of selected wavelengths gives a relative measure of the energy contained in the corresponding part of the spectrum.

(ii) Wien's (displacement) law. The variation of wavelength of maximum energy emittance (λ_{max}) with temperature of radiator is given by

$$\lambda_{max} \, T \, = \, 2.898 \times 10^{-3} \, \text{m K}^{-1}.$$

(iii) Stefan–Boltzmann law. The amount of energy emitted in unit time from unit area of a black body is proportional to the fourth power of its absolute temperature, i.e.

$$E \, = \, \sigma T^4 \text{ where } \sigma \text{ (Stefan's constant)} \, = \, 5.670 \times 10^{-9} \, \text{mW cm}^{-2} \, \text{K}^{-4}.$$

(iv) Kirchoff's law. The ratio of emissive to absorptive power, for a particular wavelength and temperature, is the same for all types of body and is numerically equal to the emissive power of a black body (whose absorptive power is, by definition, unity). Thus, if a body at a given temperature strongly absorbs radiation of a certain wavelength, it also radiates strongly this wavelength, provided it is present in the radiation spectrum for the temperature. The wavelength dependence is crucial here: for example fresh snow absorbs very little direct (short-wave) solar radiation but acts very nearly as a black body to long-wave radiation from the atmosphere.

The spectrum of solar radiation outside the earth's atmosphere, inferred from ground observations or measured directly by rockets and satellites, extends from the X-ray region through the ultraviolet, visible (0.4–0.7 μm) and infra-red to the radio-wave region. About half the total solar energy is in the form of visible light. The spectrum observed at the ground is sharply cut off in the near ultraviolet at about 0.29 μm, owing to the complete ABSORPTION of shorter wavelength, containing about 5 per cent of the total energy, by gases in the high atmosphere; there is also selective absorption of higher wavelengths by atmospheric constituents.

Application of the inverse square and Stefan–Boltzmann laws, with the assumption of a SOLAR CONSTANT of 139.6 mW cm^{-2} shows that the sun has a black-body radiation temperature of about 5800 K; while a sun 'colour' temperature of about 6100 K is obtained by insertion of the value $\lambda_{max} = 0.474 \, \mu$m into Wien's formula. Detailed examination of the extraterrestrial solar spectrum in terms of Planck's law reveals, however, that there is no close fit at all points of the spectrum with a perfect radiator of any given temperature, probably because the radiation originates at different levels in the sun's atmosphere and therefore at different temperatures.

The radiant energy delivered by the sun to the fringe of the earth's atmosphere, about 7 per cent greater at PERIHELION (early January) than at APHELION (early July), is 139.6 mW cm^{-2} at the earth's mean distance (solar constant). The mean flux perpendicular to the earth's total (illuminated plus non-illuminated) surface is about 34.9 mW cm^{-2}. About 40 per cent of this incident radiation is diffusely reflected to space without change of wavelength (ALBEDO of earth–atmosphere system), about 15 per cent heats the atmosphere by direct absorption by constituents, and the remaining 45 per cent is absorbed at the earth's surface as both direct ('sun') and diffuse ('sky') radiation. The ratio of sky to sun radiation increases with cloudiness and latitude and is greater in winter than in summer; in middle latitudes it averages 30–40 per cent.

Most of the solar energy absorbed at the earth's surface is transferred as heat to the atmosphere initially by conduction, then by turbulence and convection and by evaporation; radiation in the waveband appropriate to the temperature of the surface is smaller but also important. The total amount of long-wave radiation emitted to space by the atmospheric gases and direct from the earth's surface (terrestrial radiation) equals, on balance, the 60 per cent of solar radiation which is effective in heating the earth and atmosphere; the black-body radiation 'planetary temperature' appropriate to this radiation balance is about 250 K. Study of the conditions of balance within the earth–atmosphere system shows that, except in higher latitudes in winter, the earth's surface radiates less heat than it absorbs (heat source) and the troposphere everywhere radiates more heat than it absorbs (heat sink); in the stratosphere local radiative equilibrium prevails.

Direct absorption of solar radiation has been found by estimation and measurement to cause a daily heating of about 0.1 °C in the low stratosphere and of 0.1–0.6 °C (depending on season and water vapour content) in the lower troposphere in middle latitudes. In the high stratosphere, solar heating effects are much greater — notably at about 50 km, associated with absorption by oxygen and ozone. The daily net cooling rate, associated with the divergence of terrestrial radiation, is about 1–2 °C throughout a cloudless troposphere and low stratosphere; cooling is much reduced below a cloud layer but is increased at the top of such a layer.

See also LONG-WAVE RADIATION.

radiation balance. The resultant flux of the solar and terrestrial RADIATION through a horizontal surface. It is considered positive if the flux downwards exceeds that upwards. It is also termed 'net radiation'.

In general, the radiation balance at the earth's surface is positive by day and negative by night. The average annual surface balance is everywhere positive. It is highest in low latitudes and, in a given latitude, is greater over an ocean than over a continent. The positive balance for the surface of the globe is estimated to average about 10 W m^{-2}). Radiative equilibrium for the earth and atmosphere system as a whole is achieved by a negative balance of the same amount in the atmosphere. The radiation balance at the outer limit of the atmosphere is portrayed in Figure 38 which is derived from satellite observations and shows the measured meridional profiles of planetary albedo, absorbed solar energy, and infra-red loss; the well known feature of excess energy absorption at low latitudes and net energy loss at middle and high latitudes is clearly demonstrated. See also ENERGY BALANCE.

radiation chart. A chart (for example, that of W.M. Elsasser) which, by providing a graphical method of numerical integration of the equations of radiative transfer in the atmosphere, allows the calculation of the upward and the downward fluxes of radiation at any level, the vertical distribution of temperature and humidity being known.

radiation fog. A common type of FOG which forms overland on nights characterized by light wind, clear sky, and moist air in the lower levels of the atmosphere. The first two conditions lead to the formation of a RADIATION INVERSION. Since, however, loss of water from air in contact with cold ground proceeds rather more quickly than loss of heat, some turbulent interchange of air, with associated adiabatic cooling, and downward transfer of water vapour by EDDY DIFFUSION, are required to produce condensation in an appreciable layer. The presence of hygroscopic nuclei, as in industrial areas, facilitates fog formation by allowing condensation to occur in unsaturated air. In the British Isles fog is often caused in part by radiation and in part by advection processes — see ADVECTION FOG.

radiation inversion. An INVERSION of temperature through an atmospheric layer extending upwards from the earth's surface, such a condition developing in the course of a RADIATION NIGHT over a land surface as a result of radiational cooling of the surface. The depth of the inversion layer increases in the course of the night by

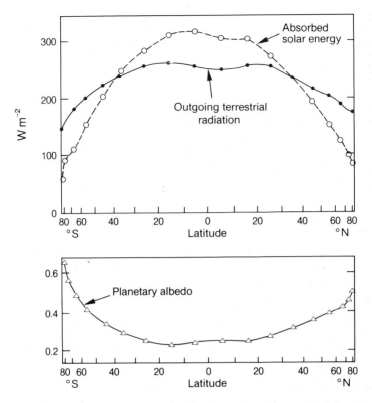

FIGURE 38. Average components of the earth's radiation budget as deduced from satellite observations, 1961–66, by Vonder Haar and Suomi [58].

turbulent transfer of heat downwards from the warmer air above the inversion and by direct radiative cooling of the lowest layers of the atmosphere. The inversion layer is typically a few tens of metres deep by dawn but notable cooling may extend to a few hundred metres above the surface.

radiation night. A night characterized by small amounts of cloud and light surface winds, on which there is marked radiative cooling of the ground and — by turbulent mixing of heat downwards or by direct radiative cooling — the surface layers of air. The absence of significant amounts of cloud reduces the long-wave downward radiation from the atmosphere which otherwise tends to compensate for the upward long-wave radiation from the ground. Light winds lead to little or no turbulent mixing which confines the cooling to a shallow layer and produces low minimum surface temperatures.

Other factors which favour low minima on radiation nights are relatively dry air (reducing the downward long-wave radiation), low heat conductivity of the ground (e.g. dry or snow-covered soil) and long hours of darkness.

radiation pressure. The pressure exerted on a body by the ELECTROMAGNETIC RADIATION incident on it. That exerted by solar radiation which is incident on the earth's surface is minute relative to atmospheric pressure.

radiative equilibrium. State of balance between the absorption and emission of RADIATION.

Turbulent transfer, evaporation and condensation are the processes which chiefly inhibit the occurrence of the state of radiative equilibrium. The state is therefore of little or no significance at the earth's surface and in the troposphere. At higher levels, evaporation and condensation are unimportant and the effect of turbulent transfer on the temperature distribution is, in general, secondary to that of radiation processes. Thus, for example, approximate radiative equilibrium prevails in the upper part of the ozone layer where ultraviolet radiant energy is strongly absorbed by oxygen and ozone in certain spectral lines and bands and is shared by collision with other atmospheric gases. The air temperature is thus raised to a level at which the terrestrial radiation emitted by the gas mixture (mainly by the constituents ozone, carbon dioxide and water vapour) is in balance with the incoming radiation. See also RADIATION BALANCE.

radiatus (ra).   One of the CLOUD VARIETIES. (Latin for having rays.)
    'Clouds showing broad parallel bands or arranged in parallel bands, which, owing to the effect of perspective, seem to converge towards a point on the horizon or, when the bands cross the whole sky, towards two opposite points on the horizon, called radiation point(s).
    This term applies mainly to CIRRUS, ALTOCUMULUS, ALTOSTRATUS, STRATOCUMULUS and CUMULUS' [2, p. 21]. See also CLOUD CLASSIFICATION.

radioactive carbon (radiocarbon).   Carbon-14 (^{14}C) is a radioactive isotope which is important in geophysics. Bombardment of the atmosphere by cosmic rays produces neutrons which react with nitrogen (^{14}N) to form ^{14}C at a rate which is greatest at about 10 km. Measurements indicate that about 2.4 ^{14}C atoms are produced in this way per second per square centimetre of the earth's surface. ^{14}C has a radioactive HALF-LIFE of about 5500 years and reverts to ^{14}N by emission of a BETA PARTICLE.
    Radiocarbon is distributed, like the vastly more plentiful non-radioactive carbon, throughout the earth–atmosphere system. Only 1 or 2 per cent of ^{14}C is stored (as radioactive carbon dioxide) in the atmosphere. About 90 per cent is contained in the carbonaceous materials of the oceans which are in exchange equilibrium with carbon dioxide. The remaining 8 or 9 per cent which is retained by plant and animal life gives rise to the technique of CARBON DATING.
    Local artificial injections of ^{14}C into the atmosphere have been caused by nuclear explosions. The ^{14}C introduced in this way has been used as a tracer element of atmospheric motion, by the measurement of the ^{14}C content of air samples collected at various levels, and of vegetation.

radioactive fall-out.   The descent to the earth's surface of radioactive material produced in a nuclear explosion.
    A distinction is made between two classes of fall-out: (i) 'close-in fall-out' (or 'local fall-out'), i.e. material that descends close to 'ground zero' (point on earth's surface at or above which the device is exploded) and (ii) 'delayed fall-out' (or 'world-wide fall-out'), i.e. material that reaches the earth's surface after a long delay, most of it far from 'ground zero'. Meteorological processes are important in both types of fall-out.

(i)   *Close-in fall-out*
    If a nuclear explosion occurs at a level sufficiently low for the associated FIREBALL to intersect the earth's surface (technically, a 'ground-burst'), fused ground material is sucked into the rising cloud and made radioactive. The average size of cloud particle is, in these circumstances, relatively large and the average TERMINAL VELOCITY correspondingly great. The largest (fastest-falling) particles reach the earth's surface about 20 minutes after the explosion, the smaller particles proportionately later and farther downwind. Since the particles start their descent from a range of levels within the cloud, they are subject to horizontal movement by mean vector winds extending through various layers (all terminating at the surface) as well as to horizontal diffusion

and so give rise to a fall-out pattern on the ground, the edges of which are likely to be irregular if there are strong vertical currents or precipitation in the region. The pattern is long and narrow if the upper winds are strong and have little vertical shear, and much more nearly circular if the upper winds are light and have pronounced shear. For a large thermonuclear ground-burst device (5–10 megatons), the height reached by the top of the cloud is of the order 25–30 km.

The radioactive dust which settles on the ground emits GAMMA RADIATION which is relatively intense downwind from ground zero along that line which corresponds to the direction of the mean vector wind in the layer from the surface to the most active part of the cloud (the so-called 'hot line'), and which is relatively weak towards either edge of the plume. After fall-out is complete at any place within the plume, the intensity of emitted radiation decreases at an approximately exponential rate, reckoned from the time of the burst. The limits of close-in fall-out are arbitrarily selected on the basis of a minimum measured dose-rate of radiation at a fixed time interval after the explosion; the selected limits are such that it is only particles of diameter greater than about 50 μm, falling in the period up to some 36 hours after the explosion, which are involved. Close-in fall-out comprises more than half the total radiochemical energy of a ground-burst weapon.

(ii) *Delayed fall-out*
If a nuclear explosion occurs at such a level that the fireball does not intersect the surface of the earth (an 'air-burst'), the size of the particles is very small and terminal velocities are low. There is then negligible close-in fall-out (except, perhaps, in the event of rain at the time of the explosion), but a delayed fall-out, partly by deposition of radioactive dust and mainly by WASH-OUT with rain. If the original cloud is confined to the troposphere, most of the delayed fall-out occurs within a period of weeks. If, however, the material reaches the stratosphere, most of the fall-out is delayed for months or years (depending on the height, season and latitude of injection). The term 'residence half-time' or 'storage half-time', signifying the time required for one half of the material to be deposited, is used with respect to delayed fall-out.

During the period when the material is stored in the atmosphere, it acts as a 'tracer element' in respect of air movement. However, other meteorological factors — notably the distribution and intensity of rainfall — play an important part in determining the eventual pattern of deposition of the material on the earth's surface.

Radioactive fall-out has been found to cause long-period changes in atmospheric electrical conductivity — see ATMOSPHERIC ELECTRICITY.

radioactivity. The property of spontaneous disintegration of unstable, into more stable, elements, accompanied by the emission of ALPHA PARTICLES, BETA PARTICLES, or GAMMA RADIATION. It is a property possessed by some of the naturally occurring elements; those in the earth's crust play a significant part in the IONIZATION of air at low levels over land areas.

radio direction-finder. An instrument, also termed 'radiogoniometer', for determining the azimuth from which radio waves are received. The term is sometimes used when elevation also is determined, being then a RADIO-THEODOLITE.

radio direction-finding (RDF). Measurement of the direction of arrival of radio waves. It is also termed 'radiogoniometry'. See RADIO DIRECTION-FINDER.

radio duct. See ANOMALOUS RADIO PROPAGATION.

radiometer. An instrument for measuring the radiant flux per unit area falling on a surface (irradiance) or its time integral, the radiant energy per unit area (irradiation). Radiometers may be designed to respond to a wide band of wavelengths (e.g. solar

237

radiation incident on the earth's surface) or to a very narrow band (e.g. the radiation corresponding to part of an infra-red CO_2 absorption line). The latter are employed in meteorological satellites (see SATELLITE, METEOROLOGICAL) in connection with REMOTE SOUNDING OF THE ATMOSPHERE and especially for measurement of minor atmospheric constituents and the variation of temperature with height; essentially, they combine a device for measuring radiant energy in combination with a diffraction grating or an optical cell which selects the particular narrow waveband required, and much technical skill and ingenuity has gone into producing a satisfactory combination of accuracy, lightness and reliability.

See BOLOMETER, NET PYRRADIOMETER, PYRANOMETER, PYRGEOMETER, PYRHELIO-METER.

radio navigation aids. Modern navigation systems depend on measuring the difference in distance from the navigator of two stations of known position which transmit radio signals of regular and precisely known form. Points of constant time-difference fall on hyperbolic surfaces with the stations at the foci, and for navigation at or near ground level one needs to consider only the lines of intersection of these surfaces with the surface of the earth. Often, a system of two or more 'slave' stations is used with a single master; the cycle of transmission begins at the master station, and the arrival of the signal at the slave station triggers the slave into producing its own signal which, however, has its phase adjusted to that of the master. A navigator receives signals from master and slaves and by accurate measurement of phase differences can calculate his position. (This calculation is largely performed automatically by the receiving equipment.)

Various practical systems are in use, and their stability is now so great that successive fixes of position of an upper-air balloon or dropsonde can be used to yield satisfactory measurements of upper winds: the balloon or dropsonde receives the relevant radio signals and retransmits them to a processing unit at a ground station or on a ship. The Meteorological Office employs the 'long-range aid to navigation' (or LORAN) and 'Omega' systems for these purposes.

radiosonde. A small radio transmitter, carried upwards by a balloon, by means of which measurements of pressure, temperature and humidity may be obtained from the upper atmosphere. A typical radiosonde contains an element sensitive to temperature such as a THERMISTOR, an ANEROID capsule to measure pressure, and an element sensitive to relative humidity such as a hygristor or GOLD-BEATER'S SKIN. The responses of these measuring devices are used to modulate in rotation, by means of suitable transducers, the carrier frequency of the radio transmitter. Various reference signals are also broadcast to aid the analysis of the data as received on the ground and to provide continuous correction to any drift in calibration. The main source of error in radiosondes is the variation in the balance of radiation emitted by the instrument and received from sunlight. Typically, the height attained by the balloon is about 24 km (30 mb), but larger balloons often go above 30 km (10 mb). Wind velocities may also be obtained by radar tracking of a target carried by the balloon, or by using RADIO NAVIGATION AIDS.

radio-theodolite. An apparatus for determining the direction (angles of elevation and azimuth) from which radio waves reach a receiver. It consists essentially of a receiver coupled to an aerial which can be rotated about horizontal and vertical axes. Used with a RADIOSONDE, it gives upper-level winds with a lower degree of accuracy than is obtained by the radar method ('radar wind') in which slant range is measured.

radio waves. ELECTROMAGNETIC RADIATION of wavelengths greater than about 1 mm, i.e. beyond the rather indefinite upper limit of INFRA-RED RADIATION.

The radio band of waves is subdivided according to frequency, and ranges from 'extremely high' to 'very low' frequency bands. Those of shortest wavelength (highest frequency) are termed 'microwaves' and are used, for example, in RADAR. For the locating of lightning sources (see SFERICS FIX) special radio receivers, tuned to receive long waves of about 30 km, are used.

radio wind. A wind in the upper atmosphere determined by radio means (see RADIO-THEODOLITE).

radon. One of the INERT GASES, radon is emitted by radioactive materials in the earth's crust and comprises a minute constituent of air at low levels. Its atomic mass is 222 and atomic number 86. The ALPHA PARTICLES which are emitted by the decay of radon (together with those emitted by its isotopes THORON and ACTINON) are responsible for part of the IONIZATION of the air at low levels over land.

rain. Liquid PRECIPITATION in the form of water drops of appreciable size (by convention, of diameter greater than about 500 μm, the limiting size of DRIZZLE drops).

For synoptic purposes, rain (other than in showers) is classified as 'slight', 'moderate', or 'heavy', for rates of accumulation less than 0.5 mm h^{-1}, 0.5–4 mm h^{-1}, and greater than 4 mm h^{-1}, respectively. See also RAINDROPS.

rainbow. A rainbow may be seen when the sun shines upon raindrops. The drops may be at any distance from the observer from a few metres to several kilometres. When sunlight falls upon a drop of water, the light is refracted on entering the drop, is reflected from the far side, and emerges with further refraction from the near side. The light which is reflected in this way does not come out in all directions but only in directions lying within about 42° from the direction of the sun. The reflected light is most intense near this limit. Accordingly an observer looking towards the raindrops receives a certain amount of light from all directions within 42° from the shadow of his head but most light along rays which make about 42° with the central line. The limiting angle depends on the colour of the light and as white light is compounded of light of different spectral colours the observer sees a number of concentric arches of different colours, generally with violet to the inside and red to the outside.

Some of the light falling on a drop does not emerge until after it has been reflected twice. None of the twice-reflected light which reaches an observer makes an angle of less than 50° with the line to the centre of his shadow. The colours of the outer bow formed in this way are in the reverse order to those of the inner bow. The space between the inner and outer ('primary' and 'secondary') bows appears darker than the space inside the primary bow or beyond the secondary bow. Several 'supernumerary' bows may also appear within the primary bow.

The coloration of a rainbow depends on the size of the drops. Drops larger than 1 mm in diameter yield brilliant bows about 2° in width, in which the limiting colour is distinctly red. With drops about 0.3 mm in diameter the limiting colour is orange and inside the violet there are bands in which pink predominates. With smaller drops 'supernumerary' bows appear to be separate from the primary bow. With still smaller drops about 0.05 mm in diameter the rainbow degenerates into a white FOGBOW with faint traces of colour at the edges. The variations of colour with drop size, and also the appearance of supernumerary bows, are due to DIFFRACTION.

Rainbows are not infrequently observed by moonlight but as the human eye cannot distinguish colour with faint lights the lunar rainbow appears to be white.

rain day and wet day. Terms at one time in use for periods of 24 hours starting at 0900 GMT on which respectively 0.2 mm (0.01 in) or more, or 1.0 mm (0.04 in) or more, of rainfall was recorded. The numbers of days satisfying these two criteria are still tabulated for statistical purposes, but the actual terms are no longer used by the Meteorological Office.

raindrops. Liquid-water drops of diameter greater than about 500 μm. The term is sometimes used to include also DRIZZLE drops, extending then to a lower limit of about 200 μm.

Surface-tension forces act on small raindrops to minimize the surface-to-volume ratio and so make them spherical. The lower surface of larger drops is, however, appreciably flattened by aerodynamic forces. Raindrops of greater (equivalent) diameter than about 6 or 7 mm break up into smaller drops when falling at their terminal velocity.

Both 'median volume diameter' (defined as the drop diameter such that half the total water is contained in larger drops) and drop concentration (drops m^{-3}) tend to increase with rate of rainfall. There is, however, considerable variability and dependence, in particular, on type of rainfall. The following are typical values: at a rate of rainfall of 0.5 mm h^{-1} there are about 250 drops m^{-3} and drops of about 1 mm diameter contribute most of the water (the average drop size being smaller), while at a rainfall rate of 25 mm h^{-1} there are about 1200 drops m^{-3} and drops of about 2 mm diameter contribute most of the water.

rainfall. The total liquid product of PRECIPITATION or CONDENSATION from the atmosphere, as received and measured in a RAIN-GAUGE. SNOW, SLEET and HAIL, in addition to RAIN, make up much the greater part of the total 'rainfall', as defined above. There are also small additions due to the deposition of DEW, HOAR-FROST and RIME on to the collecting surface of the rain-gauge. One inch of rainfall is equivalent to about 100 tons of water per acre (1 mm is equivalent to about 1 kg m^{-2}).

Rainfall is classified into three general types: OROGRAPHIC RAIN, CYCLONIC RAIN and CONVECTIVE RAIN. These types, discussed under their individual headings, are by no means mutually exclusive. Other terms, such as 'frontal rain', are sometimes also employed.

rainfall station. A STATION at which the only regular measurements made are those of rainfall. The large majority of the total of about 4800 stations in the United Kingdom which makes returns of measured rainfall to the Meteorological Office are of this category.

rain-gauge. An instrument for measuring RAINFALL.

In its ordinary form a funnel is used to collect the rain, and a tube, made of fairly narrow bore in order to minimize evaporation, leads the collected water to a receiving vessel. The rim of the funnel should have a sharp edge, bevelled outside and falling away vertically inside. The vertical walls should be sufficiently deep and the shape of the funnel sufficiently steep to prevent rain splashing in and out. In the United Kingdom the funnel has usually been made of copper with a brass rim, generally 5 inches, but occasionally 8 inches, in diameter, and mounted in an open situation with the rim horizontal and 12 inches above ground level.

The Meteorological Office 'standard' rain-gauge has a number of interchangeable units used with a choice of aperture area of 150 or 750 cm^2. The material is glass-fibre reinforced polyester resin with stainless steel for outlet pipes. The height of the rim above the ground is standardized at 30 cm. The associated units are collecting bottles, measuring flasks, tipping-bucket switches, totalizer counters, incremental recorders and magnetic recorders.

In self-recording gauges, such as the HYETOGRAPH and the 'tilting-siphon rain-gauge', the collected rainfall is usually made to raise a float to which is attached a pen which records on a chart wound on a clock-driven drum. Alternatively, as for example with the tipping-bucket switch, the record shows the intervals of time which have elapsed between the falling of successive small amounts, generally 0.2 mm, of rain. See also EXPOSURE, TIPPING-BUCKET SWITCH.

rain shadow. An area with a relatively small average RAINFALL owing to sheltering by a range of hills from the prevailing rain-bearing winds. In the United Kingdom the phenomenon is noticeable to the east of Wales in rainfall maps for months in which unusually strong westerly winds have predominated.

rain splash. RAINDROPS impinging on a surface and breaking up into smaller drops. When the surface bears, for example, fungal spores, these spores can be dispersed by the smaller drops. After wind, rain splash is perhaps the most important method by which plant pathogens are spread.

rainy season. A period of a month or more, recurring every year, which is characterized in a given region (generally tropical or subtropical) by relatively large amounts of precipitation for the region. Thus, for example, the period of the south-west MONSOON is the rainy season (or 'wet season') in most parts of south-east Asia, while winter is the rainy season in regions with a MEDITERRANEAN-TYPE CLIMATE.

random forecast. A type of FORECAST, sometimes used as a basis of comparison in the assessment of the success attained in forecasts made by more conventional methods, which is based on chance selection of values of the meteorological elements concerned.

randomization. The arrangement of material in a statistical experiment with the general object of ensuring that the initial conditions do not favour one outcome rather than another; in such a case the actual PROBABILITY of occurrence of any combination is equal to the CHANCE EXPECTATION. This object is generally achieved by numbering the items in any convenient way, and then placing them in the order of items taken from a table of RANDOM NUMBERS.

randomness. A basic statistical concept implying an absence of plan or pattern, or of any tendency to favour one consequence rather than another. See, for example, PROBABILITY.

random numbers. Sets of numbers, usually produced by mechanical or electronic means or by mathematical congruential methods, which have been tested and found to have the statistical property of RANDOMNESS. Complete and exhaustive testing for randomness is impossible, and the important thing in any application is that the numbers should behave as random numbers in the respects necessary for that application.
   Random numbers are used in RANDOM SAMPLING and other processes of RANDOMIZATION, and in MONTE CARLO METHODS. They are usually produced by use of a mathematical algorithm on a computer (see PSEUDO-RANDOM NUMBERS).

random sampling. The chance selection of one or more items from a much larger group or 'population'. The object of such sampling is generally to determine, within defined limits, the average characteristics of the entire population.

range. The difference between the extreme values in a group of measurements. As a measure of variability, range has the disadvantages of systematic dependence on size of sample and of large differences between different samples of the same size.
   The range may be used to obtain a quick estimate of the STANDARD DEVIATION of a large sample. Thus, for example, an estimate is obtained by dividing the items into groups of 9, finding the range in each group, averaging the results and dividing by 3. The dividing factor changes with group size.

rare gases. An alternative for INERT GASES.

ravine wind. A wind which blows through a ravine or narrow valley penetrating a mountain barrier because of the existence of a pressure gradient directed from one side of the barrier to the other. Such winds may attain great strength because of FUNNELLING. An example is the ravine wind at Genoa, caused by a pressure difference between the Po valley and the Gulf of Genoa.

Rayleigh number. A non-dimensional parameter (Ra) which is critical in the static stability of fluids. It is defined by the equation

$$Ra = g \, \Delta T \alpha h^3 / va$$

where g is the gravitational acceleration, ΔT the initial temperature difference between the bottom of the fluid and the fluid at height h, α the coefficient of expansion, v the kinematic VISCOSITY, and a the thermometric conductivity (see CONDUCTIVITY, THERMAL) of the fluid. If the fluid state is such that Ra is less than a critical value, which depends to some extent on the nature of the fluid boundary conditions (free or rigid), any tendency for convection is damped out by viscosity and conductivity.

Rayleigh scattering. SCATTERING of ELECTROMAGNETIC RADIATION effected by spherical particles of radius less than about one-tenth the wavelength of the incident radiation. Two important cases arise in meteorology:

(i)   scattering of incident solar radiation by air molecules in a manner which explains the BLUE OF THE SKY, and
(ii) scattering of radar waves by raindrops in the atmosphere. Solid particles of appropriately limited radius also conform to this type of scattering.

According to Lord Rayleigh's theory of molecular scattering, the scattering coefficient (β) is given by

$$\beta = \frac{32\pi^3(n-1)^2}{3N\lambda^4}$$

where n is the refractive index of the gas, N its number of molecules per unit volume, and λ the wavelength. The wavelength dependence of n is so slight that to a close approximation molecular scattering varies inversely as the fourth power of λ. Thus, molecular scattering of white sunlight is such that the scattering of blue light is about five times greater than that of the red light contained in the incident beam.

Rayleigh's theory of molecular scattering includes also the angular distribution of intensity of scattered light (symmetrical about a plane normal to the incident beam with maxima in the 'forward' and 'backward' directions) and of its state of POLARIZATION.

The radar reflectivity of raindrops of a size which satisfies the condition for Rayleigh scattering is given by

$$\frac{\pi^5}{\lambda^4} \left(\frac{n^2 - 1}{n^2 + 2} \right)^2 \Sigma N d^6$$

where λ is the radar wavelength, n the refractive index of the particle, N the number of particles per unit volume and d their diameter.

RDF. Abbreviation for RADIO DIRECTION-FINDING

Réaumur scale. A scale of temperature, now obsolete, introduced in 1731 by the French physicist Réaumur. On it the freezing-point of pure water is $0°$, and the boiling-point $80°$ at normal pressure.

recombination. The various processes by which positive and negative IONS, or positive ions and ELECTRONS, recombine to form neutral particles. The rate of recombination is expressed by a 'recombination coefficient' with dimensions L^3T^{-1}.

recovery factor. A thermometer exposed in a rapidly moving air stream will usually record a temperature different from the free-stream stagnation temperature (T_s). If the indicated temperature is T_r and the true air temperature is T_a, then the recovery factor (r) is defined by the equation

$$r = \frac{T_r - T_a}{T_s - T_a}.$$

More detailed discussion is given in WMO *Technical Note* No. 39 [56].

recurrence tendency. A recurrence tendency in a TIME SERIES is a feature which, though not strictly periodic, implies a greater-than-random frequency of separation or relatively high (or low) values in the series by a specific 'recurrence interval' of time.

A well known example in geophysics is the 27-day recurrence tendency in geomagnetic disturbance. This feature would not be revealed by PERIODOGRAM analysis of a geomagnetic disturbance time series, since it is associated with a solar cause which, after persisting for a few solar rotation periods of about 27 days, dies out before reappearing, after a variable time interval, very probably out of phase with the solar cause previously in operation. A SINGULARITY in meteorology is an example of a type of recurrence tendency which, though yet only quasi-periodic in nature, is not subject to such changes of phase.

recurvature (of tropical storm). See TROPICAL CYCLONE.

red flash. See GREEN FLASH.

reduction. In meteorology, the substitution of computed values for those directly observed, the purpose being to eliminate the effect of some particular factor or factors.

The reduction process is most commonly used to eliminate the effects of varying height on observed surface values of air temperature and pressure and is termed 'reduction to sea level'. Isobars drawn on surface synoptic charts, often also isotherms drawn on climatological charts, refer to values reduced in this way.

Mean temperature values are sometimes reduced to sea level by addition to observed values at the rate of $1\,°C$ per 165 metres or $1\,°F$ per 300 feet of station elevation. Observed pressures are reduced to sea level by application of the ALTIMETER equation; the normal assumption in the latter case is that the mean temperature of the 'missing' air column is the same as the screen temperature, and involves negligible errors except for stations at heights greater than about 300 m above sea level.

reflection. The return to the original medium of the RADIATION incident on a boundary between two media; 'total reflection' is said to occur when all the incident radiation is returned. Reflection is said to be 'specular' (or 'regular') if the reflecting boundary has irregularities which are small relative to the wavelength of the radiation, 'diffuse' if they are large relative to the wavelength. The two laws of reflection — (i) incident ray, reflected ray and normal to the reflecting surface at the point of incidence lie in the same plane, and (ii) angle of incidence (that is, angle between incident ray and normal) equals angle of reflection (angle between reflected ray and normal) — apply to specular reflection. Radiation which is diffusely reflected, on the other hand, emerges in many different directions which are unrelated to the angle of incidence.

In meteorology, the reflecting power (ALBEDO) of surfaces is of fundamental importance in the heat balance that is achieved, either locally or in the earth–atmosphere system as a whole. The albedo has some important dependence on wavelength and on the angle of incidence of radiation.

Reflection of light waves plays an important (in some cases the sole) part in some atmospheric optical phenomena; the reflection of radio waves (radar) has a number of meteorological applications.

refraction. The change of direction to which energy waves (light, sound or radio waves) are subject on passing through a medium of varying density (gradual bending) or through a boundary separating media of different densities (sudden bending).

The two laws of refraction state that:

(i) The incident and refracted rays and the normal to the surface of separation of two media at the point of incidence lie in the same plane.
(ii) Snell's law. The ratio of the sine of the angle of incidence (angle between incident ray and normal) to the sine of the angle of refraction (angle between refracted ray and normal) is a constant for any two media.

See also REFRACTIVE INDEX.

Among the many atmospheric optical phenomena in which refraction plays at least the major part are the delay of apparent SUNSET relative to the sunset time indicated by geometry, the apparent flattening of the sun or moon close to the horizon, HALO PHENOMENA, RAINBOW, GREEN FLASH and MIRAGE. The associated colouring is due to the fact that the amount of bending suffered by light waves is dependent on wavelength.

ANOMALOUS AUDIBILITY and ANOMALOUS RADIO PROPAGATION are examples of the refraction of sound and radio waves, respectively.

refractive index. The refractive index (n) of a medium is a non-dimensional measure of the degree of the REFRACTION of energy waves passing through the medium. It is given by the ratio of the velocity of an electromagnetic wave in a vacuum (c) to its velocity in the medium (v), that is

$$n = c/v.$$

Changes (continuous and discontinuous) of refractive index of the air with height are associated with lapse rates of temperature and humidity and cause various atmospheric optical effects and anomalous propagation of radio waves. Wavelength dependence of n causes dispersion of visible light and coloration of various of the optical phenomena.

The variation of n for dry air with density ρ is represented by

$$(n^2 - 1)/\{\rho(n^2 + 2)\} = \text{constant}.$$

Since $n \approx 1$ the simple relation $(n - 1)/\rho = \text{constant}$ is normally assumed. The equation implies that n generally decreases with height and that rays which pass through the atmosphere, in a direction other than a normal, acquire curvature towards the denser part of the medium. For a ray travelling nearly horizontally in a horizontally stratified atmosphere the ray curvature is proportional to dn/dh (height gradient of refractive index).

For dry air n is, for practical purposes, given by the same formula for light and radio waves, i.e.

$$n - 1 = 79\left(\frac{p}{T}\right) \times 10^{-6}$$

where p is the pressure (mb) and T the temperature (K).

For moist air at optical wavelengths, n is given by

$$(n - 1)(Tp_0/T_0p) = 0.0002918 - 0.000035r/(1 + r)$$

where T_0 and p_0 are the 'standard' values 273 K and 1013 mb, respectively, and r is the humidity mixing ratio.

For moist air at radio wavelengths, n is given by

$$n - 1 = \frac{79}{T}\left(p - \frac{e}{7} + \frac{4800e}{T}\right) \times 10^{-6}.$$

In the atmosphere $(n - 1)$ is of the order 300×10^{-6}; for convenience this is normally expressed as 300 M units, signifying 300 millionths.

For the standard values of T and p and for wavelength $= 0.5893$ μm (sodium D line) $n = 1.0002918$ for dry air and 1.000257 for water vapour. At 15 °C and for sodium D line $n = 1.333$ for liquid water and 1.31 for ice.

See also DIELECTRIC CONSTANT, MODIFIED REFRACTIVE INDEX.

refractometer. In meteorology, an instrument which employs a microwave radio technique for measuring the REFRACTIVE INDEX of the atmosphere. The instrument, carried in an aircraft, gives continuous recording of the resonant frequency of a cavity exposed to the ambient air, corresponding to changes of the quantity N, defined by $(n - 1) \times 10^6$ where n is the refractive index of the air.

regelation. Ice at a temperature near its MELTING POINT may be melted by the application of excess pressure, owing to the reduction of the melting-point effected by such pressure. The re-solidification of the ice which accompanies the removal of the excess pressure (as in the making of snowballs) is known as 'regelation'.

Regional Associations. The Regional Associations of the WORLD METEOROLOGICAL ORGANIZATION each comprise those members of the Organization the networks of which lie in or extend into one of the six meteorological Regions of the world — Africa, Asia, South America, North and Central America, south-west Pacific, and Europe. Each Association normally meets at intervals not exceeding four years.

Regional Telecommunication Hub (RTH). A centre of the global telecommunication system with international responsibilities for collection, exchange and distribution of observational and processed information. See COMMUNICATIONS, METEOROLOGICAL.

regression equation. An approximate relation, generally linear, connecting two or more quantities, derived from the CORRELATION coefficient.

relative contour. An alternative for THICKNESS line.

relative humidity. The relative humidity (U) (per cent) of moist air is defined by

$$U = 100\,\frac{e}{e'_\text{w}}$$

where e is the VAPOUR PRESSURE of the air and e'_w the saturation vapour pressure with respect to water at the same temperature. To a close approximation the corresponding ratios of the MIXING RATIO or of SPECIFIC HUMIDITY may be used. The actual relationship in terms of mixing ratio, for example, is

$$U = 100 \frac{r}{r'} \cdot \frac{0.62197 + r'}{0.62197 + r}$$

where r and r' are mixing ratio and saturation mixing ratio, respectively.

Relative humidity may be measured indirectly from wet- and dry-bulb temperature readings, with the aid of humidity tables or a humidity slide-rule, or directly, as with a hair hygrometer. At temperatures below 0 °C, relative humidity is evaluated with respect to supercooled water and not with respect to ice.

Relative humidity has a marked systematic diurnal variation opposite in phase to that of temperature; that is to say, it has a daily maximum around dawn and a minimum in the afternoon. It has a less well marked annual variation, more especially in afternoon hours, also of opposite phase to that of temperature.

relative isohypse.   An alternative for THICKNESS line.

relative vorticity.   See VORTICITY.

relaxation of a trough.   A decrease in the amplitude of a trough. When the term is used in a westerly synoptic situation in middle latitudes it also implies a poleward movement of the isopleths which define the trough (e.g. contours or thickness lines). It is the opposite of MERIDIONAL EXTENSION of a trough.

remote sensing.   The measurement of various quantities, not by exposure of an instrument at the point from which the measurements are required, but by observations made at a distance. Remote sensing may be 'active' (e.g. when a radar or laser beam is emitted by a transmitter and the reflected or scattered beam is collected and analysed by a suitable receiver), or 'passive' (e.g. when a receiver collects and analyses naturally emitted radiation from a distant source).

remote sounding of the atmosphere.   A method of determining, by REMOTE SENSING from a METEOROLOGICAL SATELLITE, the vertical distribution through the entire depth of the atmosphere of quantities such as temperature, wind, water vapour, etc. The possibility of making remote soundings of temperature depends on the fact that the atmosphere contains two gases — carbon dioxide (CO_2) and molecular oxygen (O_2) — which are almost uniformly mixed up to great heights (at least 90 km) and have strong absorption bands at wavelengths for which long-wave thermal emissions from the earth and atmosphere are appreciable. It is possible to construct RADIOMETERS, capable of being carried in satellites, which have resolving power good enough to discriminate between wavelengths that are absorbed differently by different parts of an absorption band; one can, in other words, select wavelengths that are strongly, weakly, or moderately absorbed. It is clear that any radiation received by a satellite at a wavelength that is strongly absorbed must come from the top of the atmosphere, at a wavelength that is weakly absorbed from the bottom of the atmosphere (because that is where the bulk of the emitting gas lies), and at a moderately absorbed wavelength from somewhere in between. More precisely, it may be shown that the intensity, I_v, of radiation of frequency v received by a satellite-mounted radiometer looking vertically downwards is given by

$$I_v = \int_0^\infty B_v(t) \frac{d\tau_v(z,\infty)}{dz} \; dz \; + \; B_v(T_s) \, \tau_v(0,\infty)$$

246

where $B_\nu(t)$ is the intensity of black-body radiation of frequency ν emitted at temperature T, $\tau(z,\infty)$ the TRANSMISSION COEFFICIENT between height z and the top of the atmosphere, and T_s the temperature of the earth's surface. It is convenient to replace z by $y = -\ln p$ giving

$$I_\nu = \int_0^\infty B_\nu(T)\, k(y)\, dy\, +\, B_\nu(T_s)\, \tau_\nu(0,\infty)$$

where $k(y)$ is known as the weighting function.

The shape of $k(y)$ depends on the absorption coefficient of the atmosphere for the particular frequency (or narrow band of frequencies) being observed and thus on the detailed design of the radiometer. In general, however, $k(y)$ is small at very high and very low levels with a maximum at some intermediate level. Practical soundings are made by using a radiometer with n different channels which respond to n broadly different levels in the atmosphere (see Figure 39), and the problem is to discover the structure of the atmosphere that will yield the observed measurements of radiation.

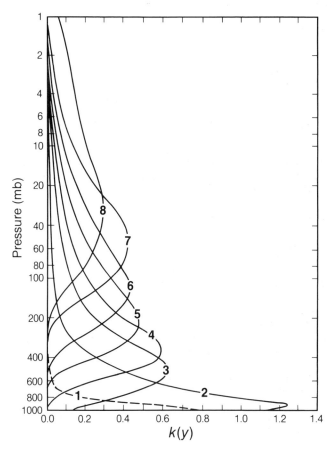

FIGURE 39. Weighting functions $k(y)$ (where $y = -\ln p$) plotted against atmospheric pressure (vertical logarithmic axis). These functions approximately describe the relative sensitivity of the eight radiance observations to temperature variations in various layers of the atmosphere for the satellite infra-red spectrometer (SIRS).

247

This 'retrieval problem' has, in fact, no unique solution: there is an infinite number of temperature structures that will fit the observations. However, it is possible to derive a most probable solution, by methods of advanced mathematical statistics, that can be shown to be near enough to the true structure to be of great practical value. A major difficulty arises from the presence of clouds which obstruct infra-red radiation. Techniques have been developed for retrieving temperature profiles when only broken cloud is present, and the O_2 bands can also be used because non-precipitating clouds are largely transparent at millimetre wavelengths.

The remote sounding of concentrations of minor constituents of the atmosphere may be effected by observations of the intensity of characteristic emissions from these constituents. Water vapour, which is distributed throughout the atmosphere, may be studied by vertical sounding. Other constituents (such as CH_4, N_2O, CO, NO, HNO_3) which are important in the chemistry of the high atmosphere, are best studied by LIMB SCANNING which makes a long atmospheric path available and avoids interference from the variable background of the earth's surface and tropospheric clouds.

As to conditions at the surface of the earth, measurements of sea surface temperature by satellite radiometer are comparable in accuracy with those made by conventional means, and surface winds may be estimated from SCATTEROMETER observations. Methods of estimating sea surface pressure by measuring total atmospheric absorption of radiation have been suggested, but no practical operational system has yet been developed.

report, meteorological. See WEATHER REPORT.

representativeness. A representative air-mass property is one that is typical of the air mass as a whole and so may be useful in AIR-MASS ANALYSIS. Some of the surface meteorological elements, in contrast to the same elements measured in the upper atmosphere, are readily changed by purely local influences and so are not representative; surface temperature is an example of such an element.

Réseau Mondial. An annual publication, now discontinued, of the Meteorological Office. It contained climatological data for all parts of the world on the basis of two stations per 10-degree square of latitude and longitude. The last volume published (in 1957) was that for 1934.

reshabar or **rrashaba.** A name, meaning 'black wind', given to a strong, very gusty, north-easterly wind which blows down certain mountain ranges in the southern Kurdistan region. It is dry, comparatively hot in summer and cold in winter.

residence half-time. See RADIOACTIVE FALL-OUT.

residual. The difference between an individual observation and the mean of a series, or the difference between an individual observation and the value derived from the adopted values of the constants which have been obtained by a discussion of the observations.

Thus an observed quantity may be known to be a function of variables x, y, z, and constants a, b, c, of the form of $ax + by + cz = l$. If a number n of observed values of l are given for known values of x, y and z, there will be n equations to determine the three constants a, b and c. The equation will not in general be satisfied for any one observation, and the value of $l - (ax + by + cz)$ for that observation is its residual.

resistance thermometer. A type of electrical thermometer in which the thermal element is a substance whose electrical resistance varies with temperature. Such a thermometer can be made with a very short LAG coefficient and is capable of highly accurate measurements.

resonance. If a periodic force is applied to a system, resonance is said to occur when the period of the force comes into close accord with a FREE PERIOD of the system, resulting in an increase in the amplitude of vibration of the system. Resonance was previously advanced as an explanation of the large magnitude of the semi-diurnal atmospheric pressure wave but is not now favoured. See ATMOSPHERIC TIDES.

resultant. The sum of a number of directed quantities or vectors. See VECTOR.

retrograde system. In synoptic meteorology, a pressure system which reverses its direction of movement.

An atmospheric wave is said to be retrograde if it moves in a direction opposite to that of the flow in which it lies.

return period. See EXTREME VALUE ANALYSIS.

return stroke. See LIGHTNING.

returning polar maritime air. See POLAR AIR.

reversal (spring, autumn). The stratospheric circulation of middle and high latitudes is dominated in winter by strong westerly winds while the summer circulation is characterized by weaker easterlies. The summer regime is of considerably shorter duration than the winter regime. The changes of circulation type which usually occur in late April and early September in the northern hemisphere are called the spring and autumn reversals.

reversing layer. The lower part of the atmosphere of the SUN, comprising a layer of relatively cool gas extending about 1000 km outwards from the PHOTOSPHERE.

revolving storm. A term synonymous with an intense TROPICAL CYCLONE.

Reynolds number. An important parameter in the flow pattern of fluids, defined by the non-dimensional quantity $Re = \bar{u}d/v$, where $\bar{u}$ is a characteristic fluid velocity, d a characteristic length, and v the kinematic viscosity of the fluid.

In 1883 Reynolds showed experimentally that turbulent, as opposed to laminar, flow is not sustained for a value of Re less than about 2000. A critical value of Re similarly exists for the onset of TURBULENCE as the speed of fluid flow past a smooth body increases.

Reynolds stresses. Fundamental stresses (τ), also termed 'eddy shearing stresses', which operate within a turbulent fluid to transport momentum. τ has the dimensions $ML^{-1}T^{-2}$.

If u', v', w' are turbulent components of velocity (instantaneous departures from average) in rectangular coordinate directions x, y, z, respectively, and ρ is the fluid density at a point, then the additional shearing stresses in the x, y plane are $-\overline{\rho u'w'}$, $-\overline{\rho v'w'}$ in the x and y directions, respectively; corresponding shearing stresses act in the other two planes. (Overbars represent time averages.)

Non-zero values of τ depend on the existence of correlation between the component eddy velocities in the above products as, for example, a general association of gusts (u' positive) with descending air (w' negative). τ is generally assumed constant within about the lowest 25 metres of the atmosphere.

RHI. Abbreviation for range–height indicator. See RADAR METEOROLOGY.

ribbon lightning. A name applied to a markedly tortuous LIGHTNING discharge from cloud to ground.

Richardson number. An important parameter in atmospheric TURBULENCE, defined as the non-dimensional quantity

$$Ri = \frac{g}{T} \frac{\{(\partial T/\partial z) + \Gamma\}}{(\partial u/\partial z)^2} = \frac{g}{\theta} \frac{\partial \theta/\partial z}{(\partial u/\partial z)^2}$$

where $\partial T/\partial z$, $\partial \theta/\partial z$ and $\partial u/\partial z$ are the vertical height gradients of air temperature, potential temperature and wind speed, respectively.

An equivalent 'flux form' of the parameter is

$$Ri = -\frac{gH}{c_p T \tau (\partial u/\partial z)}$$

where H is the flux of sensible heat.

Richardson [57] investigated the rate of consumption of energy which is implied in vertical motion of fluid elements in relation to the rate of energy production associated with vertical wind shear. On certain assumptions he derived the criterion that turbulence will persist in a fluid if $Ri < 1$ and will subside if $Ri > 1$. Whilst subsequent measurements have not yet verified as critical the value $Ri = 1$, or any other single value, Ri is employed as a fundamental stability parameter in the study of atmospheric turbulence and dynamical meteorology.

ridge. A ridge (of high pressure), also termed a WEDGE, is an extension of an ANTICYCLONE or high-pressure area shown on a SYNOPTIC CHART, corresponding with a ridge running out from the side of a mountain (see Figure 40). It is the converse of a TROUGH of low pressure and is generally associated with fine, anticyclonic-type weather.

Maximum curvature of isobars occurs along the 'axis' of a ridge. One in which such maximum curvature is relatively small is termed a 'flat' ridge and tends to be a faster-moving isobaric feature than one in which the curvature is great.

rime. Deposit of white, rough ice crystals which forms when supercooled water droplets of fog come into contact with a solid object at a temperature below 0 °C. The deposit grows out on the windward side of the object. The phenomenon seldom occurs at low levels in the British Isles because supercooled fogs are uncommon at these levels. It occurs, however, much more frequently on mountain tops. It is often popularly confused with HOAR-FROST. (See also ICE FORMATION ON AIRCRAFT.)

Ringelmann shades. A scale of shades, varying in degree of blackness, which is used by an observer to form a subjective comparison with the blackness of SMOKE emitted by a chimney and so afford an estimate of the concentration of solid material which is being emitted.

A numerical measure of the average smoke content of air over a period of time is obtained by measuring photoelectrically the reflectance of a stain made on white filter-paper by the passage of a measured quantity of air and comparing with the measured reflectance of the various Ringelmann shades.

roaring forties. A nautical expression used to denote the prevailing westerly winds of temperate latitudes (40–50° S) in the oceans of the southern hemisphere.

rocket lightning. A very rare and unexplained form of LIGHTNING in which the speed of propagation of the lightning stroke is slow enough to be perceptible to the eye.

rocketsonde. The instrumented payload which is ejected from a METEOROLOGICAL ROCKET. It is equipped to measure one or more meteorological variables. Temperature

FIGURE 40.　Ridge of high pressure over the British Isles at 1200 GMT on 7 January 1988.

251

sensors are most frequently carried but pressure and density gauges and ozone sensors are also flown.

The sonde descends attached to a parachute which is tracked by radar for wind measurement. A typical temperature sensor is a very fine tungsten wire, a change of resistance of which causes a change of frequency of the note generated by an audio-oscillator. The carrier frequency of the sonde's transmitter is frequency-modulated with this note. The transmitted signal is received at a ground station equipped with the standard RADIOSONDE reception equipment. Thus, wind and temperature measurements from about 65 to 15 km are obtained.

rocket sounding. Exploration of the earth's atmosphere up to heights of several hundred kilometres by means of instruments carried by a rocket. Among the geophysical measurements obtained from such soundings are those of atmospheric pressure, air temperature, density and composition, winds, solar radiation, electrical properties, and the earth's magnetic field. The data are obtained mainly by radio telemetering, or by recovery of a photographic record after descent by parachute.

rockoon sounding. Exploration of the earth's atmosphere by a small instrumented rocket which is carried aloft by a large balloon to a predetermined height where it floats until fired by means of a radio relay operated from the ground at the moment when some particular event, e.g. a SOLAR FLARE or AURORA, is seen to occur.

roll cumulus. An obsolete term for a form of STRATOCUMULUS in which long, parallel rolls of cloud alternate with clear spaces.

Rossby diagram. An AEROLOGICAL DIAGRAM, devised by C-G. Rossby for air-mass identification, in which the abscissa is MIXING RATIO and the ordinate is partial POTENTIAL TEMPERATURE.

Rossby number. A dimensionless parameter (Ro) expressing the ratio of the acceleration terms to the Coriolis terms in the EQUATIONS OF MOTION. It may be written

$$Ro = U/Lf$$

where U is a characteristic speed, L a characteristic length, and f the CORIOLIS PARAMETER.

Rossby radius of deformation. If a localized sudden disturbance is applied to an ocean or atmosphere in which the undisturbed large-scale flow is geostrophic, the Rossby radius of deformation is the effective distance to which the disturbed flow will eventually extend before geostrophic adjustment is complete. It was originally defined by Rossby[59] as the radius of the CIRCLE OF INERTIA corresponding to the velocity of long waves in a homogeneous incompressible ocean of depth D on a rotating earth, i.e. $f^{-1}(gD)^{1/2}$. It has been generalized by Charney[21] and others and may be more usefully defined as $\lambda = H(\omega_g/\omega_i)$ where H is the scale of vertical motion, ω_i the frequency of horizontal inertia oscillations ($=f$), and ω_g the frequency of vertical oscillations. ω_g is given by $(g/H)^{1/2}$ for a homogeneous ocean, or by the BRUNT–VÄISÄLÄ FREQUENCY (N) for gravity waves in a stably stratified atmosphere. Gill[60] has given a detailed discussion.

Rossby wave. In synoptic meteorology, a smooth, wave-shaped contour pattern on an isobaric chart with a wavelength of about 2000 km, relating more especially to the middle or high troposphere. Some four or five such waves typically extend across a hemispherical chart.

Assuming a sinusoidal long-wave pattern of wavelength L in a BAROTROPIC atmosphere with no VISCOSITY, and in which absolute VORTICITY is conserved, Rossby obtained the formula

$$c = U - \frac{\beta L^2}{4\pi^2}$$

where c is the wave speed (positive, eastward; negative, westward), U the mean zonal wind speed and β the Rossby parameter, i.e. the northward rate of change of the CORIOLIS PARAMETER. The formula accounts for progressive, stationary and retrograde waves ($c > 0$, $c = 0$, $c < 0$, respectively) but not for waves which may progress faster than the zonal current ($c > U$).

rotating fluids — laboratory studies. Experimental studies of fluids contained in axisymmetric vessels of different shapes which are subjected to rotation about the axis of symmetry. The vessel is usually either a circular pan bounded by a vertical outside wall (usually called a 'dish-pan' after the American word for 'washing-up bowl') or an annulus bounded by inner and outer cylindrical surfaces. The fluid may be maintained at constant temperature, have heat applied to it at the boundaries, or have heat generated internally by application of an alternating electrical current. The consequential motions are studied by means of arrays of immersed thermometers and by making the motions visible by the addition of tracers which can be photographed. Certain analogues of atmospheric motions are observed including patterns of long waves and meandering 'jet streams', but the real purpose is to gain deeper understanding of the essential nature of rotating fluid flow, and experimental work is pursued in close association with theoretical research and computer simulations.

See VACILLATION.

rotor. A large, closed EDDY with a horizontal axis which is produced in the lee of a range of mountains or hills crossed by a stable airstream. Such closed eddies are sometimes formed under LEE WAVES of large amplitude, the surface wind under the wave crest being reversed. Horizontal dimensions are 3 to 10 km and vertical dimensions 1 to 3 km. A very turbulent 'rotor cloud' forms in the upper part of the closed eddy when the air is sufficiently moist, with an apparent motion round the axis of the eddy which is usually below the base of such a cloud.

Examples of rotors occur in connection with the HELM WIND ('helm bar' cloud) and in the lee of the Sierra Nevada in California.

roughness length. A quantity (z_0), also called the 'roughness coefficient' or 'roughness parameter', which enters as a constant of integration into the form of the LOGARITHMIC VELOCITY PROFILE appropriate to 'fully rough' flow near a surface. z_0 is proportional to the average height of the roughness elements of the surface; wind-tunnel measurements of flow over grains of sand suggest the value $z_0 = \epsilon / 30$, where ϵ is the average height of the obstacles.

The equation which defines z_0 is

$$\frac{u}{u_*} = \frac{1}{k} \log_e \frac{z}{z_0}$$

where u is the mean velocity at distance z from the boundary, u_* the FRICTION VELOCITY, and k VON KÁRMÁN'S CONSTANT. See also AERODYNAMIC ROUGHNESS. It is necessary to measure u at a height much greater than z_0 but much less than the depth of the boundary layer.

RTH. Abbreviation for REGIONAL TELECOMMUNICATION HUB.

running means. An alternative for MOVING AVERAGES.

runoff. That portion of the rainfall over a DRAINAGE AREA which is discharged from the area in the form of a stream or streams.

run-of-wind anemometer. See ANEMOMETER.

runway visual range. The maximum distance in the direction of take-off or landing at which the runway or the specified lights or markers delineating it can be seen from a position above a specified point on its centre line at a height corresponding to the average eye-level of pilots at touchdown. (A height of approximately 5 m is regarded as corresponding to the average eye-level of pilots at touchdown.)

In practice, runway visual range cannot be measured directly from the position specified in the definition but is an assessment of what a pilot would see from that position.

S

St. Elmo's fire. 'A more or less continuous, luminous electrical discharge of weak or moderate intensity in the atmosphere, emanating from elevated objects at the earth's surface (lightning conductors, wind vanes, masts of ships) or from aircraft in flight (wing tips, propellors, etc.)' [2, p. 127].

This discharge can also occur on an aircraft where the static charge has been induced by the impact of ice or snow crystals, rain, dust, or sand. The build up of static electricity on an aircraft in flight, whether by friction or by induction, is loosely termed 'precipitation static'.

The phenomenon is usually bluish or greenish in colour, sometimes white or violet. It is accompanied by a crackling sound and occurs when the electrical field in the neighbourhood of the object becomes very strong, as when a CUMULONIMBUS cloud is overhead. The phenomenon is also termed 'corposant' (holy body).

St. Luke's summer. A period of fine weather which is popularly supposed to occur about the time of St. Luke's day, 18 October.

St. Martin's summer. A period of fine weather which is popularly supposed to occur about the time of St. Martin's day, 11 November.

St. Swithin's day. A well known example of British weather lore to the effect that if rain falls on St. Swithin's day (15 July) then, in the same locality, each of the next 40 days will also have some rain. Rainfall records lend no support to this tradition.

Saffir–Simpson hurricane scale. A scale used by the United States National Weather Service for assessing the potential for wind and storm-surge damage from a hurricane in progress. The scale ranges from 1 (minimal) to 5 (catastrophic — winds of 136 kn or more) [61].

salinity. A measure of the salts dissolved in a given solution. The salinity of a natural water surface, such as sea water, is usually expressed in parts per thousand by weight. Thus a salinity of 35 per mille (written $35^0/_{00}$) indicates that there are 35 g of salts in 1000 g of sea water. Since the total dissolved solids are difficult to determine directly with accuracy, salinity is derived in practice by applying factors to the specific gravity or to the halide content, which can be exactly measured (dissolved solids are present in constant ratios).

The salinity value of $35^0/_{00}$ is a rough average for surface ocean water. Salinity varies systematically by a few per cent with latitude and is subject also to small casual and systematic time variations.

Depression of the freezing-point temperature and of the temperature of maximum density increase with increase of salinity: for a salinity of $35^0/_{00}$ the freezing-point temperature is $-1.9\,°C$ and the temperature of maximum density is $-3.5\,°C$ (compared with $0\,°C$ and $4\,°C$, respectively, for pure water).

sand pillar. A rarely used alternative for DUST WHIRL.

sandstorm (duststorm). 'An ensemble of particles of dust or sand energetically lifted to great heights by a strong and turbulent wind' [2, p. 120].

Surface visibility is reduced to low limits; the qualification for a synoptic report is visibility below 1000 m.

SAR. Abbreviation for SYNTHETIC APERTURE RADAR.

sastruga (also **zastruga**). A Russian term for the irregularity or wave formation caused by persistent winds on a snow surface. The size varies according to the force and duration of the wind, and the state of the snow surface in which it is formed.

satellite, meteorological. Satellites launched from the earth have, since October 1957, yielded much important geophysical information both directly (by means of automatic instruments) and indirectly (by inferences drawn from the precise path of the satellite).

A satellite of mass m, in a circular orbit at distance R from the earth's centre, has a critical velocity (v) such that the CENTRIFUGAL FORCE is exactly balanced by the force of GRAVITY acting on it. The balance equation is $mv^2/R = mg(a/R)^2$, whence $v^2 = ga^2/R$.

Meteorological satellites are normally launched into one of two sorts of orbit — *sun-synchronous* (or *polar-orbiting*), or *geostationary*. In the first, the satellite orbits in a plane inclined at a small angle to the earth's axis which precesses slowly in space so that the orientation of the orbit is fixed relative to the sun. Sun-synchronous satellites orbit at heights of about 1000 km and cross the equator about 15 times a day in each direction, always at the same local time (normally noon and midnight). Geostationary satellites orbit in the equatorial plane at such a height (36 000 km) that they remain stationary relative to the ground below. All meteorological satellites carry sensitive radiometers which transmit cloud pictures and measurements of terrestrial and solar radiation in various wave bands

Other direct meteorological measurements made by satellites are those of earth albedo and cloud distribution. Air density at great heights and the shape and size of the earth have been calculated from accurate tracking of the orbit. Other geophysical measurements have included those of the earth's magnetic field, cosmic rays, meteor impact, and properties of the ionosphere.

saturated soil. A saturated soil, or waterlogged soil, is one in which all the soil pores, including those which in a healthy soil contain air, are filled with water.

saturation. A moist air sample is said to be saturated, with respect to water or to ice, if its composition is such that it can coexist in neutral equilibrium with a plane surface of pure condensed phase, water or ice, at the same temperature and pressure as the sample.

To a high degree of approximation the capacity of air to hold water in the form of vapour depends only on temperature; at temperatures below 0 °C, however, an important factor in the equilibrium conditions is the phase (liquid or solid) of the water substance (see VAPOUR PRESSURE). A sample of air is said to be 'supersaturated' if it contains more than enough water vapour to saturate it at its existing temperature. Owing to the presence of condensation nuclei, an appreciable degree of supersaturation with respect to water is rarely observed in the atmosphere.

saturation (or saturated) adiabatic (or adiabat). Line on an AEROLOGICAL DIAGRAM representing the saturation adiabatic lapse rate. See ADIABATIC.

saturation deficit. The difference between the actual VAPOUR PRESSURE of a moist air sample at a given temperature and the saturation vapour pressure corresponding to that temperature.

savanna. The term applied to a type of tropical CLIMATE, with a wet and a dry season, in which the most common form of vegetation is the tall tropical grass 'savanna'.

256

scalar. A scalar quantity is one that is completely specified by its magnitude, expressed in a given system of units (as opposed to a directed or VECTOR quantity such as wind velocity). Meteorological examples of a scalar quantity are pressure, temperature, and divergence of wind velocity.

scale analysis. A method of estimating the relative magnitudes of different terms in the EQUATIONS OF MOTION, EQUATION OF CONTINUITY, and the THERMODYNAMIC EQUATION for different types of atmospheric motion so that valid and practically useful simplifying approximations to these terms may be substituted wherever possible. The complete set of equations describing the atmosphere are of great generality and have solutions as diverse as those for high-frequency sound waves and long meteorological waves in the stratosphere. By choosing the best approximations for the type of disturbance of interest (e.g. cumulus convection, or synoptic-scale depressions and anticyclones), one may be able both to eliminate unwanted solutions entirely (e.g. sound waves) and to simplify the computations.

The procedure is to prescribe various scales of variation:

a horizontal length scale	L,
a vertical depth scale	D,
a horizontal velocity scale	U,
a vertical velocity scale	W and,
a horizontal pressure fluctuation scale normalized by the density	$\Delta P / \rho$

(L and U define a time-scale (T). The reason for normalizing ΔP by ρ is to provide a scale value that varies only slowly with height.)

For example, the eastward (y) component of the EQUATIONS OF MOTION for mid-latitude synoptic systems may be analysed as follows:

	A	B	C	D	E	F
	$\dfrac{du}{dT}$	$-2\Omega v \sin\phi$	$+2\Omega w \cos\phi$ +	$\dfrac{uw}{a}$	$-\dfrac{uv\tan\phi}{a}$	$=\dfrac{1}{\rho}\dfrac{\partial p}{\partial x}$
Scale factor	$\dfrac{U^2}{L}$	fU	fW	$\dfrac{UW}{a}$	$\dfrac{U^2}{a}$	$\dfrac{\Delta P}{\rho L}$
Magnitude (m s^{-2})	10^{-4}	10^{-3}	10^{-6}	10^{-8}	10^{-5}	10^{-3}

where we have assumed $U = 10$ m s^{-1}, $W = 10^{-2}$ m s^{-1}, $L = 10^6$ m, $D = 10^4$ m, $\Delta P / \rho = 10^3$ m^2 s^{-2} and $\phi = 45°$ ($f \approx 10^{-4}$ s).

It is clear that the first approximation to this equation is found by considering terms B and F (which together yield the geostrophic wind) and that the only other term worth including is A.

Similar arguments may be used to show that on the synoptic scale the HYDROSTATIC APPROXIMATION is valid, and that the VORTICITY EQUATION

$$\frac{d}{dt}(\zeta+f) = -(\zeta+f)\,\mathrm{div}_H\mathbf{V}\left(\frac{\partial w}{\partial y}\frac{\partial u}{\partial z} - \frac{\partial w}{\partial x}\frac{\partial v}{\partial z}\right) + \left(\frac{\partial p}{\partial x}\frac{\partial \alpha}{\partial y} - \frac{\partial p}{\partial y}\frac{\partial \alpha}{\partial x}\right)$$

may be written more simply as

$$\left(\frac{\partial}{\partial t} + u\frac{\partial}{\partial x} + v\frac{\partial}{\partial y}\right)(\zeta+f) = -f\,\mathrm{div}_H\mathbf{V}.$$

Scale analysis is used extensively in modern dynamical meteorology and in developing new numerical forecast models.

scale height. At a height z in a horizontally stratified compressible atmosphere the pressure scale height (h_z) is given by

$$1/h_z = -p_z^{-1}\,(\mathrm{d}p_z/\mathrm{d}z) = g_z/R_zT_z$$

where $R_z = R^* M_z$. In the earth's lower atmosphere M_z (the mean molecular mass) and g_z can be regarded as constant so that h_z depends only on the variation of temperature with height.

Similarly the density scale height (H_z) is given by

$$1/H_z = -p_z^{-1}\,(\mathrm{d}p_z/\mathrm{d}z) = 1/h_z(1 + \mathrm{d}h_z/\mathrm{d}z).$$

For the case in which h_z is constant with height see EXPONENTIAL ATMOSPHERE. H and h are used in discussions of the BOUSSINESQ APPROXIMATION and buoyancy oscillations of short amplitude.

SCAPE. Abbreviation for SLANTWISE CONVECTIVE AVAILABLE POTENTIAL ENERGY.

scatter diagram. A graphical point plot of corresponding pairs of associated values of two variables (ordinate and abscissa). A diagram on which, for example, the points are closely grouped round a line inclined to both axes indicates a near-linear relationship between the two variables. A diagram on which the points are randomly distributed indicates no relationship between the variables.

scattering. The process by which some of the electromagnetic RADIATION incident on particles, of molecular size upwards, which are suspended in a medium, is dispersed in all directions. The scattering process is one which gives rise to a diminution of the intensity of an incident beam of radiation; the measure of this effect — the 'scattering coefficient' (β) — is defined by the equation (analogous to the case of ABSORPTION)

$$I = I_0 e^{-\beta x}$$

where I_0 is the intensity of incident radiation and I the intensity after a path of length x through the scattering (non-absorbing) medium. The 'scattering cross-section' of a scattering particle is the area normal to a beam of radiation which would intercept the same amount of radiation as that actually scattered by the particle.

In meteorology, radiation which has been subject to scattering is generally termed DIFFUSE RADIATION. Such radiation may have been scattered once or more than once — so-called single or multiple scattering, respectively; the terms primary, secondary, tertiary scattering are also used, as appropriate, while in some usages secondary scattering signifies multiple scattering.

Scattering is a complex phenomenon which depends mainly on the ratio of the size of the scattering particle to the wavelength of the incident radiation but depends also on the refractive index, shape and composition of the scattering particle. Atmospheric scattering is usually classified as either molecular scattering or that effected by haze particles or water droplets. See RAYLEIGH SCATTERING, MIE SCATTERING, POLARIZATION.

scatterometer. An active REMOTE SENSING device, carried in an artificial satellite, which emits a radar beam and measures the radiation scattered back to the satellite by capillary waves on the surface of the sea. The radar beam falls on a long narrow strip of sea and, from the measurements of Doppler shift and received power, estimates can be made of the surface vector wind. Scatterometer measurements are liable to ALIASING errors.

scintillation (of stars).   Rapid variations of apparent brightness ('twinkling') of stars, much more marked in stars near the HORIZON than in those near the ZENITH. Variations of colour may also occur at altitudes less than about 50°. The phenomenon is caused by small variations of REFRACTIVE INDEX of air associated with atmospheric inhomogeneities, mainly in the low atmosphere. A similar effect is visible at times in terrestrial objects, for example the shimmering of objects near the earth's surface on a hot day.

scirocco.   A warm, southerly wind in the Mediterranean region. Near the north coast of Africa the wind is hot and dry and often carries much dust. After crossing the Mediterranean, the scirocco reaches the European coast as a moist wind and is often associated with low stratus.

Scotch mist.   A combination of thick mist and drizzle, so called because it is most commonly experienced in the hillier districts of much of Scotland; at times it also affects low-lying districts of the west and north. In its most typical form it is associated with a moist stream of tropical maritime air.
In the uplands of the Devon—Cornwall peninsula the same phenomenon, which is very frequent there, is known as 'mizzle'.

scud.   A mainly nautical term for ragged fragments of low cloud, often moving rapidly in a strong wind below rain clouds. The meteorological term is stratus fractus.

SEA.   Abbreviation for SUDDEN ENHANCEMENT OF ATMOSPHERICS.

sea-breeze.   See LAND- AND SEA-BREEZES.

sea disturbance.   The degree of sea disturbance is reported in a 'state of sea' code in which the scale number increases from 0 to 9 according to the average wave height. The specifications are: 0, glassy; 1, rippled; 2, smooth; 3, slight; 4, moderate; 5, rough; 6, very rough; 7, high; 8, very high; 9, phenomenal. Scale number 5, for example, corresponds to waves of average height from 2.5 to 4 m, and scale number 9 to an average height of over 14 m.

sea fret.   A local name in parts of north-east England for a sea fog in coastal districts. This is especially a spring and summer feature. See ADVECTION FOG, HAAR.

sea ice.   See ICE.

sea level.   Owing to waves, swell, tides and varying atmospheric static pressure, the actual LEVEL of the sea is constantly changing. A 'mean sea level' at any place may be determined by averaging observed heights of sea level at hourly intervals by methods designed to eliminate, as efficiently as possible, those tidal oscillations with periods up to and including a lunar day as well as the random oscillations due to waves, swell and atmospheric pressure. Mean sea level is approximated by 'mean tide level' which is the average of observed high- and low-water heights; the difference for annual mean values is, for meteorological purposes, negligible. The length of period required to obtain a suitable mean value varies considerably from place to place because of local variation of the amplitude of fluctuation about the mean position.
The present datum mean sea level — often referred to simply as 'sea level' — used in the United Kingdom, with reference to which contour levels on Ordnance Survey maps

are shown, is based on observations at Newlyn in Cornwall, on the edge of the Atlantic Ocean. This datum is 0.13 ft (40 mm) below the Liverpool datum previously used. The permanent land-survey datum is not the mean-sea-level datum itself but is referred to permanent bench-marks in the neighbourhood of the tidal gauge. Land-survey reference planes of other countries do not all depend on mean sea level. That used in Ireland, for example, refers to a particular low-water datum in Dublin Bay and is estimated to be about 8 ft (2.4 m) lower than that of Newlyn.

World sea level, as given by the average of observations in many places, varies in response to:

(i) changes of average temperature of the oceans in depth, with accompanying expansion or contraction, and

(ii) melting or accretion of ice-caps and glaciers.

There is evidence that a world-wide ('eustatic') rise of almost 100 m occurred between 15 000 and 5000 years ago because of the melting of the ice sheets of the last glaciation, and that subsequent changes have been relatively slight, the present level being lower than that 5000 years ago by a few metres. On a shorter time-scale, it is considered that a eustatic fall between about 1680 and 1850 was followed (up to at least 1930) by a more rapid rise to about the 1680 level; in the early years of the present century the rise was at the rate of about 10 mm per decade.

Mean world changes of sea level may be locally masked because of local ('isostatic') changes in response to movements of the land and in the ocean floor and by modification of coastline. Thus, for example, sea level in the Thames Estuary is apparently some 3 m higher than in Roman times. This is a local effect produced mainly by down-warping of the land — possibly a form of compensatory movement as Scotland and Scandinavia continue to rise by isostatic recovery from the depression caused by the former load of ice.

sea smoke. An alternative for ARCTIC SEA SMOKE.

seasons. In meteorology, the manner of the division of the year into seasons for climatological purposes varies with latitude. In middle latitudes the normal division corresponds to that of the 'farmer's year'; in the northern hemisphere the divisions made are: autumn — September, October, November; winter — December, January, February; spring — March, April, May; summer — June, July, August.

In the tropics the terms 'winter' and 'summer' lose their higher-latitude significance and a division into seasons is usually made in terms of rainfall amount or, in places, the associated wind direction — thus, 'dry season' and 'rainy season' or 'north-east monsoon' and 'south-west monsoon' in India. In the continental subtropical regions the natural seasons are usually defined in terms of temperature (cold and hot), or rainfall (dry and rainy), or both.

In countries of temperate climate such as the British Isles, the seasonal temperature changes progress much more gradually than in continental regions of the same latitude. In polar regions, the transition from winter to summer and vice versa is so sudden that spring and autumn largely disappear.

sea temperature. The normal methods of measuring sea temperature are:

(i) to draw water in a specially designed bucket from the ship's side, forward of all ejection pipes, and to read the temperature of the sample with a specially designed thermometer ('bucket method'), or

(ii) to read the temperature of the engine-room intake water ('condenser-intake method').

260

The water temperatures so measured are, respectively, a mean value in the surface layer of depth about 1 ft (0.3 m), and a value at a depth of several feet.

A third method is to measure the temperature of the ship's hull, in a forward unheated compartment, with a thermometer which is thermally insulated from the air in the compartment. Each of these measurements approximates to a 'sea surface temperature' except when the sea is calm.

Radiation thermometry and infra-red photography, employed from low-flying aircraft or from satellites, are able to give a clear indication of the boundaries of water masses of different temperatures.

The mean annual sea-surface temperature exceeds 27 °C (81 °F) over a broad belt of the equatorial region, and is somewhat less than −1 °C (30 °F) in the polar regions. The run of the isotherms varies in the two hemispheres and in the different oceans. The seasonal range of temperature is about 6 °C (11 °F) in both polar and equatorial regions, and is greater in middle latitudes, where for the most part it lies between 6 and 16 °C (11 and 29 °F). The greatest range, some 30 °C (54 °F) or more), is found in small areas, extending to the coast of the western North Atlantic and western North Pacific Oceans. The diurnal variation of sea surface temperature is very small, 0.5 °C (1 °F) or less. See also N.S.R.T.

secondary cold front. The development of a TROUGH or troughs of low pressure within the cold AIR MASS lying in the rear of a deep depression is relatively common. On those occasions on which a trough appears to mark the line of advance of colder air (owing to rather different recent histories of the air masses on either side of the trough) the trough line is termed a 'secondary cold front'.

secondary depression. A secondary depression is one which, forming within the region of circulation of another depression, is, at least on initial formation, of higher central pressure than the other ('primary') depression. The formation of the closed circulation which defines the 'secondary' is preceded by a widening of the isobars in the region concerned.

Most secondaries of middle latitudes form at fronts — see COLD-FRONT WAVE, WARM-FRONT WAVE, COLD-OCCLUSION DEPRESSION, and WARM-OCCLUSION DEPRESSION — but non-frontal secondaries also form, more especially in an unstable AIR MASS, as, for example, the POLAR-AIR DEPRESSION. In general, a secondary deepens at the expense of the primary depression; on some occasions, mainly with the cold-front wave, the secondary deepens to such an extent as to absorb the original primary depression. (See Figure 41.)

secular trend. In statistics, a persistent tendency for a variate to increase or decrease with the passage of time, apart from irregular variations of shorter period. A secular trend is generally revealed more clearly if the data are smoothed.

seepage gauge. See PERCOLATION.

seiche. A tidal oscillation of the waters of inland lakes, very variable in period and amplitude. Among probable causes of the phenomenon are winds, earth tremors, and atmospheric oscillations of the type revealed by a MICROBAROGRAPH. Temperature seiches — abrupt changes of temperature below a lake surface, with associated wave motion in the layer of transition — have also been observed.

seismology. The science concerned with the measurement and analysis of earth tremors and, indirectly, with the nature of the earth's interior revealed by such analysis.

A direct link with both oceanography and meteorology is provided by the study of the quasi-regular small oscillations (MICROSEISMS) which appear on seismograms. See also EARTHQUAKE, MICROSEISMS.

FIGURE 41.   Secondary depression centred off the Irish coast at 1200 GMT on 3 February 1988.

seistan wind. A strong northerly wind which blows in summer in the province of Seistan, in eastern Iran. It continues for about four months and is, therefore, known as the 'wind of 120 days'. It sometimes reaches hurricane force.

semi-geostrophic equations. See GEOSTROPHIC MOMENTUM APPROXIMATION.

sensible heat. See HEAT, ENTHALPY.

serein. Fine rain falling from an apparently clear sky. In this rare phenomenon, the cloud droplets are presumably evaporated when the larger precipitation drops are formed.
   Much more commonly, in conditions of strong vertical wind shear between cloud and ground, precipitation arrives when the sky overhead is clear and the shower cloud is visible at a lower angle of elevation.

serial correlation. An alternative for AUTOCORRELATION.

sferics fix. The estimated location of a LIGHTNING flash. The older method, whereby this location was deduced by combining observations of its direction (azimuth) made at several stations, has been replaced in the Meteorological Office by the automatic time difference (ATD) method: the times of arrival of the electromagnetic wave radiation from the flash at a central control station and several automatic slave stations are measured very accurately, and from the difference between at least three of these arrival times the location of the flash is calculated. The range of the system is about 3000 km.

shade temperature. The temperature of the air indicated by a thermometer sheltered from precipitation, from the direct rays of the sun and from heat radiation from the ground and neighbouring objects, and around which air circulates freely. A standard shelter such as the THERMOMETER SCREEN is intended to satisfy these conditions.

shadow of the earth. A steely-blue segment darker than the rest of the sky rises from the eastern horizon just after sunset, encroaching on and soon obliterating the COUNTERGLOW. This is the shadow thrown by the solid earth on the atmosphere; all light received by an observer from that part of the atmosphere within the earth's shadow has been scattered more than once. The edge of the shadow weakens as TWILIGHT progresses and, except in a very clean atmosphere, is indistinguishable well before its passage through the zenith. A similar shadow, descending to the western horizon, occurs just before sunrise.

shamal. A hot, dry north-westerly wind which blows with special persistence in summer over Iraq and the Persian Gulf. It is often strong during the daytime but decreases at night.

Shaw week. A unit of time introduced by the Meteorological Office in January 1935 for agrometeorological purposes; it fell out of use in the 1950s.
   The unit was based on a division of the year into four quarters each centred on a SOLSTICE or EQUINOX. The 'Shaw year' began on 6 November; the 7-day periods therefore start on a different day of the week in successive years. Account was taken of the one or two days additional to 52 weeks in normal or leap years, respectively.

shear. See WIND SHEAR.

shear-gravity wave. A wave disturbance which forms at the boundary between two atmospheric layers of different densities that are moving with different speeds. Theory, based on an incompressible atmosphere, specifies a critical wavelength (of about 10 km for typical atmospheric values of density and wind discontinuities) below which such waves are unstable and above which they are stable. Wind shear is therefore inadequate by itself to account for the development of frontal wave depressions.

shear hodograph. See HODOGRAPH ANALYSIS.

shearing stresses. An alternative for REYNOLDS STRESSES.

shear instability. That type of dynamical instability which arises at the boundary between two atmospheric layers moving with different speeds or in a layer containing WIND SHEAR. Also called Kelvin–Helmholtz instability.

sheet clouds. See LAYER CLOUDS.

sheet lightning. The popular name applied to a 'cloud discharge' form of LIGHTNING in which the emitted light appears diffuse and there is an apparent absence of a main channel because of the obscuring effect of the cloud.

shelter-belt. A term which is sometimes used synonymously with WIND-BREAK, but which is more usually now employed in those cases where protection against wind is provided by a belt of trees. Where the protection is afforded by shrubs or hedge, such a term as 'shelter-hedge' is often employed.

shimmer. The apparent distortion of terrestrial objects due to atmospheric inhomogeneities at low levels. It is also referred to as 'atmospheric boil'.

ship routeing. Advice to shipping companies and shipmasters of the best route to be taken, between any two ports, by a particular ship. Criteria for selection include avoidance of navigational hazards (e.g. areas of fog or ice), fuel economy, and avoidance of damage to ship or cargo. Ship-routeing advice from the Meteorological Office is provided by master mariners in close touch with the latest meteorological information including forecasts, warnings, and satellite imagery from all over the world.

shock wave. A thin layer of a medium (in particular, the atmosphere) in which the temperature, pressure, density, and velocity suddenly jump to new values. Such an effect is produced, for example, by the sudden outward movement of air particles from the site of an explosion, or by the passage of an object through air at a supersonic speed. In the former case, the passage of the shock wave is marked by a jump to high values of air pressure and temperature. This is quickly followed (at places beyond a critical distance from the explosion) by a rather longer-lived period in which the pressure and temperature fall to values lower than those which prevailed before the arrival of the wave. These phases are termed the 'compression' and 'suction' phases, respectively.

In atmospheric flight at subsonic speeds, i.e. v (speed of flight) $< V$ (speed of sound), pressure disturbances are propagated outwards through the atmosphere in all directions from the moving object as a series of non-overlapping spherical waves — see Figure 42(a). In contrast, the spherical disturbance waves emitted at successive time intervals by an object moving at a supersonic speed ($v > V$) intersect and are contained within a solid cone behind the object, the air in advance of the object being unaffected by the motion — see Figure 42(b). The semi-angle (α) of the cone ('Mach angle') is given by

$$\sin \alpha = \frac{Vt}{vt} = \frac{V}{v} = \frac{1}{M}$$

(a)   (b)

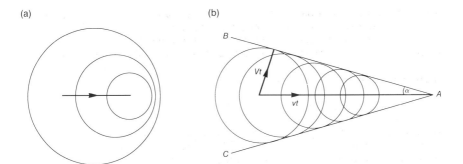

FIGURE 42.   Pressure disturbances showing as (a) a series of non-overlapping spherical waves at subsonic speeds, and (b) overlapping spherical waves emitted at successive time intervals by an object moving at supersonic speed.

where t is time and M the MACH NUMBER. Lines AB and AC in Figure 42(b) are the two-dimensional representation of so-called 'Mach lines' within which the cone is contained. The position of the lines remains unchanged with respect to the moving body. Density of intersection of successive waves and, therefore, pressure disturbance are greatest along these lines. The waves propagated by the body advance in a direction normal to the surface of the cone with the speed of sound and are heard as a sharp report, coinciding with the arrival of the 'shock wave'.

short-wave radiation.   In its common meteorological usage, solar radiation received near the earth's surface of maximum intensity at about 0.5 μm and confined within the approximate limits 0.29 and 4 μm. The term is used in contrast to the LONG-WAVE RADIATION emitted at terrestrial temperatures.

short-wave (radio) fade-out.   When a SOLAR FLARE occurs, short radio-waves, which are normally received after reflection by the F-LAYER, suffer extra absorption in the sunlit hemisphere D-LAYER, with an associated sudden fade-out of reception of the waves.

Showalter index.   An INSTABILITY INDEX (I) usually derived by assuming appropriate adiabatic ascent of an air parcel, originally at 850 mb, to the 500 mb level and subtracting the temperature (°C) so attained from the 500 mb environment temperature.
 Thunderstorms become increasingly likely the further I decreases (algebraically) below a threshold value of about +4.

shower.   In weather reports, solid or liquid PRECIPITATION from a CONVECTION cloud is designated a shower and is distinguished in such reports from the precipitation, intermittent or continuous, from layer clouds. Showers are often characterized by short duration and rapid fluctuations of intensity. HAIL invariably implies a shower, while DRIZZLE very seldom does.
 For synoptic purposes, rain showers are classified as 'slight', 'moderate', 'heavy' or 'violent' for rates of accumulation of about 0–2, 2–10, 10–50, or greater than 50 mm h^{-1}, respectively.

shred cloud.   A term sometimes applied to the cloud species FLOCCUS.

Siberian anticyclone.   A cold ANTICYCLONE which is a feature of the winter sea-level mean pressure distribution. In the January mean pressure chart for 1900–80, for example, it is represented by a centre over east central Eurasia, with central pressure of nearly 1036 mb.

265

sidereal period. The sidereal period of a planet or the moon is the time required for the body to make a complete circuit of its orbit, with respect to the stars. See also DAY.

significance. In statistics, the property which distinguishes those results which should receive further attention from those which may be ignored.

The smaller the CHANCE EXPECTATION of a statistical result, the more confidently is significance attached to it. It is convenient to speak of a result as significant at the 5 per cent or the 1 per cent level, for example, when its chance expectation is less than 1/20 or 1/100, respectively. Results with a chance expectation greater than 1/20 are not described as significant, though they may be suggestive. The use of these standard significance levels is a matter of practical convenience, since it allows the use of published tables and facilitates the comparison of the results of different authors; other significance levels may, however, be quoted.

A significant result is not necessarily important since it may mean no more than that the data contain a copying mistake; a result whether significant or suggestive must be considered in relation to the background before its importance can be estimated. See also PROBABILITY.

significant-weather chart. A form of forecast weather chart for a stated period showing fronts and centres of depressions and anticyclones together with details of weather, cloud, icing and turbulence according to agreed criteria. For fuller details reference should be made to the relevant ICAO document [62].

silver iodide. A substance used, in the form of a fine smoke which is released from the ground or from the air, as an ice-nucleating agent in CLOUD SEEDING experiments.

silver thaw. An expression of American origin. After a spell of severe frost, the sudden setting in of a warm, damp wind may lead to the formation of ice on objects which, being still at a low temperature, cause the moisture to freeze upon them and give rise to a 'silver thaw'. See also GLAZE.

similarity theory of turbulence. A dimensional approach in which characteristic scales of velocity, length, etc. are formed from the physical quantities which are reasoned to determine the properties of flow. Vertical gradients of mean quantities (wind, temperature, etc.) and statistical properties of TURBULENCE (variances, spectral properties) are then expressed in dimensionally appropriate combinations of these scales and universal functions of non-dimensional parameters.

simoom. A hot, dry, suffocating wind or whirlwind which occurs in the deserts of Africa and Arabia. Most frequent in summer, it usually carries much sand and is short-lived (less than about 20 minutes).

single-observer forecasting. Local weather forecasting which is based purely on observation of the weather elements for the same locality. The application of experience in synoptic meteorology, combined with physical reasoning, can on some occasions produce reasonably accurate forecasts for a few hours ahead. The observations chiefly used in such forecasting are those of pressure and pressure tendency, wind velocity at surface and higher levels, cloud, and temperature.

singularity. An annual RECURRENCE TENDENCY in a meteorological element during a group of successive calendar dates. An example is a period of unusual warmth or cold, departing markedly from the otherwise smooth mean annual variation of temperature, such as a BUCHAN SPELL.

266

The precise definition of singularities and their verification as reliable features of climate are often difficult and controversial. Their application in long-range weather forecasting appears to have some, though very limited, usefulness.

sink. See SOURCE.

sinusoidal pattern. That synoptic pattern (of THICKNESS), approximating to a sine curve, which is formed by alternate thermal TROUGHS and thermal RIDGES of about equal amplitude (see Figure 43). The theory of thermal DEVELOPMENT implies the tendency for CYCLOGENESIS (C) and ANTICYCLOGENESIS (A) in the regions indicated. When the 'wave' amplitude is small, THERMAL STEERING predominates and surface depressions move through the pattern without much deepening. With marked deepening and subsequent occluding of an associated depression, the pattern becomes distorted beyond recognition.

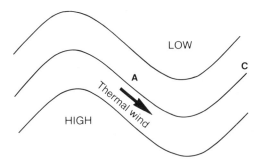

FIGURE 43. Sinusoidal pattern.

siphon barometer. A U-shaped mercury BAROMETER in which the areas of the upper and lower mercury surfaces are nearly equal. The instrument is adapted, as in the float barograph (see BAROGRAPH), to give a continuous record of pressure variation.

sirocco. Alternative spelling for SCIROCCO.

site. In order to secure observations comparable with those at other stations, the site of a meteorological STATION has to be carefully selected in accordance with certain rules which are set out in the *Observer's handbook* [1, pp. 177–180]. A RAIN-GAUGE requires a certain amount of protection from the wind, but for other outdoor instruments the more open the site, the better. A compromise is usually effected. The latitude, longitude and height of the ground on which the rain-gauge stands are used to define the position of a station. See also EXPOSURE.

SI units. Abbreviation for 'Système International d'Unités'. This system has been adopted by the General Conference of Weights and Measures and has been endorsed by the International Organization for Standardization.

SI is a coherent system based on six primary units, namely:

Quantity	Unit	Symbol
length	metre	m
mass	kilogram	kg
time	second	s
electric current	ampere	A
temperature	kelvin	K
luminous intensity	candela	cd

Derived units, related to the basic units by definition, have been given special names in some instances. For example, the unit of force in the SI system is called the NEWTON.

Six's thermometer. A U-shaped thermometer, designed by J. Six, in which the positions of two iron indexes, subsequently reset by means of a magnet, indicate the maximum and minimum temperatures attained since the previous setting.

skewness. A measure of asymmetry in a FREQUENCY DISTRIBUTION. Skewness is regarded as positive when the longer tail lies to the right of the MEAN and negative when it lies to the left. Thus, for example, most rainfall distributions are positively skew while distributions of minimum temperature tend to be negatively skew.

The coefficient of skewness (γ_1) of a distribution is defined as $\gamma_1 = \mu_3/\sigma^3$, where μ_3 is the third moment about the mean and σ the STANDARD DEVIATION.

skill score. A measure of the accuracy of a series of weather forecasts, usually in comparison with RANDOM FORECASTS or PERSISTENCE FORECASTS. For example, such an index (S) may be given by

$$S = \frac{\text{number of forecasts correct} - \text{number correct by chance}}{\text{total number of forecasts} - \text{number correct by chance}}.$$

skin-friction coefficient. An alternative for DRAG COEFFICIENT.

skip distance. The distance, measured along the earth's surface, between the point at which the ground signal from a radio-wave-transmitting source is just undetectable by a receiver of normal sensitivity and that at which reception of the transmitted waves is first possible after reflection by the IONOSPHERE. This distance corresponds to a minimum angle of incidence of the waves on the ionosphere; if this angle is not attained, the energy passes on to outer space.

The 'zone of silence' corresponds to the skip distance in the case of sound waves — see AUDIBILITY.

sky light. An alternative for DIFFUSE RADIATION.

sky radiation. An alternative for DIFFUSE RADIATION.

sky, state of. Fraction of the sky obscured by cloud on a scale of 0 (cloudless) to 8 (sky entirely covered by cloud). In the BEAUFORT NOTATION letters, b is used for a total cloud amount 0–2/8, bc for a total cloud amount 3/8–5/8 and c for a total cloud amount 6/8–8/8. For a uniform thick layer of cloud completely covering the sky o is used. The letter u, for an ugly, threatening sky, may be used with any of the preceding letters to indicate the general appearance of the sky.

slant visibility. An alternative for OBLIQUE VISIBILITY.

slantwise convection. See CONDITIONAL SYMMETRIC INSTABILITY.

slantwise convective available potential energy (SCAPE). See CONDITIONAL SYMMETRIC INSTABILITY.

sleet. The term has no agreed international meaning. In the United Kingdom it is used to describe precipitation of snow and rain (or drizzle) together, or of snow melting as it falls.

In American terminology, sleet is often used to signify ICE PELLETS.

slice method.   The estimation of vertical STABILITY in the atmosphere by a method which takes some account of the compensating downward motion induced in the environment by upward rising parcels of air. Since the environment is assumed to warm at the dry ADIABATIC lapse rate during its descent, greater stability is in general deduced by this method, as compared with the PARCEL METHOD.

sling psychrometer.   An alternative for WHIRLING PSYCHROMETER.

sling thermometer.   A thermometer mounted on a frame pivoted about a handle so that it can be whirled in the hand, thus providing 'ventilation'. If the bulb is shielded from direct solar radiation, satisfactory readings of air temperature can thus be obtained in a simple and inexpensive manner. A pair of thermometers, dry- and wet-bulb, similarly used, constitute a 'sling' or 'whirling' psychrometer. See also PSYCHROMETER.

small circle.   Any plane which does not pass through the centre of a sphere cuts the surface of the sphere in a 'small circle'.

small hail.   Precipitation in the form of showers of translucent ice particles, almost always spherical but sometimes having conical tips. Their diameter may attain and even exceed 5 mm.
   Small hail consists of SNOW PELLETS in a thin layer of ice which has formed from the freezing either of water droplets intercepted by the pellets or of water resulting from the partial melting of the pellets.

Smith index for barley mildew.   Defined by the empirical formula, using values for a particular day,

$$I = 3T + 0.5\,W + S$$

where T is the maximum temperature (°C), W the wind speed (kn) at 1200 GMT and S the number of hours of sunshine. A critical day occurs when $I > 64$.

Smith period.   Two types are defined:

(i)   For apple scab. A simplified form of MILLS PERIOD in which a minimum relative humidity of 90% following rain is substituted for leaf wetness.
(ii)   For potato blight. Two or more consecutive days when the minimum (dry-bulb) temperature is not less than 10 °C and there are at least 11 hours in each day with relative humidity of at least 90%. (A 'near miss' occurs when one or both of the two consecutive days has only 10 hours of high relative humidity.)

smog.   A term, being a contraction for 'smoke fog', which signifies a FOG in which smoke, or other form of atmospheric pollutant, besides playing an important part in causing the fog to form and to thicken (for example, by acting as condensation nuclei) has unpleasant or dangerous physiological effects. A noteworthy smog occurred in the London area in early December 1952.
   Photochemical smog is a form of air pollution characterized by haze, eye irritation and plant damage. It is produced in stagnant or slowly moving air by the reaction, in the presence of sunlight, of certain hydrocarbons and oxides of nitrogen arising mainly from motor-car exhaust; OZONE and other oxidizing agents are produced. It occurs regularly in Los Angeles.

smoke.   The visible product of incomplete combustion; in Great Britain the main source is coal burning. Coal smoke comprises mainly carbon and hydrocarbon particles of very small size (about 0.1 μm) which remain in the air, on average, for 1 to 2 days.

Atmospheric smoke concentration is measured by a 'smoke filter' method in which the weight of smoke deposited on a white filter-paper by a known volume of air is inferred from the measured reflectance of the smoke stain. Typical annual mean values for 1969 were:

   (i)   Country air — 25 to 30 μg m^{-3}.
   (ii) City air — a wide range of values varying from 250 μg m^{-3} for a few highly industrialized towns to, more typically, 75 to 100 μg m^{-3} for most cities.

A general decrease in concentration has occurred since then. See ATMOSPHERIC POLLUTION, RINGELMANN SHADES.

smoothing. A process of performing space or time averaging of data to suppress local or short-period variations. The use of efficient filters or weighted averages enables the larger-scale or more lasting components to be preserved with little distortion, while the use of inefficient filters (which include unweighted averages) entails a risk of emphasizing insignificant oscillations compared with significant features. See also FILTERING.

smudging. A method of frost protection (see FROST, PROTECTION AGAINST) consisting of the production, by combustion or by chemical means, of a smoke pall over a confined area, e.g. vineyard. The resulting reduction in outgoing terrestrial radiation results, in suitably calm conditions, in a decrease in the rate of fall of temperature of the ground in the locality.

Snell's law. See REFRACTION.

snow. Solid PRECIPITATION which occurs in a variety of minute ICE CRYSTALS at temperatures well below 0 °C but as larger SNOWFLAKES at temperatures near 0 °C. 'Granular snow' consists of opaque grains, rather flattened in shape and generally less than 1 mm in diameter.

For synoptic purposes, snow (or a snow shower) is classed as 'slight', 'moderate', or 'heavy' for a rate of accumulation of snow (in the absence of drifting or melting) less than 0.5 cm h^{-1}, 0.5–4 cm h^{-1}, and greater than 4 cm h^{-1}, respectively.

See also SNOWFALL.

snow, day of. In British climatology, any period of 24 hours ending at midnight GMT upon which snow is observed to fall is regarded statistically as a 'day of snow'.

See also SNOWFALL.

snowdrift. When a strong wind blows, there is a strong tendency for falling snow, or fallen snow on the ground, to accumulate not in open places but in any sheltered place, as in the lee of natural or artificial obstacles, there forming 'drifts'.

snowfall. The depth of fresh snowfall is normally measured with a graduated ruler. Its measurement as RAINFALL (i.e. its water content) may be made in a suitable SNOW-GAUGE, or by melting the snow caught in a normal rain-gauge, or by collecting and melting samples of fresh snow which has fallen in the open. Thirty centimetres of freshly fallen snow has about the same water content as 25 mm of rainfall.

The amounts of snow which fall over the British Isles are measured as rain; separate statistics of snowfall amount are therefore not available. Separate records are, however, kept of the numbers of days of snowfall and of SNOW LYING, and also of snow depth.

The average annual number of days with snow falling on low ground up to about 60 m above mean sea level increases with increasing latitude and from west to east, and ranges from less than 5 days in southern Cornwall to over 35 days in north-east Scotland. At places above 60 m the average number of days increases by approximately one day for every 15 m of elevation; above 300 m the increase is greater than this, the corrections being about +20, +52, +90, and +140 days at 300, 600, 900 and 1200 m, respectively.

The average annual number of days with SNOW LYING is one of the most variable of meteorological elements over the British Isles. Factors which influence it include monthly mean temperature, frequency of snowfall, quantity of snowfall and the character of the station and its surroundings, such as its height, aspect, and distance from sea. Values range from less than 5 days per year in southern and western coastal districts to over 100 days per year in the Grampians. In some winter months, such as February 1942 and February 1947, snow lying has been reported on every day of the month over large areas of Britain, whereas in February 1943 the majority of stations reported no single day of snow lying.

Falls of undrifted snow of depth exceeding 150 mm on level ground at low altitudes occur somewhere in England in one-third to one-half of winters; the average number of days in 10 years with snow lying to such depths ranges from less than 1 day in south-west coastal districts to between 30 and 40 days in eastern England.

snowflakes. Aggregates of ICE CRYSTALS which occur in an infinite variety of shape and form. At very low temperatures the flakes are small and their structure is simple. At temperatures which are close to freezing-point the individual flakes may be composed of a very large number of ice crystals (predominantly star-shaped) and the flakes may then have a diameter up to several inches (1 in = 25.4 mm).

snow-gauge. A device for the retention and measurement of SNOW. In the Hellmann–Fuess snow-gauge the snow is caught in a receiver supported on a balance, the displacement of which is continuously recorded so that an autographic record of snowfall (and of the fall of rain and hail also) is obtained. Most snow-gauges are, however, merely rain-gauges fitted with jackets or other devices to make them suitable for collecting solid precipitation, and for melting it before taking the reading.

snow grains. Precipitation of very small white and opaque grains of ice. These grains are fairly flat or elongated; their diameter is generally less than 1 mm.

snow-line. The lower limit in altitude of the region of perpetual snow. In high polar latitudes the snow-line is at sea level; in northern Scandinavia it is at about 1200 m, in the Alps at about 2400 m, in the Himalayas at about 4500 m. These figures are only a guide, as the height of the snow-line varies on the north and south sides of a mountain and from one mountain to another in the same latitude or region. It has no direct relation to the mean annual temperature, depending more on the summer temperature, but many other factors exert an influence, such as amount of snow in winter prevailing winds, exposure and steepness of the slopes, etc.

snow lying. This expression is used for occasions when one-half or more than one-half of the ground representative of the STATION is covered with snow. The ground representative of the station is defined as 'the flat land easily visible from the station and not differing from it in altitude by more than 30 metres'. British statistics of snow lying refer only to occasions when this state of affairs exists at the hour of morning observation.

See also SNOWFALL.

snow pellets. Precipitation of white and opaque ice particles, which are spherical or sometimes conical; their diameter is in the range 2–5 mm.

snow rollers. Cylinders of snow, formed and rolled along by the wind.

snow survey of Great Britain. A survey started in 1937/38 by the British Glaciological Society, with an annual report covering each snow season published in the *Journal of Glaciology*. The survey lapsed during the Second World War but was restarted with the report for 1946/47 and continued by the Society until 1952/53.

Starting with the report for 1953/54 the work was taken over by the Meteorological Office; reports were first published in the *Meteorological Magazine*, in *British Rainfall* from 1956/57, and as a separate publication from 1968/69.

sodium. A very minor constituent of the atmosphere, estimated to total about 1 tonne (1000 kg) in weight. The sodium D line (5893 Å) is observed in the NIGHTGLOW emission spectrum and is conspicuous in the TWILIGHT GLOW. The sodium is thought to have maximum concentration at about 85 km.

soil moisture. The moisture content of soils is generally expressed as the percentage ratio of the mass of water to that of dry soil, but may be expressed also in terms of inches of water per given depth of soil. Soil moisture is of obvious importance for the growing of plants and is of direct meteorological interest in affecting the thermal conductivity of soil and so the rate at which heat is conducted upwards to or downwards from the atmosphere. In certain circumstances, soil moisture affects the rate of EVAPORATION from the soil and of TRANSPIRATION from vegetation. Conversely, the rates of evaporation and transpiration are much affected by the meteorological variables temperature, sunshine, humidity and wind.

Among the measures of soil moisture which are employed are SATURATED SOIL, FIELD CAPACITY, and WILTING POINT. The varying force with which soil retains contained water is termed the CAPILLARY POTENTIAL.

soil moisture deficit. The amount of rainfall or irrigation required to restore soil to its FIELD CAPACITY.

soil temperature. See EARTH TEMPERATURE.

soil thermometer. A thermometer for measuring the temperature of the earth at various depths. The depths used in the United Kingdom are 5, 10, 20, 30 and 100 cm. Two types of thermometer are used depending upon the depth at which the temperature is measured. For depths up to 30 cm the thermometers are unmounted and unsheathed with a bend in the stem between the bulb and the lowest graduation. The bend allows the bulb to be at 5, 10 or 20 cm when the vertical part of the stem is sunk into the ground with the horizontal part of the stem in contact with the surface. Measurements at these depths are made under a bare soil surface.

For depths of 30 cm or more the thermometers are enclosed in glass tubes and their bulbs embedded in wax to make them insensitive to sudden changes. This allows them to be drawn to the surface and read before their temperature changes appreciably. At these depths temperatures are measured under a grass surface with the thermometers suspended in steel tubes sunk through the surface of the grass plot.

solar activity. See SUN.

solar constant. The solar radiation flux at a surface normal to the sun's beam outside the earth's atmosphere at the earth's mean distance from the sun.

Measurement of the solar constant has been made at high-level observatories since 1902, mainly by the Smithsonian Institution. Direct solar radiation intensity is continuously measured and is related to an absolute standard; atmospheric attenuation is allowed for by measurement, at various solar zenith angles, of relative flux over a wide band of selected wavelengths and extrapolation to 'zero path length'. The generally accepted value of the solar constant is 139.6 mW cm^{-2}.

There has been much controversy as to whether measured day-to-day variations and longer time-average variations of the solar constant are real or spurious. As measuring technique has developed, the values have more nearly approached a constant value (for a given absolute standard). New measurements by satellite-borne radiometers have shown some real variation, but this seems to be at most 0.1 per cent. This is in marked contrast to the large percentage solar-cycle variations known from geophysical evidence to occur in the far ends of the solar spectrum where, however, the amount of energy involved is very small.

solar corpuscular streams. Streams of charged corpuscles (particles) which are emitted at high speed from disturbed regions of the sun are affected by the earth's magnetic field in the event of their crossing the earth's path and, in turn, affect this field, causing geomagnetic and ionospheric storms. It is generally held that these particles do not penetrate the earth's atmosphere below about 70 km (above which height their interaction with the atmosphere gives rise to visible AURORA) and that no relation between the incidence of particle streams and lower atmosphere meteorological phenomena has yet been shown. See also GEOMAGNETISM.

solar cycle. The relative sunspot number varies in a quasi-periodic manner, with successive maxima separated by an average interval of about 11 years — the so-called 'solar cycle'. If reversal of sunspot magnetic-field polarity in a given solar hemisphere in successive 11-year periods is taken into account, the complete solar cycle may be considered to average some 22 years. See SUNSPOT.

solar day. See DAY.

solar flare. A solar explosion, unpredictable in nature and up to a few hours in duration, from a restricted region of the CHROMOSPHERE above certain types of SUNSPOT. Flares are classified on an ascending scale from 1 to 3+, on the visual basis of intensity of emitted light and solar area covered. In the sunlit hemisphere, short-lived 'sudden ionospheric disturbances' start almost simultaneously with the visual appearance of a great flare, and are attributed to the arrival in the high atmosphere of a flood of ionizing radiation released by the flare; the extra ionization is thought to occur mainly in the D-LAYER.

An intense and world-wide magnetic storm follows a great flare, which is near the centre of the sun's disc, with a frequency which is much in excess of CHANCE EXPECTATION. It is thus inferred that a SOLAR CORPUSCULAR STREAM is ejected almost radially from the flare region, the delay of about 20 hours corresponding to the slower speed of travel of the solar particles compared with that of the wave radiation.

solarimeter. See PYRANOMETER.

solar radiation. See RADIATION.

solar-radiation thermometer. An alternative for BLACK-BULB THERMOMETER.

solar spectrum. See RADIATION.

solar–terrestrial relationships. The relationships between the (variable) physical state of the SUN ('solar activity') and the (variable) particle and wave radiations emitted by the sun on the one hand, and the resulting physical effects produced in the earth's atmosphere on the other. Meteorological events observed in the troposphere and stratosphere appear to depend very little, if at all, on observed solar variability. The latter is, however, very important in certain types of COSMIC RADIATION, and in studies of the AURORA, GEOMAGNETISM and the IONOSPHERE.

solar wind. Term for the motion of interplanetary gas outwards from the sun towards the earth near which it interacts with the earth's magnetic field. It is generally assumed that the strength of this 'wind' increases with increasing solar activity (see SUN).

solenoids. The intersection in a BAROCLINIC atmosphere of surfaces of constant pressure with surfaces of constant specific volume (isobaric and isosteric surfaces, respectively) forms three-dimensional 'isobaric–isosteric solenoids'; since isosteric surfaces are also isopycnic (constant density) the intersections may also be said to form 'isobaric–isopycnic solenoids'. 'Unit solenoids' are formed by the intersection of surfaces separated by one unit of pressure and specific volume, or of pressure and density.

The existence of such solenoids in the atmosphere tends to produce a so-called 'direct circulation'; in the absence of the CORIOLIS FORCE the rate of production of circulation is proportional to the concentration of unit solenoids and is so directed as to cause the lighter air to rise and the denser air to subside.

soliton. A type of solitary wave of finite amplitude which, although it exists as a solution of a non-linear equation or system, propagates without change of form. This is because of a balance between the non-linearity (which tends to steepen the wave front) and DISPERSION (which tends to spread the wave front). One soliton may interact strongly with another soliton, but after the interaction both retain their forms almost as if the principle of superposition were valid.

In meteorology, solitons have been proposed as analytic solutions of the equations of motion for Rossby-type waves in a zonal flow and of small-scale internal gravity waves. Redekopp [63] has shown that long ROSSBY WAVES in zonal shear flow are described by the classical equation of soliton propagation, the Korteveg–deVries (KdV) equation of form

$$\frac{\partial y}{\partial t} + c_0 (1+\epsilon y) \frac{\partial y}{\partial x} + c_0 \beta \frac{\partial^3 y}{\partial x^3} = 0.$$

A general review is given by Miles [64].

solstice. The time of maximum or minimum DECLINATION of the sun when, for a few days, the altitude of the sun at noon shows no appreciable change from day to day. The summer solstice for the northern hemisphere (winter solstice for southern hemisphere) occurs on about 22 June, when the sun is farthest north of the equator; the winter solstice for the northern hemisphere (summer solstice for southern hemisphere) occurs on about 22 December, when the sun is farthest south of the equator.

sounding. A direct or indirect measurement of the vertical distribution of some physical or chemical property of the atmosphere. See also BALLOON SOUNDING, METEOROLOGICAL ROCKET.

sound waves. Sound passes through a medium by means of longitudinal waves whose velocity of propagation depends on the temperature and nature of the medium.

The passage of sound waves at a point in the atmosphere is associated with air pressure fluctuations about a mean value. Where these fluctuations are small relative to the mean pressure, the velocity of the waves is given by

$$V = (\gamma RT)^{\frac{1}{2}}$$

where T is the absolute temperature.

V in air at 0 °C is about 332 m s^{-1} or 760 mile h^{-1}. See also WAVE MOTION, MACH NUMBER.

source. In hydrodynamics, a 'source' is a point, line, area or volume within a fluid at which fluid is continuously created and from which it moves equally in all directions; the source 'strength' is measured by the rate of production of fluid. The converse is a hydrodynamic 'sink' to which fluid converges and at which it disappears.

In an analogous way a 'heat source' in a thermodynamic system is that part of the system in which heat is continuously generated and from which it is transferred to be continuously dissipated at a 'heat sink'. The large-scale atmospheric heat source is at low levels in low latitudes, the heat sinks are at low levels in high latitudes and at high levels in all latitudes.

southerly buster. A name given in south and south-east Australia to a sudden change of wind, usually from a north-westerly direction to a southerly direction, which is accompanied by a sudden fall in temperature. This change of direction occurs behind a cold front, and if the rise of pressure is considerable the southerly wind is violent. The arrival of the southerly wind is usually marked by a long crescent-shaped roll of cloud. The temperature sometimes falls as much as 20 °C in half an hour. These storms are sometimes accompanied by thunder and lightning. They are similar to the PAMPERO of South America and the LINE SQUALL of middle latitudes. They are most prevalent from October to March.

southern oscillation. A global-scale fluctuation of atmospheric mass and motion fields with opposing centres of action over Indonesia and the south-eastern tropical Pacific. One extreme case is associated with the EL NIÑO SOUTHERN OSCILLATION (ENSO). The southern oscillation was documented by G.T. Walker who devised an index based on the observed seasonal distribution of pressure, and to a lesser extent temperature and rainfall, over a large and predominantly oceanic region of lower latitudes.

A 'North Atlantic oscillation' and a 'North Pacific oscillation', derived in a similar manner for these regions, were also employed by Walker. A variety of additional regional oscillations such as 'Pacific/North American' (PNA) (often associated with the ENSO) have been diagnosed in recent studies.

space charge. In ATMOSPHERIC ELECTRICITY an excess, within any specified portion of the atmosphere, of positive over negative IONS, or vice versa — positive or negative space charge, respectively.

The downward movement of positive electric charge, and upward movement of negative charge, in response to the existing POTENTIAL GRADIENT, implies a positive space charge in fair-weather regions where the field is directed downwards. This space charge is greatest at low levels where the field is greatest. Large space charges of either sign are, however, measured in association with precipitation elements.

Spanish plume. A plume of potentially warm air that flows from Spain towards France forming a capping inversion. This inversion traps moisture below it by restraining small-scale convection which would otherwise spread the moisture up through the TROPOSPHERE. The moisture is continuously augmented by evapotranspiration over France leading to very high values of wet-bulb potential temperature in the lower layers. The interaction of this very moist low-level flow with a trough approaching from the west, combined with the eventual erosion of the capping inversion, often leads to the development of violent thunderstorms over northern France and southern England [43]. See also MEXICAN PLUME.

The term 'Spanish plume' is used in the Meteorological Office to describe the spreading to the British Isles of widespread thunderstorms that have developed over Iberia and western France. The phenomenon is associated with the slow eastward movement of an upper trough towards Biscay and Iberia which interacts with a mass of very warm air in the lower troposphere over Iberia.

specific heat. The specific heat of a substance is the heat required to raise the temperature of unit mass of it by one degree. The dimensions are $L^2T^{-2}\theta^{-1}$.

The specific heat of a substance is to some extent dependent on the temperature (see CALORIE). The specific heat of water is 4187 J kg^{-1} K^{-1}; the specific heat of ice is 1450 J kg^{-1} K^{-1} at -90 °C, 1884 J kg^{-1} K^{-1} at -30 °C and 2106 J kg^{-1} K^{-1} at 0 °C.

The specific heat of dry air at constant pressure (c_p) is 1005 J kg^{-1} K^{-1} and at constant volume (c_v) 716 J kg^{-1} K^{-1}. The ratio (γ) of the specific heats (i.e. c_p/c_v) is 1.40. The specific heats of water vapour at constant pressure (c_{pv}) and at constant volume (c_{vv}) are, respectively, 1846 J kg^{-1} K^{-1} and 1386 J kg^{-1} K^{-1}. Admixture of water vapour increases the specific heat of air, to a degree negligible for most purposes, as follows:

$$c_{pm} \text{ (moist air)} \approx c_p (1 + 0.9r)$$
$$c_{vm} \text{ (moist air)} \approx c_v (1 + 1.0r)$$

where r (HUMIDITY MIXING RATIO) is expressed in kilograms per kilogram.

specific humidity. The specific humidity (q) of moist air is the ratio of the mass (m_v) of water vapour to the mass ($m_v + m_a$) of moist air in which m_v is contained, m_a being the mass of dry air, i.e.

$$q = \frac{m_v}{m_v + m_a} .$$

Since m_v is much smaller than m_a, specific humidity for a given sample is almost identical with HUMIDITY MIXING RATIO.

specific volume. The volume occupied by unit mass of a substance, at a specified temperature and pressure. It is the inverse of DENSITY.

spectrobolometer. See BOLOMETER.

spectrophotometer. An instrument which measures the intensity of radiation of a given wavelength. In meteorology, such an instrument is used mainly in the measurement of OZONE — see DOBSON SPECTROPHOTOMETER.

spin-down time. The time taken by an atmospheric or oceanic vortex to lose most of its vorticity under the influence of EKMAN PUMPING in the BOUNDARY LAYER. In mathematical formulations it is defined as the time in which the vorticity is reduced to e^{-1} of its initial value where e is the base of natural (Napierian) logarithms.

spin-up time. The time taken for a fluid circulation to be set up by the application of a constant stress at a boundary. Examples are oceanic circulation generated by the wind and the motion of a liquid in a cylinder which is suddenly set into rotation.

spissatus (spi).   A CLOUD SPECIES. (Latin for thickened.)
'CIRRUS of sufficient optical thickness to appear greyish when viewed towards the sun' [2, p. 18]. See also CLOUD CLASSIFICATION.

splintering.   The splintering or fragmentation of ICE CRYSTALS, more especially of the delicate branched form of crystal or the ejection of small ice particles during the freezing of supercooled water droplets is considered to be an important source of the multiplication of ice crystals within a cloud. The electric charge separation associated with splintering has been proposed as a factor in the rapid rate of charge separation which occurs within a THUNDERSTORM cloud.

split cold front.   See UPPER COLD FRONT.

spontaneous nucleation.   An alternative for HOMOGENEOUS NUCLEATION.

spring.   See SEASONS.

squall.   A strong wind that rises suddenly, generally lasts for some minutes, and dies comparatively suddenly away. It is distinguished from a GUST by its longer duration.
The term is often used in such a sense as to include the precipitation, thunderstorm, etc., which are a common accompaniment of the sudden increase of wind. The following definition of 'squall' was adopted in April 1962 by the Third Session of the Commission for Synoptic Meteorology of the WMO:
'A sudden increase of wind speed by at least 8 m s^{-1} (16 kn), the speed rising to 11 m s^{-1} (22 kn) or more and lasting for at least one minute. Note: When Beaufort scale is used for estimating wind speed, the following criteria should be used for the reporting of squalls: a sudden increase of wind speed by at least 3 stages of the Beaufort scale, the speed rising to Force 6 or more and lasting for at least one minute.'

squall line.   The name originally given to what is now known as a COLD FRONT.
The use of the term is now generally confined to violent convective phenomena extending along a line or belt which is non-frontal in nature — see INSTABILITY LINE.

stability.   A system which is subjected to a small disturbing impulse is said to be in stable, neutral, or unstable equilibrium, according to whether it returns to its original position, remains in its disturbed position, or moves farther from its original position, respectively, when the disturbing impulse is removed.
Investigation of the STATIC STABILITY of the atmosphere is made most simply by the PARCEL METHOD, in which an assessment is made of changes of kinetic energy of a test parcel of air, displaced vertically and adiabatically with respect to its environment as represented by an ascent curve on an AEROLOGICAL DIAGRAM. The environment is termed stable, neutral, or unstable (at defined points or in defined layers) according as the kinetic energy of the parcel decreases, remains constant, or increases, respectively.
The following rules apply to the most general case of a moist but unsaturated test parcel which is subject to a smaller or larger vertical ascent through a moist but unsaturated environment (see Figure 44). If γ is the existing LAPSE of temperature:

   (i)   'Absolute stability' exists if $\gamma < \Gamma_s$.
  (ii)   'Absolute instability' exists if $\gamma > \Gamma_d$.
 (iii)   'Conditional instability' exists if $\Gamma_d > \gamma > \Gamma_s$.

Case (iii) is subdivided into two classes defined by the vertical distribution of humidity in the environment curve, as follows:

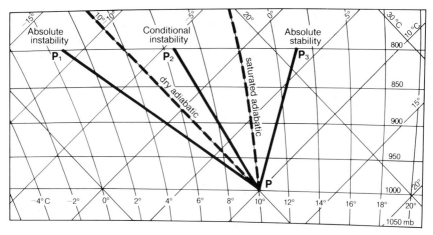

FIGURE 44. Stability and instability in relation to the lapse rate of temperature. Absolute stability occurs when the observed lapse rate is less than the saturated adiabatic as P–P₃, absolute instability occurs when the observed lapse rate exceeds the dry adiabatic as P–P₁ and conditional instability occurs when the observed lapse rate lies between the dry and saturated adiabatics as P–P₂.

(a) The case is one of 'stability' if none of the pseudo wet-bulb potential temperature (θ_{sw}) lines (with sufficient accuracy, the WET-BULB POTENTIAL TEMPERATURE (θ_w) lines) corresponding to possible test parcels intersects the environment curve — see PSEUDO WET-BULB TEMPERATURE.

(b) The case is one of 'latent instability' if one or more θ_w lines intersects the environment curve. The latent instability is termed 'real latent' if the 'negative area' (that lying between the ascent curve of the test parcel and the environment curve, and to the left of the environment curve) is less than the 'positive area' (to the right of the environment curve); the latent instability is termed 'pseudo-latent' if the converse is true. (The 'positive area', being proportional to energy, is sometimes called the 'convective available potential energy' (CAPE).)

The term 'convective instability', or 'potential instability', is applied to the case in which a layer of air will become unstable on being lifted bodily (as over high ground) until it is saturated; the criterion for this case is that θ_{sw} (with sufficient accuracy, θ_w) decreases with increasing height through the layer.

Estimation of the static stability by means of the parcel method and, in particular, calculations by this method of available kinetic energy in conditions of instability with associated convection, cloud formation and precipitation, are liable to serious error by the neglect of such factors as the mixing of rising parcels with the environment, the compensating downward motion induced in the environment, and the additional energy released by the cooling of the environment by the evaporation of precipitation into it. The method is, nevertheless, capable of producing useful results. See also DYNAMIC STABILITY.

stability categories. A classification of the state of the lower atmosphere, related to its ability to diffuse wind-borne material, which is derived from routine measurements of ordinary meteorological variables. Stability categories were originally introduced by Pasquill in 1958 and have been subsequently modified in various ways. They are specified alphabetically from A (most unstable) to G (very stable). An account is given by Pasquill and Smith [65].

278

standard. A prescribed measure or scale of any kind, such as a unit or scale of reference. The legal magnitude of a unit of measure or weight.

A 'primary standard' instrument (ABSOLUTE INSTRUMENT) is often used in the CALIBRATION of some 'substandard' or 'secondary standard' instruments which, in turn, may be used to calibrate many instruments for field use.

standard atmosphere. Hypothetical ATMOSPHERE, corresponding approximately to the average state of the real atmosphere, in which the pressure and temperature are defined at all heights. Such an atmosphere is adopted internationally as the basis for the calibration of altimeters, evaluation of aircraft performance, etc.

The principal standard atmospheres now in use are:

(i)   The ICAO STANDARD ATMOSPHERE to 20 km and its extension to 32 km.
(ii) The *U.S. standard atmosphere, 1976* [4], extends the ICAO standard atmosphere upwards from 32 km; values are given up to 1000 km. The *U.S. Standard atmosphere supplements, 1966* [66] gives seasonal and latitudinal supplementary atmospheres for the STRATOSPHERE and MESOSPHERE, linking at 120 km with THERMOSPHERE models which vary with solar activity. These values are widely used in the western hemisphere. Supplementary data on other atmospheric parameters are included.
(iii) The COSPAR CIRA 1972 international reference atmosphere 30–700 km gives reference values plus winds seasonally and latitudinally in the stratosphere and mesosphere and variations with the solar intensity in the upper thermosphere.
(iv) The Russian 1964 GOST atmosphere.

standard atmospheric pressure. A concept used in some physical definitions. It is the pressure exerted by a column of mercury 760 mm high, of density $13\,595.1$ kg m^{-3}, subject to gravitational acceleration of 9.80665 m s^{-2}), and equals $101\,325$ N m^{-2} or 1013.25 mb.

See also STANDARD DENSITY, STANDARD GRAVITY.

standard density. A conventional value of the density of mercury, adopted for the sake of uniformity in the conversion of pressure readings from units of pressure to units of height, or vice versa. The value adopted by the WMO is the density at 0 °C, equal to $13\,595.1$ kg m^{-3}. See also STANDARD GRAVITY, BAR.

standard deviation. The standard deviation (s) of a group of measurements is the root-mean-square departure of the observations from their MEAN. It is the square root of the VARIANCE and is the most generally useful measure of their dispersion. It is defined by

$$ s = \left\{ \frac{1}{n} \sum_{j=1}^{n} (x_j - \bar{x})^2 \right\}^{\frac{1}{2}} $$

where $x_1, x_2, ..., x_n$ are measurements of the same kind.

When n is small, the standard deviation of the sample is an underestimate of that of the POPULATION from which it is derived; a better estimate of the population standard deviation (σ), sometimes termed the STANDARD ERROR, is given by

$$ \sigma = \left\{ \frac{1}{n-1} \sum_{j=1}^{n} (x_j - \bar{x})^2 \right\}^{\frac{1}{2}}. $$

standard error. The MEAN values of samples, each of size n, drawn at random from a given POPULATION of STANDARD DEVIATION σ have a normal or near-NORMAL (FREQUENCY) DISTRIBUTION about the population mean, and a standard deviation of $\sigma/\sqrt{n}$. This quantity may be regarded as a measure of the dependability of an individual sample mean and is termed the 'standard error of the mean' (σ_m). For small n, a better estimate of σ_m is $\sigma/\sqrt{(n-1)}$.

The standard error of the sum of n independent values drawn from different populations of standard deviation σ_1, σ_2, etc. (σ_s) is $\sigma_s = \sqrt{(\sigma_1^2 + \sigma_2^2 + ..., + \sigma_n^2)}$; when drawn from the same population of standard deviation 1, this reduces to $\sigma\sqrt{n}$.

The standard error of the difference of two independent values, or means, of standard deviation σ_1 and σ_2, (σ_d) is $\sigma_d = \sqrt{(\sigma_1^2 + \sigma_2^2)}$; when the means are drawn from the same population of standard deviation σ, this reduces to $\sigma\sqrt{2}$.

The provision concerning randomness of selection, implying independence of selected values, is very important in geophysical statistics owing to time persistence and geographical coherence of many physical quantities. See PERSISTENCE.

standard gravity. A conventional value of the gravitational acceleration, adopted for the sake of uniformity. The value adopted by the WMO is 980.665 cm s^{-2} (9.80665 m s^{-2}). It is to this value, not to the value of 980.616 cm s^{-2} (the best determined value of gravity at sea level in 45° latitude), that pressure data in height units (millimetres or inches) refer. (The value 980.665 cm s^{-2} is a previous best determined value.)

See also GRAVITY, BAR, STANDARD DENSITY.

standard temperature (of barometer). That temperature at which, under STANDARD GRAVITY, the indicated reading of a mercury barometer is correct. At any temperature of an ATTACHED THERMOMETER other than the standard temperature, a 'temperature correction' must be applied to the barometer reading to take account of differences between the density of the mercury and dimensions of the metal scale of the barometer and those values assumed in the CALIBRATION of the instrument. It is also known as the 'fiducial' temperature.

The WMO resolved that, with effect from 1 January 1955, the standard temperature for all mercury barometers should be 0 °C.

standard time. Time referred to the mean time of a specified MERIDIAN. The meridian of Greenwich is the standard for western Europe. The standard meridian for other countries is generally so chosen as to differ from Greenwich by an exact number of hours or half-hours. See also ZONE TIME.

standard vector deviation. The standard vector deviation of a group of vectors is a measure of the scatter of the vector end points about the point which represents the end of the vector mean, all the vectors emanating from a common origin. The standard vector deviation (σ) is defined as the root-mean-square vector deviation; it is most readily calculated from the equation

$$\sigma^2 = \Sigma|\mathbf{V_i}|^2/n - |\mathbf{V_m}|^2$$

where $|\mathbf{V_i}|$ is the modulus (magnitude) of an individual vector and $|\mathbf{V_m}|$ the modulus of the vector mean.

For the wind vector, σ is found to increase with height in the troposphere in accordance with the approximate relation $\rho\sigma = $ constant. The average value of σ at a pressure level of 500 mb over the British Isles is about 30 knots in summer and 40–45 knots in winter.

See also CONSTANCY OF WINDS.

standing wave. In meteorology, an air wave which is (almost) stationary with respect to the earth's surface. Such waves are commonly associated with airflow over mountains. See MOUNTAIN WAVE, LEE WAVES.

Stanton number. The reciprocal of the PRANDTL NUMBER.

starshine recorder. An alternative for NIGHT-SKY RECORDER.

state, equation of. An alternative for GAS EQUATION.

state of ground. See GROUND, STATE OF.

state of sea. See SEA DISTURBANCE.

state of sky. See SKY, STATE OF.

static. See ATMOSPHERICS.

static stability. See STABILITY.

static stability, equation of. An alternative for HYDROSTATIC EQUATION.

station. In meteorology, a location at which regular weather observations are made. Among the classes of station are the SYNOPTIC STATION, CLIMATOLOGICAL STATION, AGRO-MET STATION, HEALTH RESORT STATION, RAINFALL STATION, OCEAN WEATHER STATION and AUTOMATIC WEATHER STATION.

station index number. A group of three figures used in synoptic messages to signify the particular station, within a given block area the boundaries of which coincide in most cases with national frontiers, at which the observation was made.

statistical inference. The process by which one proceeds from factors derived from samples of measurements which are available for analysis to reach conclusions about the statistical POPULATIONS of which these measurements form part, or about other samples of the same kind. Conclusions drawn by statistical inference are never absolute, but are always regarded as valid with a greater or lesser degree of PROBABILITY, depending on the nature and statistical SIGNIFICANCE of the evidence.

statistical theory of diffusion. Treatment of DIFFUSION involving the statistical properties of TURBULENCE (VARIANCE, correlation (see CORRELATION COEFFICIENT) and EDDY SPECTRUM) based on the velocity changes experienced by a moving particle, in contrast to the velocity changes evident at a fixed point in the flow.

statistics. This word is used in the following three different, but related, ways:

(i) Numerical measurements made in a precisely defined way on all members of a given group or on several related groups.

(ii) Various numerical quantities derived from these measurements which sum up the general characteristics of the group in a compact form (such as STANDARD DEVIATION and MEAN) or describe some general numerical relationship between two similar groups (e.g. CORRELATION COEFFICIENT).

(iii) The science of such measurements and derived quantities including the study of the homogeneity of different groups and their interrelationships, and of how and to what extent general conclusions may be drawn from particular samples.

See PROBABILITY, RANDOM NUMBERS, FREQUENCY, POPULATION, NORMAL (FREQUENCY) DISTRIBUTION, SIGNIFICANCE.

steam fog. An alternative for ARCTIC SEA SMOKE.

steering. In synoptic meteorology, controlling factor(s) in the direction and speed of movement of pressure systems, sometimes also of precipitation belts, thunderstorms, etc.

The forecasting problem relating to the movement of pressure systems is one of DEVELOPMENT combined with steering. The principle of THERMAL STEERING operates well when development is not very pronounced. Attempts to find by more empirical means an appropriate 'steering current', that is wind velocity at a particular 'steering level' or mean wind velocity in a particular 'steering layer', have had only limited success.

The 700 mb wind velocity is often found to give useful guidance to the movement of existing areas of precipitation including groups of showers; it is not, however, a reliable indicator of the movement of severe thunderstorms.

Stefan–Boltzmann law. The law, discovered empirically by Stefan and later shown theoretically by Boltzmann, that the total RADIATION in all directions from an element of a perfect radiator is proportional to the fourth power of its absolute temperature.

steppe. A name given to the grassy, treeless plains in south-east Europe and Siberia. The word is sometimes extended to mean similar plains and regions of semi-ARID CLIMATE elsewhere.

steradian. The unit of solid angle, being the solid angle subtended at the centre of a sphere of unit radius by a cap of unit area on the spherical surface. The whole sphere subtends a solid angle of 4π steradians at the centre of the sphere.

stereographic projection. See PROJECTION.

Stevenson screen. A standard housing for meteorological thermometers. It is similar to the 'large thermometer screen' but is smaller.

The Stevenson screen was designed by Thomas Stevenson, a civil engineer and father of Robert Louis Stevenson, the well known author. See THERMOMETER SCREEN.

Stokes's law. A solid sphere of radius r, moving with velocity v through a fluid of viscosity η, experiences a viscous drag F tending to oppose its motion, given by

$$F = 6\pi\eta r v.$$

This formula holds only for a small sphere moving with low velocity in conditions of laminar flow in a wide expanse of fluid. See also TERMINAL VELOCITY.

storage half-time. For a specified variable constituent of the atmosphere, the time required for one half of the constituent to be removed from the atmosphere. See RADIOACTIVE FALL-OUT.

storm. The term 'storm' is commonly used for any violent atmospheric phenomenon, such as a gale, thunderstorm, line squall, rainstorm, duststorm, and snowstorm. In synoptic meteorology, the term is applied to an active centre of low pressure with which are associated gales, precipitation, etc.

storm surge. A deviation, positive or negative, of the observed tide from the computed astronomical tide at the corresponding time and place. The storm surge is an essentially dynamical phenomenon; wind velocity is the main cause, with static pressure a minor contributory factor. A notable storm surge occurred in the North Sea on 31 January and 1 February 1953 and caused widespread flooding in adjacent land areas. In 1953 a unit was set up within the Meteorological Office to provide warnings to various public authorities of the likely occurrence of dangerously high storm surges. Various empirical and semi-empirical methods of forecasting have been developed which use as data the forecast wind and pressure fields over the North Sea combined with telemetered readings from a number of tide-gauges around the coasts of Scotland and the east of England; these older methods are supplemented by a mathematical model developed by the Institute of Oceanographic Sciences; this model also provides forecasts of surges along the south and west coasts.

s.t.p. An alternative for N.T.P.

stratiformis (str). A CLOUD SPECIES. (Latin, *stratus* flattened and *forma* appearance.)
'Clouds spread out in an extensive horizontal sheet or layer.
This term applies to ALTOCUMULUS, STRATOCUMULUS and, occasionally, to CIRROCUMULUS' [2, p. 18]. See also CLOUD CLASSIFICATION.

stratocumulus (Sc). One of the CLOUD GENERA. (Latin, *stratus* flattened and *cumulus* heap.)
'Grey or whitish, or both grey and whitish, patch, sheet or layer of cloud which almost always has dark parts, composed of tessellations, rounded masses, rolls, etc., which are non-fibrous (except for VIRGA) and which may or may not be merged; most of the regularly arranged small elements have an apparent width of more than five degrees' [2, p. 39]. See also CLOUD CLASSIFICATION.

stratopause. The atmospheric boundary between the stratosphere and MESOSPHERE. See STRATOSPHERE.

stratosphere. That region of the ATMOSPHERE, lying above the TROPOSPHERE and below the MESOSPHERE, in which, in contrast to these regions, temperature does not decrease with increasing height. The stratosphere therefore extends from the TROPOPAUSE to a height of about 50 km, where the temperature reaches a maximum. Subdivision of the region is sometimes made into the 'lower' (tropopause–20 km), 'middle' (20–30 km) and 'upper' (30–50 km) stratosphere.
An alternative definition of the stratosphere as that region from the tropopause to about 20 km in which temperature changes little with height, is not now favoured.
The stratosphere is a region which is characterized by relatively large amounts of ozone but by amounts of water vapour which are lower (mixing ratio of the order 10^{-2} g kg^{-1} or less) than in the high troposphere. These constituent gases, together with carbon dioxide, largely determine the radiation balance which, in general, controls the vertical temperature distribution of this region. Despite the absence of convective motion, the (lower) stratospheric region has vigorous circulations which are often clearly related to low-level pressure systems.

stratospheric oscillation. An alternative for QUASI-BIENNIAL OSCILLATION.

stratospheric warming. An alternative for SUDDEN WARMING.

stratus (St). One of the CLOUD GENERA. (Latin for flattened.)
'Generally grey cloud layer with a fairly uniform base, which may give drizzle, ice prisms or snow grains. When the sun is visible through the cloud, its outline is clearly

discernible. Stratus does not produce halo phenomena except, possibly, at very low temperatures.

Sometimes stratus appears in the form of ragged patches'[2, p. 43]. See also CLOUD CLASSIFICATION.

streak lightning. LIGHTNING discharge which has a distinct main channel, often tortuous and branching; the discharge may be from cloud to ground or from cloud to air.

stream function. At a level of non-DIVERGENCE in a horizontal air current a stream function (ψ) may be defined such that

$$\mathbf{V} = \mathbf{k} \times \bigtriangledown \psi$$

where $\mathbf{V}$ is the wind velocity vector, $\mathbf{k}$ the vertical unit vector and $\bigtriangledown \psi$ the GRADIENT of the stream function.

The wind velocity vector is normal to, and to the left of, $\bigtriangledown \psi$, that is the wind blows along the isopleths of ψ (and with low values to the left). The isopleths of ψ are therefore STREAMLINES — hence the term 'stream function'. It is a scalar quantity with dimensions $L^2 T^{-1}$.

streamline. A curve which is parallel to the instantaneous direction of the wind vector at all points along it. Isobars are streamlines only in strict GEOSTROPHIC FLOW.

Student's *t*-test. A statistical test which is used to indicate whether a given sample is derived from a given NORMAL (FREQUENCY) DISTRIBUTION, or to indicate whether two samples are derived from the same normal distribution. It is recommended when small samples are being considered and its use is described in statistical textbooks.

Stüve diagram. See AEROLOGICAL DIAGRAM.

subcooling. A seldom-used alternative for SUPERCOOLING.

subjective forecast. A forecast in which, in contrast to an OBJECTIVE FORECAST, the personal judgement of the forecaster plays a significant part. A forecast in SYNOPTIC METEOROLOGY, though based on physical and dynamical principles, is in some degree subjective.

sublimation. In chemistry, the conversion of a solid to a vapour, without melting. In meteorology, the term is applied with respect to water both in the above sense (direct evaporation from an ice surface) and in the converse sense (direct deposition of ice from water vapour).

sublimation nucleus. A type of NUCLEUS, on which direct SUBLIMATION of water vapour to ice crystals may occur.

subsidence. The word used to denote the slow downward motion of the air over a large area which accompanies DIVERGENCE in the horizontal motion of the lower layers of the atmosphere. The greatest divergence is from regions of rapidly rising pressure and the subsidence is probably of the order of 30 to 60 metres per hour in many cases. In stationary unchanging anticyclones the subsidence is due to the outward airflow at the earth's surface only, and is then very much slower. The subsiding air is warmed dynamically (see ADIABATIC) and its relative humidity therefore becomes low, occasionally falling below 10 per cent at about 1–2 km after prolonged subsidence. The downward movement and consequent warming increase with height, up to 3 km or perhaps more, so that the LAPSE rate of temperature is decreased, and INVERSIONS are often developed. The vertical velocity is zero at the horizontal ground, but turbulence often mixes up the lower layers and brings some of the warm dry air to the ground. Subsidence normally results in fine dry weather, but fog, stratus or stratocumulus clouds may occur in certain conditions.

substantial change. An alternative for LAGRANGIAN CHANGE.

subtropical high. One of the cells of high atmospheric pressure (ANTICYCLONES) which compose the quasi-permanent belt of high pressure of the 'subtropics' (i.e. that part of the earth's surface between the TROPICS and the 'temperate regions' whose equatorial boundaries are about 40° N and S). See GENERAL CIRCULATION.

sudden enhancement of atmospherics (SEA). When a SOLAR FLARE occurs, the extra IONIZATION of the D-LAYER in the sunlit hemisphere makes it a more efficient reflector of the radio waves emitted at times of lightning flashes; there is, therefore, an associated sudden increase in the recorded level of distant atmospherics.

sudden warming. A term applied to a relatively sudden temperature rise which occurs on some occasions in the stratosphere at higher latitudes, generally in late winter. The main warming, typically of about 50 K in one or two weeks but sometimes much more rapid, occurs at levels of 25 km or above; modified effects occur, however, at lower levels. The warming is thought to be associated with downward motion of air at the levels concerned since, with the lapse rate which prevails in the stratosphere, relatively slow subsidence is able to produce appreciable warming. The phenomenon is also termed 'explosive warming'.

sulphur dioxide. Gas, of chemical formula SO_2, which occurs in minute and variable concentration in the atmosphere and is of industrial and volcanic origin. In populated regions SO_2 is formed by oxidation of much of the sulphur content of coal or coke or of heavy fuel oils, on combustion. The gas is estimated to amount to about 3 per cent by weight of fuel burned. It dissolves readily in water to form sulphurous acid, and oxidizes photochemically in sunlight to sulphur trioxide which similarly becomes sulphuric acid. These acids cause damage by corrosion.

SO_2 concentration of air is measured either by finding the acidity of a sulphuric acid solution formed by the reaction between known volumes of air and hydrogen peroxide, or by exposing a prepared surface of lead peroxide to the air for a considerable time (usually a month) and measuring the yield of lead sulphate; in either case the SO_2 concentration of the air may be inferred. Typical annual mean VOLUME FRACTION values in Great Britain are, for country air 1×10^{-8}, and for city air 2×10^{-8}. See ATMOSPHERIC POLLUTION.

sultriness. In meteorology, a combination of high atmospheric temperature and humidity.

sumatra. A SQUALL which occurs in the Strait of Malacca, blowing from between south-west and north-west. There is a sudden change of wind from a southerly direction and a rise in speed is accompanied by a characteristic cloud formation — a heavy bank of cumulonimbus which rises to a great height. These squalls usually occur at night, and are most frequent between April and November. They are generally accompanied by thunder and lightning and torrential rain. There is a sudden fall of temperature at the moment the squall arrives.

summer. See SEASONS.

sun. A luminous gaseous sphere round which the earth moves in a slightly elliptical orbit at an average distance of 1.4953×10^8 km. The sun's diameter is 1.3914×10^6 km, its apparent diameter at the earth's mean distance 0.533 degree, its mass 1.9866×10^{30} kg, its mean density 1.41×10^3 kg m^{-3}. The RADIATION emitted from the sun's luminous disc (photosphere) corresponds to a black-body radiation temperature of about 5800 K, the internal gases being at a temperature of many millions of degrees. The gaseous regions above the photosphere, visible during solar eclipses, comprise the REVERSING LAYER, the CHROMOSPHERE, and finally the solar CORONA which extends outwards to a distance of several solar diameters. The Fraunhofer spectral lines due to absorption by gases in these regions show the presence of terrestrial elements. The SYNODIC PERIOD of rotation of the sun, as judged by sunspot movement, is 26.9 days in latitude 0° and 28.3 days in latitudes 30° N and S.

The energy output of the sun, both particles and waves, varies with time. This so-called 'solar activity' is associated with disturbances which are observed in the photosphere and solar atmosphere and which are in large measure interrelated. Chief among the solar disturbances are SUNSPOTS and SOLAR FLARES but they also include, for example, faculae, flocculi, prominences and outbursts of solar 'radio noise'. The sun is described as 'quiet' or 'disturbed' if the disturbances are relatively few and weak or numerous and active, respectively. The relationships between solar activity and various geophysical phenomena to which they give rise are termed SOLAR-TERRESTRIAL RELATIONSHIPS.

sun dog. A popular alternative for mock sun or PARHELION.

sun drawing water. See CREPUSCULAR RAYS.

sun pillar. A vertical column of light above (and sometimes below) the sun, most often observed at sunrise or sunset. The colour may be white, pale yellow, orange or pink. The phenomenon, which is due to the reflection of sunlight from ice crystals, may be seen over a wide area. See Figure 45.

J.C. Smart

FIGURE 45.   Sun pillar with mock sun.

sunrise, sunset. The times at which the sun appears to rise and set, in consequence of the rotation of the earth on its axis. Owing to the effect of atmospheric REFRACTION, which increases the apparent angular altitude of the sun when near the horizon by about 34', sunrise is earlier and sunset later than geometrical theory indicates. There is a further uncertainty caused by the fact that the sun has an appreciable diameter (32') so that time elapses between the first and last contacts with the horizon.

For meteorological purposes, allowance is made for normal refraction of 34' and it is assumed that, for an observer at sea level with a clear horizon, sunrise and sunset occur when the sun's upper limb contacts the apparent horizon; at such a moment the true centre of the sun's disc is 50' below the horizon.

The times of sunrise and sunset vary with latitude and with the declination of the sun. Diagrams illustrating the variations, so far as the British Isles are concerned, are given in the *Observer's handbook* [1, pp. 190–194].

The general explanation of the variety of colours that are to be seen in the sky about the time of sunrise or sunset is as follows. White light such as that from the sun may be regarded as composite, the constituents being light of all the colours of the spectrum. When the light waves meet obstacles in their course, such obstacles as the molecules of the atmospheric gases or larger obstacles such as particles of dust, the waves are broken and secondary waves proceed in all directions from the obstacles. The direct light is therefore reduced in strength and the farther the light goes through an atmosphere of such obstacles the more the strength is reduced, the energy being used up in producing scattered light. This effect is more pronounced with blue light, for which the wavelength is short, than with red light for which the wavelength is longer. Accordingly a beam of white light passing through air loses the constituents of shorter wavelength and becomes yellow, then orange, and finally red.

This accounts for the changing colour of the sun as it nears the horizon. The SCATTERING by air alone merely makes the setting sun yellow, but if there is dust in the air or even the nuclei on which water vapour is condensed then the sun becomes orange or red before it sets.

Clouds which are illuminated by the light from the setting sun are also red, whilst other clouds which are illuminated by scattered light in which the blue constituents are present are white or grey; higher clouds illuminated by light which has only passed through less dense and cleaner air may also appear white.

The colours of the sky itself can be explained in the same way. When sunlight has already travelled a great distance through the lower atmosphere it has lost the constituents of short wavelength, and in the light which remains to be scattered the longer wavelengths predominate. In the further passage of scattered light to the eye of the observer the longer wavelengths again have preference.

When we look at the sky in a particular direction we receive light which has been scattered by the atmosphere at all heights. Blue may predominate, whilst that coming from the lower levels may be red. The combination of light from both ends of the spectrum gives us purple. On the other hand in other parts of the sky the middle wavelengths may predominate and the resulting colours are green or yellow.

sunset. See SUNRISE, SUNSET.

sunshine. Direct, as opposed to diffuse, solar RADIATION.

The routine measurements of the duration of sunshine which are made for climatological purposes refer in the British Isles, as in most other countries, to so-called 'bright' sunshine. Since different instruments differ in their response characteristics to the radiation, this term has lacked precise definition — but see SUNSHINE RECORDER.

sunshine recorder. An instrument for recording the duration of bright SUNSHINE. Such instruments depend either on the heating action of the sun or on the chemical action produced by the sun's rays. The Campbell–Stokes recorder, an instrument of the former class, is in general use in the British Isles. A spherical glass lens focuses the solar image on a graduated card held in a frame of special design. The duration of sunshine is indicated by the length of the burnt track of the image. The WMO decided in 1962 to adopt the Campbell–Stokes recorder, used with record cards to specifications of the French Meteorological Office, as a standard of reference to which values of sunshine duration should be reduced in future. See also EXPOSURE.

sunspot. A relatively dark region on the disc of the SUN, with an inner 'umbra' of effective radiation temperature about 4500 K and an outer 'penumbra' of somewhat higher temperature.

Sunspot duration varies from a few hours to many solar rotation periods. Their frequency is quasi-periodic, with an average 'period' of about 11 years. In the typical sunspot 'cycle' there are at first few spots in about solar latitude 30° N and S, maximum spots in about 15° N and S after some 4½ years, and again few spots in about 8° N and S after a further 6½ years, these last spots overlapping in time the first high-latitude spots of the following cycle. There are, however, some large departures from these average figures. The mean time taken for a spot in solar latitude 0° to return to the central meridian, as seen from the earth, is 26.9 days; in latitude 30° N and S the time taken is 28.3 days.

Sunspots are vortex-like disturbances with large associated magnetic fields. There is yet no accepted theory of their formation or quasi-periodic nature. There are well established relationships of a general nature between sunspot variation and effects measured in the IONOSPHERE, GEOMAGNETISM, etc., but not as yet in meteorology.

superadiabatic lapse rate. A 'lapse rate' (rate of fall of temperature with height) greater than the dry ADIABATIC lapse rate of 1 °C per 100 m. Such a LAPSE rate does not occur within the free atmosphere, but the dry adiabatic lapse rate is often exceeded by a factor of several times near a land surface which is strongly heated by solar radiation.

supercell. A unit of convection consisting of an intense updraught and downdraught forming a single persistent three-dimensional pattern in a quasi-steady state on a horizontal scale of about 20 km; it is associated with violent thundery activity, the formation of very large hailstones, and damaging winds including tornadoes. Supercells form in strongly baroclinic zones, with the wind increasing sharply and veering with height, when very warm and moist air is overrun by cooler and drier air. They have an anomalous direction of travel relative to the mean tropospheric wind; this is usually to the right of the tropospheric winds in the northern hemisphere and to the left in the southern hemisphere. They often persist for an hour or more.

Supercell storms differ qualitatively and quantitatively from ordinary multicell storms which contain one or more convective cells each of which has a relatively short life-cycle and tends to form discrete new cells as it moves and decays. Supercell storms are rare in the British Isles, but notable examples were the Horsham storm in 1958 [67,68,69] and the Wokingham storm in 1959 [70].

supercooling. Supercooling of a liquid (sometimes termed 'subcooling' or 'undercooling') signifies the existence of a substance in the liquid state at a temperature below the normal freezing-point.

Although the supercooled state is regarded as unstable in the sense that its achievement in the laboratory requires very careful cooling of the liquid, supercooling of cloud droplets is common in the atmosphere. All clouds which extend above the 0 °C isotherm contain supercooled droplets at some stage in their history; in particular, altocumulus is predominantly a water-droplet cloud though generally at a temperature well below 0 °C and it is necessary to attain cirrus level to find clouds almost invariably in the form of ice crystals. The larger raindrops do not undergo a marked degree of supercooling.

Supercooling is fundamental in the ice-crystal process of PRECIPITATION.

superior air.   A term sometimes used, in synoptic meteorology, in respect of air at higher levels which has been made very dry by the process of SUBSIDENCE.

superposed-epoch method.   A method of statistical analysis (also called the 'n-method') which is used, for example, to investigate the possibility of a RECURRENCE TENDENCY in a given TIME SERIES, or the relationship between two synchronous time series.

Where a single series is involved, average values are calculated of the n terms $(+1, +2, ..., + n)$ of the series which follow various 'O' terms selected on the basis of an objective criterion, for example the peak value in successive equal blocks of terms; statistical SIGNIFICANCE may be looked for in the departures of the computed average values from the mean of the whole series, taking into account the standard deviation of the series.

Where the relationship between two series is involved, average values of terms $+1, +2, ..., + n$ are computed in each series, the epoch of the 'O' term being selected on the basis of a criterion applied to one of the series; the significance of the distribution of the means may be assessed by the CHI-SQUARE TEST, for example. Corroboration of a suggested relationship between the series may be sought by defining the 'O' epoch on the basis of a criterion applied to the second series and repeating the calculations on the data arranged on this basis. Nowadays, and especially where long series of data are involved, POWER SPECTRUM and cross-spectrum analysis would be preferred.

super refraction.   See ANOMALOUS RADIO PROPAGATION.

supersaturation.   See SATURATION.

surface chart.   In synoptic meteorology, an alternative for WEATHER MAP.

surface-free energy.   An alternative for SURFACE TENSION.

surface inversion.   An INVERSION of temperature through an atmospheric layer extending upwards from the earth's surface. This is frequently a RADIATION INVERSION, but it may form also as the result of a drift of air over a surface colder than itself.

surface temperature.   Unless otherwise specified, the air temperature measured in the shade at a height of between 1.25 and 2 metres, as in the THERMOMETER SCREEN.

surface tension.   Any surface of a liquid is subject to a tension, expressed in newtons per metre in SI UNITS, due to forces of attraction between the liquid molecules which act in such a way as to minimize the surface area.

The phenomenon, which is also termed 'surface-free energy' ($J\ m^{-2}$), is important, for example, in NUCLEATION processes and in SOIL MOISTURE. The surface tension of pure water increases with decrease of temperature; values at $+20\ °C$, $0\ °C$ and $-20\ °C$ are, for example, 7.27×10^{-2}, 7.57×10^{-2} and $7.91 \times 10^{-2}\ N\ m^{-1}$, respectively.

surface wind.   Generally, the WIND velocity at a height of 10 metres in an unobstructed area.

surge.   A term in synoptic meteorology which is sometimes used (first by Abercromby) to denote a substantial and general rise of atmospheric pressure over an

289

area, the rise being greater than that attributable to the movement of depressions or anticyclones in the vicinity.

In tropical meteorology, the term is used to signify a marked and sudden increase of strength of a monsoon or trade-wind current.

The term is also used of water disturbances — see STORM SURGE.

swell. Swell is wave motion in the ocean caused by a disturbance which may be at some distance away; the swell may persist after the originating cause of the wave motion has ceased or passed away. It often so continues for a considerable time with unchanged direction, as long as the waves travel in deep water. The height of the waves rapidly diminishes but the length and velocity remain the same, so that the long, low, regular undulations characteristic of swell are formed. Swell is often observed to have a wavelength greatly in excess of that of waves seen during a storm; the probable explanation is that the longer waves are then masked by the shorter and steeper storm waves. Swell observations are useful in denoting the direction in which sea disturbance has taken place due to tropical cyclones or other storms. See OCEAN WAVES.

SWF. Abbreviation for 'short-wave (radio) fade-out'.

symbols, international meteorological. See WEATHER.

synodic period. The synodic period of a planet or the moon is the interval of time between successive conjunctions (see CONJUNCTIONS, ASTRONOMICAL) of the body and the sun, as viewed from the earth. See also DAY.

synoptic chart. An alternative for WEATHER MAP.

synoptic meteorology. That branch of meteorology which is concerned with a description of current weather as represented on geographical charts and applied especially to the prediction of its future development.

The starting point for this branch of meteorology is the 'synoptic report' or 'synoptic message' containing a coded summary of the current weather at each of a large number of SYNOPTIC STATIONS (including AUTOMATIC WEATHER STATIONS). Many such reports, transmitted by a variety of telecommunication systems are, on receipt at selected centres, plotted symbolically on a 'synoptic chart' which thus provides a representation of the weather at a particular time, and generally over a large geographical area, in a synoptic (summary or condensed) form which is suitable for the purposes of synoptic meteorology, as defined above.

synoptic station. A STATION at which meteorological observations are made for the purposes of SYNOPTIC METEOROLOGY. The observations are made at the 'main synoptic hours' 0000, 0600, 1200, 1800 GMT and normally also at the 'intermediate synoptic hours' 0300, 0900, 1500, 2100 GMT, the observed elements being plotted symbolically on the WEATHER MAP.

synthetic aperture radar (SAR). An active REMOTE SENSING device designed to provide high resolution images of ocean waves. The resolution is expected to attain from 5 to 25 m allowing detection of ships' wakes. An important practical application is the detailed analysis of sea ice for navigation purposes. SAR operates by measuring and storing the amplitude and phase of the returned signal over a short time interval and using this information to produce an effective resolving power greater than would normally be obtained from the 'real' aperture of the antenna system. It operates on frequencies of about 1.25 GHz and requires considerable power.

T

tablecloth. The term applied to the orographic stratus cloud which often occurs on the windward side of Table Mountain near Cape Town, South Africa.

tail wind. See EQUIVALENT HEAD WIND.

Taylor–Proudman theorem. That GEOSTROPHIC motion of a homogeneous fluid is the same in all planes perpendicular to the axis of rotation. Such motion, being two-dimensional only, can be pictured as consisting of travelling columns each orientated parallel to the axis of rotation. The columns are called 'Taylor columns' or 'Proudman pillars'.

Technical Commissions. The Technical Commissions of the WORLD METEOROLOGICAL ORGANIZATION are each composed of experts in the various meteorological fields: Atmospheric Sciences, Aeronautical Meteorology, Agricultural Meteorology, Climatology, Hydrometeorology, Instruments and Methods of Observation, Maritime Meteorology, and Synoptic Meteorology. Meetings of the Commissions take place at least once every four years.

teleconnection. A connection between variations or anomalies at locations separated by distances greatly in excess of the normal synoptic scale. The variations at one location are usually taken to be purely atmospheric while those at the other may be either purely atmospheric or related to the boundary conditions (e.g. fluctuations in sea surface temperature). An example of the first kind is the Pacific/North American (PNA) pattern of correlation between monthly mean 500 mb geopotential heights [71]; of the second, the relationship between rainfall variations over the Sahel and global patterns of sea surface temperature. The reality of teleconnections can be established by statistical analysis, but their explanation requires physical reasoning and the employment of computer models of the GENERAL CIRCULATION.

temperature. The general idea of temperature is intuitively known to everybody from subjective responses to the feel of hot or cold bodies. It was discovered a long time ago that the physical properties of certain substances, including their size (see EXPANSION) and electrical resistance, varied continuously with temperature, and this fact made possible the construction of THERMOMETERS and the devising of various empirical TEMPERATURE SCALES. In the last century Kelvin showed how a temperature, with an associated scale, could be defined in a way independent of the working substance by making use of the laws of THERMODYNAMICS; this is known as the thermodynamic (or absolute) temperature. The thermodynamic temperature may be shown to be identical with the PERFECT GAS temperature, and this enables the thermodynamic scale to be realized, to a known degree of accuracy, in the laboratory.

temperature scales. The scales in common use are the Celsius, the Fahrenheit and the Kelvin. They are defined primarily in relation to THERMODYNAMIC TEMPERATURE, but standards for the various practical realizations of the thermodynamic scale are also prescribed. (Fortunately the differences between theoretical and practical scales, with modern technology, are negligible for meteorological purposes.)

In 1968, thermodynamic and practical units of temperature were defined to be identical and equal to $1/273.15$ of the thermodynamic temperature of the TRIPLE POINT of water, this unit being known as the 'kelvin' and denoted by 'K'.

If temperatures measured on the thermodynamic scale are denoted by T, and those on the Celsius scale by t_C, then $t_C = T - T_0$ where $T_0 = 273.15$ K; this amounts to saying that on the Celsius scale the freezing-point and boiling-point of water at standard pressure are 0 °C and 100 °C, respectively. If temperature in Fahrenheit is denoted by t_F, then

$$t_C = (t_F - 32) \times \frac{5}{9} \text{ and } t_F = \left(\frac{9 t_C}{5}\right) + 32,$$

i.e. the freezing-point and boiling-point of water are 32 °F and 212 °F, respectively.

tendency. A term used in synoptic meteorology to signify local time rate of change of an element, for example of surface pressure or of GEOPOTENTIAL (height) at a fixed pressure level. See BAROMETRIC TENDENCY.

tendency equation. The equation relating the change of pressure with time at a point at height h in the atmosphere (pressure TENDENCY) to the change in the weight of the air above is

$$\left(\frac{\partial p}{\partial t}\right)_h = -g \int_h^\infty \rho \left(\frac{\partial u}{\partial x} + \frac{\partial v}{\partial y}\right) dz - g \int_h^\infty \left(u \frac{\partial \rho}{\partial x} + v \frac{\partial \rho}{\partial y}\right) dz + (g\, \rho w)_h$$

where g is assumed constant with height.

The equation shows that the local pressure tendency at height h has contributions from three processes (the three terms, taken in order):

(i)   horizontal DIVERGENCE at heights greater than h,
(ii) horizontal ADVECTION of air of different density at heights greater than h, and
(iii) vertical motion of air at height h.

The third process does not normally operate in the surface pressure tendency $(\partial p / \partial t)_0$ because of the boundary condition $w = 0$ at the earth's (level) surface.

tenuity factor. In BALLISTICS, the ratio of the density of air having the observed pressure at the surface and temperature equal to the BALLISTIC TEMPERATURE, to the density of air at pressure 1015 mb and temperature 16 °C.

tephigram. See AEROLOGICAL DIAGRAM.

terce. A term used in the Meteorological Office for the set of values included between the two TERCILES of a FREQUENCY DISTRIBUTION or those in the tails below the lower and above the upper terciles.

tercentesimal scale. The name for the approximate absolute scale, sufficiently accurate for meteorological purposes, obtained by adding 273 to the Celsius temperature. See TEMPERATURE SCALES.

tercile. One of two values of a random variable which divide its DISTRIBUTION FUNCTION into three equal parts.

terminal velocity. The velocity of a body falling through a fluid reaches a 'terminal' value when the weight of the body is balanced by the combined upthrust and drag due to the fluid.

For very small particles, viscous drag predominates, STOKES'S LAW is obeyed, and the terminal velocity v of a sphere of radius r metres and density σ kg m^{-3} is given by

$$v = \frac{2}{9}\, g\, \frac{r^2(\sigma - \rho)}{\eta} \text{ m s}^{-1}$$

where ρ (Kg m^{-3}) is the density of the fluid, η (Ns m^{-2}) its viscosity and g (m s^{-2}) the gravitational acceleration. The expression is accurate to 1 per cent for water-drops up to 30 μm in radius. For larger drops, inertial drag becomes more important and the empirical relation $v = 0.085r - 0.15$ m s^{-1} (where r is in micrometres) fits well for drops from 30 to 500 μm in radius. Drops with radius larger than 500 μm become non-spherical when falling and the expression $v = 6.5\ \sqrt{}\ r - 0.7\,(1.3 - r)^2$ m s^{-1} (where r is in millimetres) is accurate to within 3 per cent for drops from 0.5 mm radius up to the largest raindrops (3 mm radius) which have a limiting terminal velocity of about 9 m s^{-1}. Terminal velocities at heights of 3 and 6 km exceed those at ground level by about 10 and 30 per cent respectively because of changes in air temperature and pressure.

Approximate terminal velocities are: for a single ice crystal 0.5 m s^{-1}, for a snowflake 1 m s^{-1}, and for a rimed snowflake 2 m s^{-1}.

The terminal velocity of hailstones is a function of density, shape and surface roughness as well as radius and no formula will fit all hailstones. However, the terminal velocity is approximately proportional to $\sqrt{}\ r$ and the largest hailstones (about 50 mm radius) have terminal velocities around 45 m s^{-1}.

terrestrial magnetism.   An alternative for GEOMAGNETISM.

Tetens' formula.   A relationship between vapour pressure (e) in millibars and dew point (T_d) in degrees Celsius of the form

$$e = \exp\left(1.8099 + \frac{17.27\, T_d}{T_d + 237.3}\right).$$

This empirical formula [72, pp. 297–309] is an adaptation of a form originally put forward on the basis of original observations by Gustav Magnus in 1844 [73].

tetroon.   See CONSTANT-LEVEL BALLOON.

thaw.   The transition by melting from snow or ice to water. The term is especially used to indicate the end of a spell of FROST, which in the British Isles in winter is generally associated with the displacement of a stagnant or continental AIR MASS by one of maritime origin.

theodolite, pilot-balloon.   An instrument consisting of a telescope mounted to permit rotation both in elevation and azimuth, and fitted with a right-angled prism so that the observer continues to look horizontally into the eyepiece no matter what the elevation of the balloon.

thermal.   A volume of air which possesses BUOYANCY on account of low density relative to its environment and so rises through the environment.

Thermals are produced in conditions of intense solar heating over land, with a resulting SUPERADIABATIC LAPSE RATE at low levels. Strong thermals tend to occur over regions where the earth's surface is warmer than the surrounding area; sun-facing slopes and towns are among the good thermal sources recognized by glider pilots. Cumulus clouds show the presence of thermals which, however, often exist without such evidence.

Observations have shown that the typical pattern of air motion within a thermal is one in which maximum vertical velocity occurs within the central core, with circulatory motions ('vortex rings') towards either edge of the thermal; however, the air motion in a cumulus cloud is often more complex, with regions of descending air within the body of the cloud. A thermal grows in size by mixing with surrounding air at its upward-moving head and by the mixing of air into its wake. The temperature excess and buoyancy of the thermal relative to its environment are thus progressively reduced.

thermal capacity. The thermal capacity, also called heat capacity, of a body is the product of its mass and its SPECIFIC HEAT.

thermal conductivity. See CONDUCTIVITY, THERMAL.

thermal diffusivity. See CONDUCTIVITY, THERMAL.

thermal equator. The latitude of highest mean surface air temperature. Because of the non-uniform distribution of land and sea, this does not coincide, over a year, with the geographic equator. In northern summer the thermal equator is at about latitude 15°N, and in northern winter it is about 5°S averaging about 5°N for the year as a whole.

thermal high, low. A closed centre of high (low) values of THICKNESS on a thickness chart. The centre is so called because it represents, to a close approximation, a centre of high (low) mean temperature in the isobaric layer concerned. But see also THERMAL LOW.

thermal low. A surface depression whose formation is the direct result of differential solar heating of neighbouring land and sea areas; in summer, for example, air density and therefore also surface pressure over land tend to be reduced relative to the values over the sea, because of the higher surface temperatures reached.

A thermal low may form over the British Isles during afternoon and evening in summer if the general pressure gradient is very slight; such a depression is weak, has little vertical extent and has no pronounced associated weather characteristics. Another example is the tendency for the formation of shallow depressions in winter over the relatively warm Mediterranean. The Asiatic 'monsoon low' is an example of a thermal depression on a much larger scale. See also THERMAL HIGH, LOW.

thermal precipitator. An apparatus used for sampling small particles, especially Aitken and condensation nuclei — see NUCLEUS.

Air is drawn past a heated wire at a slow, controlled speed. The particles are unable to penetrate a dust-free barrier surrounding the wire and form line deposits on, for example, cold glass surfaces located at a convenient distance from the wire; the deposits are examined and counted microscopically.

thermal Rossby number. A variant of the ROSSBY NUMBER in which the characteristic wind speed is that of the THERMAL WIND.

thermal steering. The principle of thermal steering states that surface patterns of the vertical component of vorticity (corresponding closely to surface depressions and anticyclones) are 'steered' in the direction of the THERMAL WIND in the troposphere, with a speed proportional to that of the thermal wind.

thermal wind. The thermal wind (V_T) in a specified atmospheric layer, at a given time and place, is the vertical geostrophic WIND SHEAR in the layer concerned. V_T is defined by the relationship

$$V_T = V_1 - V_0,$$

294

where V_1 and V_0 are the geostrophic winds at the top and bottom of the layer, respectively.

The term 'thermal wind' was adopted because wind shear is determined by the distribution of mean temperature in the layer concerned. The direction of the wind is such that it blows parallel to the mean isotherms (or THICKNESS lines) of the layer keeping low mean temperature (or low thickness) on the left in the northern hemisphere and on the right in the southern hemisphere. Its magnitude is given by the expression

$$V_T = \left| \frac{g}{f} \frac{\Delta z'}{\Delta n} \right|$$

where $\Delta z'/\Delta n$ is the gradient of thickness of the layer (expressed in geometrical units). See also VECTOR, HODOGRAPH ANALYSIS, POLAR VORTEX.

thermistor. A semiconductor with a large temperature coefficient of resistivity, and a linear relationship at constant temperature between current and applied electromotive force, which may conveniently be used as the basis of a THERMOMETER.

thermocline. An oceanic layer in which the rate of decrease of temperature with increasing depth is a maximum.

A thermocline is a stable region in which vertical mixing of water is strongly inhibited. A permanent thermocline exists at a depth of some hundreds of metres in low and middle latitudes. In summer, a 'seasonal thermocline' is formed at a shallower depth, especially in middle latitudes, because of the heat which is then absorbed in the surface layers and distributed by wave motion throughout a limited depth.

thermocouple. An instrument for measuring temperature. It consists basically of two wires of different metals joined at each end. One junction is kept at a fixed (known) temperature, the other put at the point where the temperature is to be measured. A thermoelectric electromotive force (e.m.f.) is generated, of magnitude proportional to the temperature difference (for two given metals); this e.m.f., or the resulting electric current in the circuit, may be used as a measure of the temperature difference between the junctions.

thermodynamic diagram. A diagram on which may be represented graphically the states of air samples (or the varying state of a single air sample) in terms of pressure, temperature, and humidity or of other functions related to these. It is also termed an 'adiabatic diagram' or, in meteorology, an AEROLOGICAL DIAGRAM, many forms of which exist.

The term 'thermodynamic diagram' is restricted by many authors to those types of adiabatic diagram in which the area enclosed on the diagram by a curve which represents a cyclic thermodynamic process is proportional in all parts of the diagram to the amount of work performed in the process.

thermodynamic equation. An equation expressing the law of conservation of energy, in meteorology usually written in terms of the POTENTIAL TEMPERATURE (θ) as follows:

$$\frac{d}{dt}(\ln\theta) = \frac{1}{\theta}\frac{d\theta}{dt} = \frac{1}{c_p}\frac{dS}{dt}$$

where dS/dt is the rate of increase of entropy due to non-adiabatic processes such as evaporation and condensation, and the effects of radiation. The term $d/dt\,(\ln\theta)$ may be written in various ways according to the choice of VERTICAL COORDINATE SYSTEM.

thermodynamics. That part of the science of heat which deals with the transformation of heat into other forms of energy, and vice versa.

thermodynamic temperatures. Definitions of thermodynamic dew-point, frost-point, ice-bulb, and wet-bulb temperatures are identical with those of DEW-POINT, FROST-POINT, ICE-BULB TEMPERATURE, and WET-BULB TEMPERATURE. The distinction in nomenclature is made in recognition of the fact that the measured values of these elements may differ slightly from the defined values because of instrumental limitations or procedure.

thermogram. The continuous record of temperature yielded by a THERMOGRAPH.

thermograph. A recording THERMOMETER. Many patterns are in use, the sensitive member being, for example, a bimetallic strip, a BOURDON TUBE, a resistance element, or a steel bulb filled with mercury.

thermohygrograph. An alternative for HYGROTHERMOGRAPH.

thermometer. An instrument for measuring temperature, from one or other of the physical changes produced in matter by heat, e.g. expansion of solids, liquids or gases, changes in electrical resistance, production of electromotive force at the junction of two different metals, etc.

In normal meteorological practice, mercury-in-glass thermometers are used. At temperatures below $-38.9\,°C$ (freezing-point of mercury), and also in measuring minimum temperatures, alcohol (freezing-point $-114.4\,°C$) is used.

See BOURDON TUBE, SOIL THERMOMETER, MAXIMUM THERMOMETER, MINIMUM THERMOMETER, RESISTANCE THERMOMETER, THERMISTOR.

thermometer screen. The standard housing for meteorological thermometers. It consists of a wooden cupboard, with hinged door, mounted on a steel stand, so that the bulbs of the wet-bulb and dry-bulb thermometers are 1.25 m (4 ft) above the ground. The whole screen is painted white. Indirect ventilation is provided through the bottom, double roof and louvered sides. Thermometers are placed within it to give a close approximation to the true air temperature, undisturbed by the effects of direct solar or terrestrial radiation.

The 'small thermometer screen' accommodates the dry-bulb, wet-bulb, maximum and minimum thermometers. The 'large thermometer screen' provides additional accommodation for a THERMOGRAPH and HYGROGRAPH. It also has another door at the back. See also STEVENSON SCREEN.

thermoneutral zone. The range of ambient temperature in which normal metabolism provides enough spare heat to maintain an essentially constant body temperature. The limits of the zone depend on the species and breed of animal and its age, sex, degree of acclimatization, how it is fed, and even the time of day. The zone is narrow for the young of a species (e.g. 29 to 30 °C for a chick) and wide for a well fed large adult (e.g. -30 to $+25\,°C$ for a cow).

thermopile. An instrument for measuring RADIATION. It consists essentially of a number of THERMOCOUPLES, either connected in series (if electromotive force is measured) or in parallel (if electric current is measured), as a means of increasing the sensitivity beyond that possible with a single thermocouple.

thermosphere. That part of the ATMOSPHERE, extending from the top of the MESOSPHERE at about 85 km to the atmosphere's outermost fringe, in which the temperature increases with increasing height.

thickness. The GEOPOTENTIAL height difference at a given place between specified pressure levels. Thickness values relating to selected standard pressure levels are obtained from radiosonde observations and are plotted on geographical 'thickness charts'. Contours are drawn, at an appropriate thickness interval, joining places of equal thickness and are termed 'thickness lines'. The analysis of such charts is termed 'thickness analysis' and has an important role in synoptic meteorology — see, for example, DEVELOPMENT.

thoron. Gas, of atomic mass 220 and atomic number 86, which is a radioactive isotope of RADON. It occurs in minute concentration in the atmosphere and plays a small part in the IONIZATION of the air at low levels.

thunder. The noise which follows a flash of LIGHTNING, caused by a sudden heating and expansion of air along the path of the lightning. The distance of a lightning flash may be roughly estimated from the interval between seeing the flash and hearing the thunder, counting one kilometre for every three seconds. The long duration of thunder compared with the associated lightning flash is explained by the different distances travelled by the sound from different parts of the flash and by echoing from neighbouring hills. Echoing causes intensity variations, which, however, also arise from the multiple and tortuous nature of many lightning strokes.

Thunder is seldom heard at distances greater than 20 km though distances up to about 65 km have been reported on occasion. Owing to refraction of sound waves in the lower atmosphere, thunder is sometimes inaudible at distances much less than 20 km, especially when the initiating lightning flash is not to ground.

thunderbolt. Popular term for a LIGHTNING discharge from cloud to ground.

thunder-cloud. Popular expression for CUMULONIMBUS, the cloud associated with THUNDERSTORMS.

thunderstorm. 'One or more sudden electrical discharges, manifested by a flash of light (lightning) and a sharp or rumbling sound (thunder)' [2, p. 126].

Necessary conditions for a thunderstorm are generally stated to be a CUMULONIMBUS cloud base lower than the 0 °C isotherm, such vertical depth of cloud as to ensure a cloud top of temperature less than about −20 °C (implying probable glaciation of the cloud top), and the occurrence of precipitation. There are, however, well substantiated observations, mainly in low latitudes, of the occurrence of lightning in clouds no parts of which were at a temperature below 0 °C. The mature thunder-cloud often has a cellular structure and there are both updraught and downdraught regions; in earlier stages of development only updraught occurs, in later stages of dissipation only downdraught. Vertical updraughts average about 6 m s^{-1} and may reach 30 m s^{-1}; downdraught velocities are about half those of updraught.

Charge separation in thunder-clouds is such as to produce a positive charge in the upper part of the cloud and negative charge in the lower part; small regions of positive charge near the base have also been observed. Initial charge separation has been attributed by various workers to mechanisms such as the selective capture of ions by water drops or ice particles, the frictional rupture of large raindrops, and surface and volume interactions of water, in its various phases, contained in the cloud. Experimental support has been obtained for most of the theories but it is believed that the dominant mechanism involves the growth of soft HAIL.

Rainfall in a thunderstorm at a particular place typically reaches peak intensity more quickly than it dies away; behaviour is, however, sometimes complicated by the effects of two or more cells. Other well marked surface effects are attributable to the arrival of the downdraught air which, cooled by evaporation of water drops, is accelerated earthwards. Surface wind suddenly becomes strong and gusty; temperature falls sharply, sometimes by 10 °C or more; relative humidity rises unsteadily to nearly 100 per cent; and pressure rises sharply, reversing a previous fall.

Thunderstorms are fairly frequently accompanied by hail. In Great Britain, snow in thunderstorms is mainly confined to exposed west and north coasts.

An adequate supply of moisture and a lapse rate of temperature in excess of the saturated adiabatic through a range of height not less than 3000 metres above cloud base are required for the development of a thunderstorm; high surface temperature is a common but not an essential condition. An initial 'trigger' action is often provided by orographic uplift or, especially, by horizontal convergence of surface air. This latter most frequently exists in a shallow depression, trough of low pressure (as at a cold front), or col. The line of convergence which often marks the farthest inland penetration of the sea-breeze may provide the necessary impulse.

In Great Britain thunder is a very variable element, the highest and lowest annual totals of thunderstorm days at many individual stations ranging from less than 5 in a quiet year to 20 or more in an active one. One consequence of this is that published maps showing the average frequency of days of thunder differ considerably in detail according to the period of records used. They agree, however, in showing that the average annual frequency is less than 5 days in western coastal districts and over most of central and northern Scotland, and 15–20 days over the east Midlands and parts of south-east England. There is relatively little seasonal variation on the western seaboard but elsewhere summer is the most thundery season.

The most thundery part of the earth is the island of Java where the annual frequency of thunderstorms is about 220 days per year. C.E.P. Brooks has estimated that the earth has a total of about 44 000 thunderstorms per day and a total of about 100 lightning discharges each second.

thunderstorm day. A local calendar day on which THUNDER is heard at a particular location.

tidal wave. A popular term for a destructive type of wave motion in seas and oceans, associated either with strong winds or with underwater earthquakes. In technical terms they are classified as STORM SURGE and TSUNAMI, respectively.

tide. The periodic rise and fall of the earth's oceans due to combined gravitational forces applied by the moon and sun. Similar, though more complex, effects occur in the earth's atmosphere — see ATMOSPHERIC TIDES.

tilt of a trough. The angular departure of a TROUGH line from a meridian. A trough whose axis is rotated anticlockwise from a meridian is said to have a positive tilt.

time. When the centre of the sun is due south of an observer, the time is called 12h or noon, local apparent time (LAT). The sun is said to TRANSIT at this time. The interval between two consecutive transits of the sun is divided into 24 equal parts, and the times where the lines of division fall are numbered 13h, 14h ... 23h, 24h (or midnight), 1h, 2h ... 12h. LAT is recorded by SUNSHINE RECORDERS and sundials.

The interval of time between successive transits of the sun is not quite constant, but goes through a cycle of changes during the year. This is because the orbit of the earth is elliptical and the earth's axis is not at right angles to its orbit. As it would be very inconvenient in daily life if the length of the day varied in this way, astronomers have invented a 'mean sun' whose apparent motion round the earth is uniform throughout the year. The apparent positions of the real and mean suns are always very close, and coincide four times a year. The moment of transit of the mean sun is 12h (or noon) local mean time (LMT). The interval between successive transits is called a DAY and each day is divided into 24 equal parts. All hours, and all days, are equal in duration. LMT is obtained from LAT by adding (or subtracting, if sign is minus) the 'equation of time' which is given very accurately in *The nautical almanac* [74] and elsewhere, and sufficiently accurately for meteorological purposes in the *Observer's handbook*

[1, p. 208]. The equation of time varies from about $-14\frac{1}{4}$ minutes in mid-February to about $+16\frac{1}{2}$ minutes in early November.

LMT at Greenwich is called Greenwich mean time (GMT). Differences in LMT between different places are determined solely by longitude differences on the scale 1 hour per 15° of longitude. Thus, LMT for any place is derived by adding to GMT, or subtracting from GMT, a correction on this scale (4 minutes of time per degree of longitude), according as to whether the place is to the east or west of the Greenwich meridian.

The inconvenient use of a clock time which varies continuously with longitude is avoided by the use of ZONE TIME. In the British Isles BRITISH SUMMER TIME (BST) is 1 hour in advance of GMT. However, GMT is adhered to throughout the year for meteorological observation purposes, other than evening reports for the Press from selected stations, which are based on clock time.

See also UNIVERSAL TIME.

time series.   In statistics, a series of values which are arranged in order of occurrence and which refer, in general, to equally spaced time intervals. Such series are common in geophysics. The values may be those which obtain at discrete intervals of time or which are averaged over successive periods of time.

Some degree of persistence normally exists between successive values in a time series, provided the time interval of separation is not too long. The presence of persistence implies that the number of statistically independent values in the series may be much smaller than the total number of values, with important implications in questions of statistical SIGNIFICANCE. See PERSISTENCE.

tipping-bucket switch.   A switch, suitably fitted to the delivery tube of a RAIN-GAUGE, consisting of a delicately balanced bucket divided into two equal compartments each of which is constructed to collect a precise quantity of rain. The compartments are presented to the delivery tube in the manner of a see-saw so that as one overbalances when the required amount of rain is collected, the other assumes the collecting position. Each tip causes a magnet to actuate a reed switch which, in turn, actuates an impulse counter or magnetic-tape event recorder.

tonne.   Metric ton, equal to 10^3 kg (0.9842 ton).

topography.   In general usage, all features of the earth's surface that might be shown on a map, including hills and mountains, lakes and rivers, woods, built-up areas, large buildings, etc. In meteorology, topography is usually understood to mean features influencing atmospheric flow including contours of height of ground above sea level (OROGRAPHY) and the various types of land usage (urban, pasture, forest) that affect ROUGHNESS LENGTH and surface exchange of energy.

tornado.   A violent whirl, generally cyclonic in sense, averaging about 100 m in diameter and with an intense vertical current at the centre capable of lifting heavy objects into the air. Uprooting of trees and the explosive destruction of buildings, due to local pressure differences that occur in the intense horizontal pressure gradient near the tornado centre, mark the paths of tornadoes. The paths vary in length from a few hundred metres to some hundreds of kilometres; associated winds in extreme cases are estimated to attain speeds of about 200 kn. Heavy rain, and generally thunder and lightning, occur with the tornado.

While the conditions required for the formation of a tornado are similar to those required for a severe THUNDERSTORM, namely great instability, high humidity, and horizontal convergence of winds at low levels, the precise conditions which cause tornadoes (rather than merely thunderstorms) are not yet known. Tornadoes are most frequent and intense in the USA, east of the Rocky Mountains, especially in the central plains of the Mississippi region where they form in unstable air of tropical origin and move towards north or north-east.

During the period 1970–84 there were on average 11 days a year in the British Isles on which known tornadoes caused damage to buildings, chiefly in the south and east of England. Because of their local nature most tornadoes are not recorded but, on 18 November 1981, 105 tornadoes were observed in association with a cold front associated with a small secondary depression.

The term 'tornado' has also been used for thunderstorm squalls in west Africa.

torrid zone. The torrid (or equatorial) zone is the region of the earth which lies between the Tropics of Cancer (23° 27′N) and Capricorn (23° 27′S).

tower of the winds. An octagonal tower, built in Athens in about the 2nd century BC. The tower carries on its sides the names of the winds associated with the eight compass points and also symbolic figures which represent the character of the winds. A description of the figures is given by Theophrastus in his treatise 'On winds and on weather signs' and is quoted by Sir Napier Shaw [75, p. 80], together with an estimate of the corresponding weather.

trace of rain. In rainfall measurement, a 'trace' is recorded if either:

   (i)   the measured fall is less than 0.05 mm and the observer knows that the water in the gauge is not the result of water draining from the sides of the can after the previous measurement, or

  (ii) the gauge contains no water but the observer knows that some rain or other precipitation has fallen since the previous observation.

tracer. A property, or a substance, which 'labels' a particular mass of air and so makes it possible to infer the three-dimensional flow of the mass over a period of time.

Wet-bulb potential temperature, ozone, dust and radioactive material are examples of tracers. A well defined source of the property or substance, leading to large differences in its concentration throughout the atmosphere, is normally required. See, for example, OZONE, RADIOACTIVE FALL-OUT.

trade winds. The trade winds (or 'tropical easterlies') are the winds which diverge from the subtropical high-pressure belts, centred at 30–40°N and S, towards the equator, from north-east in the northern hemisphere and south-east in the southern hemisphere.

The characteristics of the trade-wind belt vary considerably with both latitude and longitude. Marked steadiness of the winds is a feature only in latitude belts some 10–15° wide, centred on about 15°N and S (but varying somewhat with season), mainly in the eastern half of the tropical oceans. Fine weather prevails in the poleward and eastern sections of the belts, owing to the marked anticyclonic subsidence undergone by the trade-wind air. Towards the equator and the western oceanic regions of the belt, the stability of the air is decreased by added moisture; more cloudy, showery weather prevails in these regions, accentuated at times by horizontal convergence of air and the development of cyclones.

The name originated in the nautical phrase 'to blow trade', meaning to blow in a regular course or constantly in the same direction, afterward shortened to 'trade'. The word is allied to the words 'track' and 'tread'; its use in the sense of commerce was a later development.

See also ANTI-TRADES.

trajectory. A curve drawn to represent the actual path of an air particle over a finite time interval. Such a path is in general three-dimensional; normally, only the horizontal projection of the path is drawn because of insufficient knowledge of the relatively small vertical component of motion.

tramontana. A local name in the Mediterranean for a northerly wind. It is usually dry and cold.

transit. Transit of a heavenly body is said to occur when the body is on the MERIDIAN of the observer. The body is at its maximum elevation at the moment of transit.

transitive and intransitive systems. If all initial states of a physical system lead eventually to the same set of statistical properties for the system, then the system is called *transitive*. If, however, there are two or more sets of statistical properties, and some initial states lead to one set while others lead to another, the system is called *intransitive*.

If the system is strictly transitive, but is found to exhibit different average properties over successive long time intervals, then the system is called *almost transitive*. It has been suggested that the atmosphere is a transitive system on short time-scales, but that the atmosphere–ocean–cryosphere system may be almost intransitive on long time-scales.

translucidus (tr). One of the CLOUD VARIETIES. (Latin for transparent.)
'Clouds in an extensive patch, sheet or layer, the greater part of which is sufficiently translucent to reveal the position of the sun or moon.

The term applies to ALTOCUMULUS, ALTOSTRATUS, STRATOCUMULUS and STRATUS' [2, p. 22]. See also CLOUD CLASSIFICATION.

transmission coefficient. A quantity (τ), also called the 'transmissivity', which is the fraction of the radiation intensity incident on a medium which remains in the beam after passing through unit thickness of the medium. It is related to the ATTENUATION coefficient (σ) by the relation $\tau = e^{-\sigma}$.

transmissivity. An alternative for TRANSMISSION COEFFICIENT.

transmissometer. Automatic visibility-measuring equipment which employs a light source and a photoelectric cell to measure atmospheric opacity or transmissivity; certain assumptions are then made to convert the measurements to visibility. The alternative term 'visibility recorder' is also used.

Equipment currently in use in the Meteorological Office employs a selenium barrier-layer photocell at the focus of a condensing lens to measure the decrease in brightness of a beam from a small projector at a distance of 180 m. The equipment is designed to minimize the possibility of extraneous light reaching the photocell, and periodic adjustments are made to allow for the slow reduction in performance of the light source in the projector and of the photocell in the receiver.

Visibilities in the range 120–3600 m may be deduced from the strip-chart record provided.

transmittancy. The transmittancy (T) of a column of a medium of length r is the pure number $T = \tau^r$, where τ is the TRANSMISSION COEFFICIENT of a unit length of the medium, and r is measured in unit lengths.

transparency. The capacity of a medium for allowing RADIATION to pass. Of fundamental importance in meteorology are the different transparencies of the various atmospheric constituents in respect of radiation of a given wavelength, and of individual constituents in respect of radiation of different wavelengths. See also ABSORPTION.

transpiration. The process by which the liquid water contained in soil is extracted by plant roots, passed upwards through the plant and discharged as water vapour to the atmosphere. The process is necessary for the health and growth of the plant. The rate of transpiration during the day is about the same as evaporation from an open water surface in the same meteorological conditions, but is almost zero during night hours.

If the root system of a plant is never short of water the process corresponds to so-called 'potential transpiration'. The actual rate of transpiration falls significantly below the potential value only in a long spell of dry weather, a difference between the rates appearing sooner with the shallower-rooted plants. Values of potential transpiration may be calculated in a similar way to values of potential EVAPORATION. See also EVAPOTRANSPIRATION.

tree line. A term used to signify, on the hemispherical scale, the latitudinal limit of tree growth; on the regional scale, the higher altitude limit of tree growth within a region.

In the northern hemisphere the tree line divides the TUNDRA regions from those of BOREAL CLIMATE. Within the British Isles the tree line is generally at a height of 450 to 600 m above sea level. The tree line depends mainly on summer temperature, an approximate criterion for growth being a mean temperature in excess of 10 °C in at least two months. Such physical factors as exposure and drainage are also important.

tree ring. See DENDROCHRONOLOGY and DENDROCLIMATOLOGY.

triple point. That point on a pressure–temperature diagram which is the common meeting point of the liquid–vapour, solid–liquid, and solid–vapour lines for a given substance. These lines sharply define the conditions of pressure and temperature at which the changes of state from liquid to vapour, etc. occur (see Figure 46). The triple point thus represents the pressure–temperature conditions, unique for a given substance, at which the substance may be solid, liquid or gas. The triple point of water substance has the coordinates $p = 6.11$ mb, $T = +0.01$ °C.

The term is also used to describe the tip of the WARM SECTOR of a depression which is partly occluded. See, for example, COLD-OCCLUSION DEPRESSION and WARM-OCCLUSION DEPRESSION.

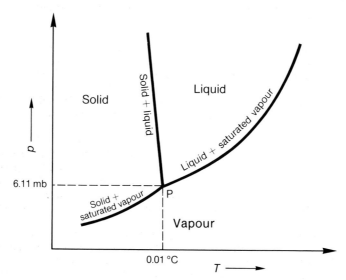

FIGURE 46   Triple point (P) for water (not to scale).

tritium. A radioactive isotope of HYDROGEN which is continuously formed in the high atmosphere by the action of COSMIC RADIATION, and has also been injected into the atmosphere by thermonuclear explosions. Tritium is used as a TRACER of atmospheric motions.

tropical air. An AIR MASS originating in low latitudes, normally in the SUBTROPICAL HIGHS at around 30–35° N and S.
Tropical air which reaches the British Isles generally originates in the Atlantic ('tropical maritime air') and is mild and moist in winter, warm and moist in summer. It is characterized on and near the south and west coasts by stability at low levels, low stratus clouds, hill fog, poor visibility, and frequent drizzle. Inland and in the east, the clouds are more broken and have a higher base, especially in summer. Occasionally, tropical air reaching the British Isles is of continental (African) origin ('tropical continental air') and is then generally associated with fine, mild (in summer, very warm) weather.

tropical climate. A type of CLIMATE which prevails in most equatorial and tropical parts of the earth and is characterized by high temperatures and high humidity throughout the year and frequent rain throughout most of the year.

tropical continental air. See TROPICAL AIR.

tropical cyclone. A CYCLONE of tropical latitudes. WMO nomenclature is as follows: 'tropical depression' when winds on the BEAUFORT SCALE are up to and including force 7; 'moderate tropical storm' for Beaufort forces 8 and 9; 'severe tropical storm' for Beaufort forces 10 and 11; and 'hurricane' for Beaufort force 12.
A tropical cyclone of moderate intensity is a 'tropical storm'; if of great intensity, a tropical cyclone in the Indian Ocean, Arabian Sea or Bay of Bengal is termed a CYCLONE, in the western Pacific a TYPHOON, in Western Australia a WILLY-WILLY and in most other tropical latitudes a HURRICANE.
Shallow tropical cyclones (depressions) are a relatively common feature of low-latitude non-desert regions and accentuate cloudiness and showeriness. They are often associated with horizontal convergence, especially at the INTERTROPICAL CONVERGENCE ZONE. Infrequently, such a depression greatly intensifies and is termed a hurricane, typhoon, or cyclone, depending on locality.
The more intense tropical cyclones are confined to fairly specific regions and seasons which, broadly, are the western sides of the great tropical oceans, beyond 5° from the equator, towards the end of the hot season or seasons. More specifically, the main oceanic regions and times are: North Atlantic (West Indies), North Pacific off the west coast of Mexico, and North Pacific westwards of 170° E (China Seas), July to October; south Indian (Madagascar to 90° E and near north-west Australia) and South Pacific (150° E eastwards to 140° W), December to March; Bay of Bengal and Arabian Sea, April to June and September to December.
The mean annual cyclone frequencies for the North Atlantic (1887–1948) are 7.3 (all intensities) and 3.5 (hurricanes); the respective mean monthly frequencies for the month of maximum frequency (September) are 2.4 and 1.3. Months of maximum frequency of cyclones of hurricane intensity are usually, though not always, those of maximum frequency of cyclones of all intensities; for example, depressions are most common in the Bay of Bengal in July but development to hurricane intensity does not occur then.
Central pressure of the more intense tropical cyclones is often about 960 mb and pressure at the periphery about 1020 mb. These values are comparable with those of a mid-latitude depression but the average tropical storm diameter is much smaller (some 800 km compared with 2400 km) and pressure gradients and winds are correspondingly greater. Very low surface pressures are sometimes attained; the lowest known value

303

reduced to mean sea level is about 870 mb, recorded at 16° 24'N, 137° 46'E (about 500 kilometres north-west of Guam) on 12 October 1979. Pressure tendencies are very large near the centre of an intense tropical cyclone.

A tropical cyclone generally moves initially towards west or north-west in the northern hemisphere and towards west or south-west in the southern hemisphere; the speed is generally about 10 knots. 'Recurvature' of the cyclone, that is change of path direction towards north-east in the northern hemisphere and towards south-east in the southern hemisphere, sometimes occurs at about latitudes 20–30°. Much more complex tracks, however, are not uncommon. After recurvature, the cyclone tends to assume the characteristics of a mid-latitude depression.

Apart from a central EYE OF STORM region, usually some 10–50 km in diameter, (though some Pacific typhoons may have larger eyes) heavy and continuous rain and multi-layer cloud occupy the central regions of the cyclone, with more showery precipitation towards the edges. Decay of a cyclone is usually rapid after passage inland.

A sufficient supply of both real and latent heat (sea surface temperature at least 27 °C) at a distance from the equator (at least some 5°) sufficient for the CORIOLIS FORCE to be active are necessary conditions for the formation of a tropical storm. The precise mechanisms which cause shallow tropical cyclones to form or, having formed, to intensify to a tropical storm or hurricane are as yet uncertain. An essentially dynamical explanation, rather than a frontal or convective explanation, is now favoured.

tropical maritime air. See TROPICAL AIR.

tropics. That region of the earth's surface lying between the Tropics of Cancer and Capricorn at 23° 27'N and S, respectively.

tropopause. The atmospheric boundary between the TROPOSPHERE and the STRATOSPHERE.

(i) The 'first tropopause' is the lowest level at which the LAPSE rate decreases to 2 °C km^{-1} or less, provided also that the average lapse rate between this level and all higher levels within 2 km does not exceed 2 °C km^{-1}.

(ii) When, above the first tropopause, the average lapse rate between any level and all higher levels within 1 km exceeds 3 °C km^{-1}, then a 'second tropopause' can occur and is defined by the criteria of (i) above. This tropopause can either be above or within the 1 km layer.

(iii) Further tropopauses may be defined similarly.

Day-to-day changes of tropopause height occur and there are also appreciable systematic seasonal changes (height greater in summer than in winter).

Detailed synoptic studies have revealed a complex, sometimes discontinuous, tropopause structure, usually in association with deep depressions, jet streams and fronts. Terms used to describe such complexities include 'tropopause funnel' (a bowl-shaped lowering of the tropopause to an unusually low level), 'folded' tropopause and 'multiple' tropopause (also termed 'laminated' or 'foliated' tropopause).

troposphere. The lower layers of the atmosphere, extending to about 16 km near the equator, 11 km in latitude 50°, 9 km near the poles, and with upper limit at the TROPOPAUSE.

The troposphere is characterized in general by a positive LAPSE rate of temperature and is the region to which precipitation and clouds (apart from certain rather rare types) are confined.

trough.   A trough (of low pressure) is a pressure feature of the SYNOPTIC CHART; it is characterized by a system of isobars which are concave towards a DEPRESSION and have maximum curvature along the axis of the trough, or 'trough line' (see Figure 47). The trough is said to be 'deep', or 'shallow', according as the maximum curvature of the isobars along the trough line is great, or small, respectively; the former corresponds to the V shape referred to in the obsolete term 'V-SHAPED DEPRESSION'. If the isobars of a depression are circular the trough line is generally taken to be the line through the centre perpendicular to the line of advance of the centre.

A FRONT is necessarily marked by a trough but the converse is not true. Those troughs which are not frontal in character are, however, also generally marked by cloudy, showery weather.

The term 'trough' is also used in meteorology to signify an elongated region of low values of any specified element, e.g. 'thickness trough', 'temperature trough'.

trowal.   A term used, mainly in Canada, to signify the projection on the earth's surface of a tongue of warm air aloft, such as may be formed during the OCCLUSION process of a depression. This feature is often found to mark a line of discontinuity of surface weather, cloud type and pressure tendency.

truncation error.   If a quantity is defined by a mathematical series it may be approximated by cutting off or truncating the series after a chosen number of terms. The error which arises by using such an approximation is termed a truncation error. The phrase is also applied to errors due to the finite length of numbers held in a digital computer.

tsunami.   A wave generated by an underwater upheaval of the earth's crust. Such a wave moves out in all directions from the point of origin and is capable of causing great destruction on arrival at a coast. Popularly called a 'tidal wave'.

***t*-test.**   See STUDENT'S *t*-TEST.

tuba (tub).   A supplementary cloud feature. (Latin for trumpet.)

'Cloud column or inverted cloud cone, protruding from a cloud base; it constitutes the cloudy manifestation of a more or less intense vortex.

This supplementary feature occurs with CUMULONIMBUS and, less often, with CUMULUS' [2, p. 24]. It is commonly known as a 'funnel cloud'. See also CLOUD CLASSIFICATION.

tundra.   Treeless lands of northern Canada and Eurasia which lie mainly just inside or just outside the Arctic Circle. In these regions the mean monthly temperature rises above freezing-point in some 2 or 3 months in summer but at a depth of about 30 cm remains below freezing-point throughout the year.

turbidity.   That property of a cloudless atmosphere which produces ATTENUATION of solar RADIATION. Measurements of atmospheric turbidity (turbidity factor) are generally concerned with the attenuation which is additional to that associated with molecular SCATTERING, the particles responsible being DUST, SMOKE, etc.

turbulence.   Whereas in LAMINAR FLOW the path of a typical fluid particle is repeatable, and usually steady and smooth, in turbulent flow the paths become irregular, with random oscillations whose detail cannot be related to the boundary conditions. A turbulent flow-field typically contains a range of motions on different time-scales and space scales (eddies) which is commonly represented as a spectrum of interacting sinusoidal oscillations (see EDDY SPECTRUM). The phenomenon is mathematically intractable, the usual method of dealing with it being to express the

FIGURE 47. Non-frontal trough extending south-west from the British Isles at 1200 GMT on
6 January 1988

statistical properties of the small-scale motions in terms of a more slowly varying 'mean flow'.

In laminar flow the transfer of momentum, heat or matter is effected by the interaction of individual fluid molecules, but in turbulent flow this transfer is effected by masses of fluid each containing very large numbers of molecules; this is a much more efficient process, so that the effective DIFFUSIVITY of a turbulent flow is usually many orders of magnitude greater than that of an equivalent laminar flow.

Most naturally occurring flows are turbulent. In meteorology, turbulent transfer effects are most commonly studied in relation to processes in the atmospheric BOUNDARY LAYER. The same ideas are, however, frequently applied to processes occurring on all time-scales and space scales: for example, the horizontal transfer properties of depressions and anticyclones. See for example, AERODYNAMIC ROUGHNESS/SMOOTHNESS, EXCHANGE COEFFICIENT, K-THEORY, MIXING LENGTH, REYNOLDS NUMBER, SIMILARITY THEORY OF TURBULENCE, STATISTICAL THEORY OF DIFFUSION, LAGRANGIAN SIMILARITY.

turbulence spectrum. An alternative for EDDY SPECTRUM.

turbulent boundary layer. A layer of air, adjacent to a fixed boundary, in which the air velocity increases from zero at the boundary to the free stream value. The REYNOLDS STRESS in such a layer greatly exceeds the molecular viscous stress.

turbulent diffusion. An alternative for EDDY DIFFUSION.

turbulent flux. An alternative for EDDY FLUX.

twilight. The pre-sunrise or post-sunset period of partial daylight. See SUNRISE, SUNSET.

Twilight is caused by the reflection and scattering of sunlight by the upper atmosphere towards an observer on the earth, when the sun is below the observer's horizon. The amount of light received progressively diminishes after sunset as the sun sinks farther below the horizon and as the sunlight is scattered by progressively higher and less dense air and is subject more and more to multiple scattering before reaching the observer.

Various stages of twilight are recognized. Thus, in the evening, 'astronomical twilight' (AT) ends when the sun's centre is 18° below the horizon, corresponding to the last trace of daylight. 'Civil twilight'(CT) ends when the sun's centre is 6° below the horizon, corresponding to the lower limit of sufficiency of daylight for outdoor activity. Intermediate is 'nautical twilight' when the sun's centre is 12° below the horizon.

While the intensity of indirect illumination from sunlight (assuming no cloud or haze) is fixed by the angular depression of the sun below the horizon, the duration of twilight (morning or evening) varies with latitude and season as shown in Table III.

TABLE III — *Duration of astronomical twilight (AT) and civil twilight (CT) in different latitudes and seasons*

	Equator		50°		60°	
	AT	CT	AT	CT	AT	CT
	Minutes					
Winter solstice	75	26	121	45	168	69
Equinox	70	24	112	37	151	48
Summer solstice	75	26	—	51	—	119

At midsummer, between the Arctic Circle and latitude 48½° N, there is a belt with no true night, twilight extending from sunset to sunrise.

The following figures are quoted from Kimball and Thiessen for the intensity of illumination, in foot-candles (1 foot-candle = 10.7639 lumens per square metre), of a horizontal surface in cloudless conditions: sun in zenith, 9600; sun on horizon, 33; sun 6° below horizon, 0.4; sun 18° below horizon, 0.0001; full moon in zenith, 0.02.

twilight arch.   The 'primary twilight arch' appears after the sun has set, as a bright, but not very sharply defined segment of reddish or yellowish light resting on the western horizon. The 'secondary twilight arch' is the slightly luminous segment near the western horizon in the last stages of TWILIGHT.

twilight flash.   An alternative for TWILIGHT GLOW.

twilight glow.   A marked intensification (also termed 'twilight flash') of the brightness of certain lines, notably the SODIUM D line at 5893 Å, in the AIRGLOW emission spectrum near the times of sunrise and sunset. Observations of the variation of intensity of the emission with solar zenith angle near these times has enabled the height distribution of sodium in the high atmosphere to be inferred.

twinkling (of stars).   See SCINTILLATION.

type.   Different distributions of atmospheric pressure are characterized by more or less definite kinds of weather. Accordingly, when a certain form of pressure distribution is seen on a chart the weather is described as being of a given type. The types are therefore defined by the shape or general trend of the isobars. Thus an 'anticyclonic' or a 'cyclonic' type denotes that an ANTICYCLONE or a DEPRESSION is the main feature of the pressure distribution; on the other hand a 'westerly' type indicates that the isobars run in more or less parallel lines over a considerable distance from west to east, having (in the northern hemisphere) the lowest pressure to the north; a 'northerly' type will have isobars running north and south with the low pressure to the east, etc.

The weather associated with each type varies with season but members of the same type nearly always have something in common; thus, the anticyclonic type usually has dry weather, the cyclonic type, wet weather; the southerly type in the northern hemisphere will in general be relatively warm and the northerly type cold. Over the British Isles, the westerly type is very persistent and often gives rise to long periods of rather unsettled weather. The easterly type gives in winter suitable conditions for severe frosts, while in summer, in at least the southern part of the British Isles, the weather is usually very warm.

typhoon.   A name of Chinese origin (meaning 'great wind') applied to the intense TROPICAL CYCLONES which occur in the western Pacific Ocean. They are of essentially the same type as the Atlantic HURRICANE and Bay of Bengal CYCLONE.

U

UDC. An abbreviation for UNIVERSAL DECIMAL CLASSIFICATION.

Ulloa's circle (or ring). An alternative for BOUGUER'S HALO.

ultraviolet radiation. ELECTROMAGNETIC RADIATION in the approximate wavelength interval from 10 to 4000 Å, i.e. in the wavelength region below visible radiation. See VISIBLE SPECTRUM.

The relatively small fraction of the total energy contained in solar RADIATION in the 'far ultraviolet' is strongly absorbed in the high atmosphere, resulting in various photochemical reactions including that of OZONE formation and a sharp cut-off, at about 2900 Å, of the solar spectrum observed at the earth's surface. The latter is, therefore, in large measure protected from the strongly actinic and biological effects which are produced by ultraviolet radiation.

Umkehr effect. An effect which is used to infer the vertical distribution of OZONE from surface measurements.

A series of measurements of the relative intensities (I), in light scattered from the zenith sky, of two selected wavelengths, one (A) strongly absorbed by ozone, the other (B) less strongly absorbed, is made when the sun is near the horizon. The ratio I_A/I_B decreases with increasing zenith angle (Z) of the sun, owing to increasing path length through the ozone, up to the point at which the 'effective' scattering height for both wavelengths (increasing with increase of Z) lies below or within the OZONE LAYER. A critical point is reached (e.g. $Z > 85°$) when most of wavelength A, but not of B, reaches the observer after being scattered from above the ozone layer; during a further increase of Z by a few degrees, the ratio I_A/I_B increases, constituting a reversal of the previous trend. The vertical distribution of ozone may on certain assumptions be inferred from the precise variation of I_A/I_B with Z.

uncinus (unc). A CLOUD SPECIES. (Latin for hooked.)
'CIRRUS often shaped like a comma, terminating at the top in a hook, or in a tuft the upper part of which is not in the form of a rounded protuberance' [2, p. 18]. See also CLOUD CLASSIFICATION.

undercooling. A seldom-used alternative for SUPERCOOLING.

undersun. A HALO PHENOMENON produced by reflection of sunlight on ice crystals in clouds. 'It appears vertically below the sun in the form of a brilliant white spot, similar to the image of the sun on a calm water surface. It is necessary to look downward to see the undersun; the phenomenon is therefore only observed from aircraft or from mountains' [2, p. 122].

undulatus (un). One of the CLOUD VARIETIES. (Latin for waved.)
'Clouds in patches, sheets or layers, showing undulations. These undulations may be observed in fairly uniform cloud layers or in clouds composed of elements, separate or merged. Sometimes a double system of undulations is in evidence' [2, p. 20]. See also CLOUD CLASSIFICATION.

Universal Decimal Classification (UDC). A method of classifying the subject matter of books and documents. It divides all knowledge into 10 classes, numbered from 0 to 9, which can each be further divided and subdivided in steps of 10 according to a decimal notation. The scheme is internationally agreed and is recommended by WMO for use in meteorological libraries. Meteorology is allotted the number 551.5 and is subdivided into main divisions as follows: 551.50 Practical meteorology (methods, data, instruments, forecasts and other applications); 551.51 Structure, mechanics and thermodynamics of the atmosphere in general; 551.52 Radiation and temperature; 551.54 Atmospheric pressure; 551.55 Wind; 551.57 Aqueous vapour and hydrometeors; 551.58 Climatology; 551.59 Various phenomena and influences.

universal time. TIME defined by the rotational motion of the earth relative to the stars which is determined from the apparent diurnal motions that reflect this rotation; because of variations in the rate of rotation, universal time is not rigorously uniform. The universal time determined directly from observations is denoted by UT0; upon correction for the variation of the meridian of the observer due to the observed motion of the poles it is denoted by UT1. When UT1 is corrected for the mean seasonal variations in the rate of rotation of the earth it is denoted by UT2. The smoothed approximation of UT2 that is distributed by most broadcast time signals is denoted by UTC (universal time coordinated) which provides a relatively uniform time measure. The numerical measure of UTC is known as sidereal time and is defined by the diurnal motion of the first point of Aries (see ZODIAC).

upbank thaw. The precedence of a THAW in a valley, sometimes by many hours, by a thaw or marked rise of temperature at mountain level in the same vicinity. The phenomenon is usually caused by the arrival at higher levels of the warm air in advance of a surface warm front; it may also be caused by the subsidence and dynamical heating of air at the higher level.

The associated inversion of the normal temperature lapse rate is a contributory cause of GLAZE.

upper cold front. The leading edge of dry air, with low wet-bulb potential temperature, that may advance ahead of a surface kata cold front (see KATAFRONT) and overrun the WARM CONVEYOR BELT. Ahead of the upper cold front the depth of warm moist air increases abruptly in association with an organized band of convection. The combination of an upper cold front with a surface cold front some way behind it is called a 'split cold front'. The existence of split cold fronts creates difficulties in attempts to apply simple classical frontal models to complex real situations.

upper-level trough. In synoptic meteorology, a line along which there exists in the upper air a TROUGH of low pressure (or the analogous contour trough on an isobaric chart), with an associated change of wind direction.

By implication, such a feature is not associated with a trough or front on the surface chart. It may, however, especially in summer, be associated with a line of convective phenomena — see INSTABILITY LINE.

upper-level (U-type) rain-band. A wide mesoscale upper-tropospheric convective rain-band associated with general frontal precipitation in the ascending parts of the WARM CONVEYOR BELT that contain upper- or middle-level convective cells, often in clusters.

upslope fog. FOG which is formed on the windward slopes of high ground by the forced uplift of stable, moist air until saturation is reached by adiabatic expansion.

upwelling. The term applied to the movement of cold water from moderate depths up to the surface, as can happen near a coast on occasions when the warmer surface water is driven from the coast by the wind.

V

vacillation. Periodic oscillation of the amplitude or shape of regular waves observed in laboratory studies of thermal convection in a fluid annulus that rotates about a vertical axis and is subjected to axisymmetric applied differential heating (i.e. when the cylindrical walls of the annulus are maintained at different constant temperatures). For vacillation to occur, the angular velocity of rotation must lie between two critical values which depend on the gravitational acceleration, the shape and dimensions of the apparatus, the physical properties of the fluid, and the nature and intensity of the applied heating.

See ROTATING FLUIDS — LABORATORY STUDIES.

valley wind. Usually, the ANABATIC WIND which blows up a valley during the day in quiet, clear conditions. The counterpart at night is the KATABATIC WIND termed 'mountain wind'.

The term 'valley wind' is sometimes also used in the same sense as RAVINE WIND.

vane. See WIND VANE.

vapour. A gas which is at a temperature below its 'critical temperature', i.e. at a temperature at which it can be liquefied by pressure alone. Water vapour is the main example of such a gas in the earth's atmosphere. Carbon dioxide and sulphur dioxide are also technically vapours at atmospheric temperatures, but in their low concentrations are not liquefied in the ranges of pressure and temperature that exist in the atmosphere.

vapour concentration. The density of water vapour (ρ_v) in a mixture of water vapour and dry air, being defined as the ratio of the mass of water vapour (m_v) to the volume (V) occupied by the mixture, i.e.

$$\rho_v = m_v / V.$$

The quantity is of the order 10 g m^{-3}. The alternative terms 'absolute humidity' and 'vapour density' applied to this quantity are not now favoured.

vapour density. An alternative for VAPOUR CONCENTRATION.

vapour pressure. In meteorology, that part of the total atmospheric pressure which is exerted by WATER VAPOUR. The vapour pressure (e) of water vapour in moist air at total pressure p and with HUMIDITY MIXING RATIO r is defined by

$$e = \frac{r}{0.62197 + r} \, p.$$

Vapour pressure is measured indirectly from wet-bulb and dry-bulb temperature readings, with the aid of a humidity slide-rule or tables (see PSYCHROMETER).

If r_w, r_i denote SATURATION mixing ratio of moist air with respect to a plane surface of pure water and ice, respectively, then the 'saturation vapour pressure' with respect to water (e'_w) and that with respect to ice (e'_i), of moist air at pressure p and temperature T are, respectively, defined by

$$e'_w = \frac{r_w}{0.62197 + r_w} \, p, \text{ and}$$

$$e'_i = \frac{r_i}{0.62197 + r_i} \, p.$$

The implied pressure dependence of e'_w and e'_i is in practice negligible; to a close approximation the saturation vapour pressure of moist air depends only on temperature, as is strictly true of pure water vapour in equilibrium with a plane (or ice) surface, in accordance with the values shown in Table IV.

In the table the values of e'_w at temperatures 0 °C and below are those which occur with respect to supercooled water. At such temperatures, the excess of e'_w over e'_i (which has a maximum of 0.27 mb at about −12 °C) is important in the formation of PRECIPITATION.

For equilibrium conditions other than those at a plane surface of pure water or ice, the values of saturation vapour pressure are changed, relative to those in the table, to a degree which is significant in the CONDENSATION process, as follows:

(i) Kelvin showed that at a given absolute temperature (T) the equilibrium vapour pressure (e'_r) over a drop of pure water of radius r is greater than the corresponding value (e'_w) appropriate to a plane surface, i.e.

$$\rho R T \log_e \frac{e'_r}{e'_w} = \frac{2\sigma}{r}$$

where ρ is the water density, R the gas constant for water vapour, and σ the surface tension of the water drop.

(ii) The saturation vapour pressure over water which contains dissolved substance is reduced, by an amount Δe, in accordance with Raoult's law, i.e.

$$\frac{\Delta e}{e'_w} = -\frac{in'}{n + in'}$$

where n and n' are the numbers of moles of water and solute, respectively, and i is a factor which varies with the concentration of the solution.

TABLE IV — *Variation of e'_i and e'_w with temperature*

T (°C)	−40	−30	−20	−10	0	+10	+20	+30
e'_i (mb)	0.13	0.38	1.03	2.60	6.11	—	—	—
e'_w (mb)	0.19	0.51	1.25	2.86	6.11	12.27	23.37	42.43

vardarac (or **vardar**). A cold northerly wind which blows through the Morava–Vardar gap in the rear of a depression and affects the Thessaloniki region of Greece; it is a type of RAVINE WIND.

variance. A statistical measure of variability, equal to the square of the STANDARD DEVIATION. The success of a forecasting technique is often measured by the percentage reduction in variance found by expressing the variance of the forecast errors as a percentage of that of the element being forecast.

variance-ratio test. See F-TEST.

vector. A quantity which requires both direction and magnitude for its complete specification. Meteorological examples are wind velocity, vorticity and pressure gradient, as opposed to such SCALAR quantities as temperature and pressure. The magnitude of a vector is termed the 'modulus' of the vector.

A vector quantity may be represented by a straight line drawn in a specific direction and of specific length. Graphical addition of two (or more) forces not acting in the same straight line or, alternatively, resolution of a single force into two (or more) 'components' is done by using the parallelogram (or polygon) law relating to vector quantities. Thus, for example, in Figure 48 $\overrightarrow{AC}$ is the vector sum or 'resultant' of $\overrightarrow{AB}$ and $\overrightarrow{BC}$, which are thus components of $\overrightarrow{AC}$. A meteorological illustration relating to this diagram is that the GEOSTROPHIC WIND at an upper isobaric level ($\overrightarrow{AC}$) may be regarded as the vector sum of the geostrophic wind at a lower isobaric level ($\overrightarrow{AB}$) and the THERMAL WIND in the isobaric layer concerned ($\overrightarrow{BC}$).

Vector mean is the vector sum divided by the number of observations. The vector sum may be obtained by graphical , or may be computed by resolution of each vector into north and east components, algebraic addition of the respective components, and recombination of the two sums into a single vector.

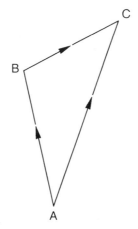

FIGURE 48.   Vector addition.

veering.   The changing of the wind direction in a clockwise rotation in either hemisphere. The opposite to backing.

velocity.   A VECTOR quantity signifying rate of change of position with time in a specified direction.

In meteorology, this term is often loosely used, e.g. in relation to motion of air or pressure systems, as being synonymous with speed, which is a SCALAR quantity.

velocity potential.   A scalar function (ϕ) which always exists in irrotational fluid motion and is defined by the equation

$$\mathbf{V} = -\nabla\phi$$

where $\mathbf{V}$ is the velocity vector. The equation implies that $\mathbf{V}$ is normal to the equipotential lines and is directed from high to low potential. See GRADIENT.

velum (vel).   (Latin for sail of ship.)

'An accessory cloud veil of great horizontal extent, close above or attached to the upper part of one or several cumuliform clouds which often pierce it.

Velum occurs principally with CUMULUS and CUMULONIMBUS' [2, p. 24]. See also CLOUD CLASSIFICATION.

vendavales. Strong, squally south-west winds in the Strait of Gibraltar and off the east coast of Spain. Associated with depressions mainly between September and March, they bring stormy weather and heavy rain.

venturi tube. A tube used in the measurement of fluid velocity, as in wind-tunnel experiments. The fluid velocity (V) is related to the measured pressure difference (Δp) at the tube entrance, relative to that at a constriction in the tube through which the fluid passes, by the equation

$$V = \left(\frac{2\Delta p}{\rho(r^2 - 1)} \right)^{1/2}$$

where ρ is the fluid density and r the ratio of the tube cross-sections at entrance and constriction.

veranillo. The two or three weeks of fine weather which break the rainy season near midsummer in tropical America.

verano. The long, dry season near midwinter in tropical America.

verification of forecasts. The process of obtaining a measure of the success of FORECASTS by relating predicted weather to actual weather. While simple comparison may serve to reveal certain features, for example a systematic bias towards optimism or pessimism, the verification process generally consists of deriving an index by one or other of a variety of methods which depend on the nature of the forecast. Such an index may then be compared with various standard indices based, for example, on a RANDOM FORECAST, a PERSISTENCE FORECAST, or a forecast based on climatic normals.

Among the many purposes of forecast verification is the evaluation of: forecasting techniques, the nature of forecasting errors, the short- or longer-period variation of the accuracy of forecasts, the economic value of forecasts, and the skill of individual forecasters.

vernier. A device for estimating fractions of a scale division when the reading to the nearest whole division is not sufficiently accurate. The vernier is a uniformly divided scale which is arranged to slide alongside the main scale of an instrument.

vertebratus (ve). One of the CLOUD VARIETIES.

'Clouds, the elements of which are arranged in a manner suggestive of vertebrae, ribs, or a fish skeleton.

This term applies mainly to CIRRUS' [2, p. 20]. See also CLOUD CLASSIFICATION.

vertical coordinate systems. In addition to the 'normal' system of CARTESIAN COORDINATES in which the vertical coordinate (z) is equal to the geometrical height above mean sea level measured along the local vertical, several other systems have been used in numerical forecast models or in studies of atmospheric dynamics, each one having certain advantages which make the equations simpler to formulate and solve under some special set of conditions. There is, however, no 'perfect' system suitable for all conditions and scales of motion. The systems that have been most widely used are as follows:

(i) *Pressure coordinates.* The vertical coordinate is p. This is probably the most widely used system. Isobaric analysis is widely preferred to constant-level analysis, and the equations for GEOSTROPHIC WIND and DIVERGENCE take a particularly simple form.

(ii) *Sigma (σ) coordinates.* The vertical coordinate is σ, given by σ = (pressure/pressure at ground level) = p/p_*. The advantage is that the lower boundary conditions for the solution of the equations of a numerical forecast model take the simple form $d\sigma/dt = 0$; i.e. the motion has no motion normal to the ground. It thus takes variations of topographic height into account very easily.

(iii) *Hybrid σ-pressure systems.* These systems employ a combination of σ and pressure coordinates in a varying proportion so that near the ground the coordinate is purely sigma and in the stratosphere purely pressure. The idea is to have the advantages of sigma coordinates at low levels without their disadvantages at high levels.

(iv) *Isentropic coordinates.* The vertical coordinate is potential temperature (θ). This system is useful for studying trajectories since large-scale flow is, to the first approximation, isentropic and takes place on surfaces of constant θ.

(v) *Log pressure coordinates.* The vertical coordinate is $\log(p_s/p)$ where p_s is a reference pressure usually at ground level or mean sea level. The coordinate is proportional to z for an isothermal atmosphere.

(vi) *HB coordinates.* These were introduced by Hoskins and Bretherton [76] who defined the vertical coordinate as

$$z' = (R\theta_0/g\kappa)\{1-(p/p_0)^\kappa\}$$

where θ_0 and p_0 are reference values. It is equal to z in an atmosphere with a dry adiabatic lapse rate throughout and, in general, $\theta dz' = \theta_0 dz$.

(vii) *Z minus H coordinates.* The vertical coordinate is $Z-H$ where H is the height of the ground above mean sea level. This system is used in the Meteorological Office mesoscale numerical forecasting model.

vertical visibility. The visual range of a dark object, of moderate angular size, viewed vertically upwards against a sky background in daylight.

This element, required in synoptic observations on occasions when the sky is obscured by fog, etc., may be measured by PILOT BALLOON and theodolite, the vertical visibility being taken as h cosec E where h is the height of the balloon and E its angular elevation at the moment of its disappearance from view.

virga (vir). A supplementary cloud feature. (Latin for rod.)

'Vertical or inclined trails of precipitation (fallstreaks) attached to the under surface of a cloud, which do not reach the earth's surface.

This supplementary feature occurs mostly with CIRROCUMULUS, ALTOCUMULUS, ALTOSTRATUS, NIMBOSTRATUS, STRATOCUMULUS, CUMULUS and CUMULONIMBUS'[2, p. 23]. See also CLOUD CLASSIFICATION.

virtual height. In radio echo sounding of the IONOSPHERE, the equivalent height of reflection of the radio waves obtained from the time delay between emission and reception of the wave on the assumption that the wave travels with the speed of light.

virtual temperature. The virtual temperature of a sample of moist air is that temperature at which completely dry air of the same total pressure would have the same density as the given sample.

The following closely approximate relations hold between absolute virtual temperature (T_v) and absolute air temperature (T):

$$T_v \approx T/\left(1 - \frac{3}{8}\frac{e}{p}\right).$$

'Virtual temperature increment' $\equiv T_v - T \approx 0.61qT \approx 0.61rT$, where e is the vapour pressure, p the total pressure, q the SPECIFIC HUMIDITY (kg kg^{-1}) and r the HUMIDITY MIXING RATIO (kg kg^{-1}).

As a further, but generally sufficient, approximation

$$T_v - T \approx q/6 \approx r/6$$

where q and r are expressed in grams per kilogram.

viscosity. That property of a fluid whereby it resists deformation. In a fluid in which different layers move with different velocities, molecular viscous forces operate so as to tend to make the velocities more uniform; for two layers a short distance apart, both parallel to the direction of flow, the viscous stress per unit area (τ) is proportional to the velocity gradient, the constant of proportionality being the coefficient of (dynamic) viscosity (μ), i.e. $\tau = \mu\partial u/\partial z$. The ratio of the dynamic viscosity to the density (ρ) of the fluid is termed the kinematic viscosity (v). Air near the earth's surface has the approximate values $\mu = 1.8 \times 10^{-5}$ kg m^{-1} s^{-1} and $v = 1.5 \times 10^{-5}$ m^2 s^{-1}.

In the atmosphere, turbulent eddies are very much more important in effecting mixing of momentum than is the molecular viscosity. By analogy with the definition of v, the vertical 'eddy viscosity' (K_M), for example, is defined by the equation

$$\tau/\rho = K_M \, \partial\bar{u}/\partial z$$

where τ is the corresponding REYNOLDS STRESS. K_M varies with height and is generally about 1 m^2 s^{-1}, i.e. some 10^5 times greater than v. See also DIFFUSIVITY.

visibility. Visibility is defined as the greatest distance at which an object of specified characteristics can be seen and identified with the unaided eye in any particular circumstances, or, in the case of night observations, could be seen and identified if the general illumination were raised to the normal daylight level. Lower visibilities are expressed in metres or yards, higher visibilities in kilometres or miles. Reports generally refer to a visibility based on all directions; where there is marked variation with direction, the lowest visibility is recorded for synoptic purposes, with an appropriate entry in a 'special phenomena' group.

'Visibility objects' by day are ideally confined to black or nearly black objects which appear against the horizon sky. Night visibility objects comprise mainly unfocused lights of moderate and known intensity at known distances. Conversion of such night observations to daylight scales involves an assumption of the different values of contrast threshold appropriate to the visibility objects by day and night. Various types of VISIBILITY METER are also used for observation by night.

Visibility, though to some extent dependent in its measurement on extraneous physiological and physical factors, is an element which is governed mainly by the atmospheric EXTINCTION COEFFICIENT associated with solid and liquid particles held in suspension in the atmosphere; the extinction is primarily caused by SCATTERING, rather than by ABSORPTION, of the light. While visibility is an element which is characteristic, in a general way, of an air mass — it is, for example, broadly much better within air masses which originate in high latitudes and move equatorward than in those which originate in low latitudes and move poleward — local variations of visibility associated with precipitation, atmospheric pollution and other factors prevent its use as a reliable air-mass indicator. See also METEOROLOGICAL OPTICAL RANGE, CONTRAST THRESHOLD OF THE EYE, KOSCHMIEDER'S LAW, OBLIQUE VISIBILITY, VERTICAL VISIBILITY.

visibility meter. A class of instruments designed to measure VISIBILITY by the determination of the EXTINCTION COEFFICIENT and/or the SCATTERING of light by the atmosphere. In an instrument measuring the latter of these properties the assumption is made that the reduction of visibility due to direct ABSORPTION is negligible.

Visibility meters are not ABSOLUTE INSTRUMENTS and require calibration in terms of the daylight visibility scale. Their use in practice tends to be limited to night observations in places where suitable night visibility points are not available.

visibility ratio. An alternative for LUMINOSITY.

visibility recorder. An alternative for TRANSMISSOMETER.

visible spectrum. That part of the ELECTROMAGNETIC RADIATION spectrum, between about 0.4 and 0.7 μm, to which the human eye is sensitive. Within the visible spectrum the wavelength increases through the range of colours violet, indigo, blue, green, yellow, orange and red. Of the total solar RADIATION intensity 41 per cent is contained within this part of the electromagnetic spectrum. See also LUMINOSITY.

volcanic dust. Material from volcanic eruptions penetrating the STRATOSPHERE affects the earth's radiation balance by reflecting and absorbing solar radiation, and by absorbing terrestrial radiation. However, if the material remains in the TROPOSPHERE, it is rapidly removed by scavenging by rainfall, and the effects of the eruption on the weather are confined to the immediate locality. The effects of a stratospheric penetration depend on the quantity and type of material injected, the area over which the material spreads, and its residence time in the stratosphere.

A variety of compilations of the dates, locations, types of materials, and degrees and duration of stratospheric penetration of past volcanic eruptions have been made, based on solar radiation measurements, historical reports, measurements of the thickness, composition and distribution of volcanic ash layers, and measurements of the sulphate content of annual deposition layers in polar ice-caps. Volcanic dust is known to have spread in the stratosphere as a veil covering more than half the surface area of the globe in some instances, and to have persisted in observable quantities for up to three years. The climatology of stratospheric winds ensures that material from tropical eruptions tends to spread to affect the extratropics within a year, but that material from extratropical eruptions remains confined to the higher latitudes of the same hemisphere before eventually settling out. Volcanic dust veils in the stratosphere are associated with atmospheric optical effects such as BISHOP'S RING.

The major volcanic eruptions of the past appear to have been followed by widespread intervals of cooler climate lasting from one to several years, though the association remains open to doubt. Recent volcanic eruptions have been the object of more detailed research, and measurements showed abnormally warm conditions in the tropical stratosphere following the eruptions of Agung (Indonesia) in 1963 and El Chichon (Mexico) in 1982. The latter eruption disrupted the global observing system by its marked effects on the radiation received by satellite-borne instrumentation designed to monitor surface, tropospheric and stratospheric temperatures.

volume fraction. See GASEOUS CONCENTRATION, UNITS OF

von Kármán's constant. A non-dimensional quantity (k) in the equation which defines the nature of the wind structure in the low atmosphere in adiabatic lapse rate conditions — see LOGARITHMIC VELOCITY PROFILE. k is found experimentally to have a value close to 0.4.

vortex. A fluid flow which possesses VORTICITY. A 'vortex line' is one drawn from point to point of a fluid such that it coincides at all points with the instantaneous direction of the axis of rotation of the fluid. A 'vortex tube' is the surface which contains all the vortex lines which intersect a closed small curve within the fluid. A 'vortex filament' is the fluid contained within a vortex tube. A 'vortex sheet' is a surface of discontinuity of velocity which separates two adjacent streams of a fluid and on which the vorticity is infinite.

vorticity. The vorticity at a point in a fluid is a vector which is twice the local rate of rotation of a fluid element. The component of the vorticity in any direction is the CIRCULATION per unit area of the fluid in a plane normal to that direction. The dimension is T^{-1}.

Vorticity is a three-dimensional property of the field of motion of a fluid. In large-scale motion in the atmosphere the vorticity component of chief significance is that which occurs in the horizontal plane (i.e. rotation about the vertical axis); the other components are, however, significant in some dynamical problems.

In vector notation the vorticity of a velocity vector ($\mathbf{V}$) is written as curl $\mathbf{V}$ or rot $\mathbf{V}$ or $\nabla \times \mathbf{V}$. In CARTESIAN COORDINATES the vertical (z) component of 'relative vorticity' (i.e. rotation in a horizontal plane, evaluated from winds measured relative to the rotating earth) is

$$\text{vorticity}_z = \zeta = \left(\frac{\partial v}{\partial x} - \frac{\partial u}{\partial y} \right).$$

Similarly, $\quad \text{vorticity}_x = \xi = \left(\frac{\partial w}{\partial y} - \frac{\partial v}{\partial z} \right)$

and $\quad \text{vorticity}_y = \eta = \left(\frac{\partial v}{\partial z} - \frac{\partial w}{\partial x} \right).$

The expression for the vertical component of vorticity in terms of velocity (V), radius of curvature of the streamlines (r), and differentiation along the normal to the streamlines ($\partial V / \partial n$) (see NATURAL COORDINATES), namely

$$\zeta = \frac{V}{r} - \frac{\partial V}{\partial n},$$

shows that the vertical component of vorticity may be regarded as the sum of components due to curvature (V/r) of horizontal flow and to horizontal wind shear ($-\partial V / \partial n$). Thus, for example, the contribution to vorticity about the vertical axis made by the horizontal wind shear associated with a westerly jet stream in the northern hemisphere is strongly cyclonic poleward of the jet axis and anticyclonic equatorward of the axis.

In 'solid rotation' of angular velocity ω the vorticity is 2ω. In latitude ϕ, where the ANGULAR VELOCITY OF THE EARTH about the vertical axis is $\Omega \sin\phi$, the earth has a vorticity about this axis of $2\Omega \sin\phi$ which is cyclonic in sense. Air partakes of the vorticity of the earth appropriate to its latitude, in addition to any relative vorticity it may possess. Thus, in latitude ϕ, 'absolute vorticity$_z$' is given by

$$\zeta_a = \zeta + 2\Omega \sin\phi.$$

Relative vorticity in a cyclonic sense is reckoned positive, in an anticyclonic sense negative. See also GEOSTROPHIC VORTICITY.

vorticity equation. The vorticity equation as used in meteorology relates the rate of change of the vertical component of VORTICITY to the horizontal DIVERGENCE. It is derived by eliminating geopotential (or pressure) from the equations of motion. In PRESSURE COORDINATES the vorticity equation can be written

318

$$\frac{d}{dt}(\zeta+f) = -(\zeta+f)\,\mathrm{div}_p\,V + \left(\frac{\partial\omega}{\partial y}\frac{\partial u}{\partial p} - \frac{\partial\omega}{\partial x}\frac{\partial v}{\partial p}\right).$$

$$\qquad\qquad\quad\text{(a)}\qquad\qquad\qquad\quad\text{(b)}$$

In CARTESIAN COORDINATES it is

$$\frac{d}{dt}(\zeta+f) = -(\zeta+f)\,\mathrm{div}_H\,V + \left(\frac{\partial w}{\partial y}\frac{\partial u}{\partial z} - \frac{\partial w}{\partial x}\frac{\partial v}{\partial z}\right) + \left(\frac{\partial p}{\partial x}\frac{\partial\alpha}{\partial y} - \frac{\partial p}{\partial y}\frac{\partial\alpha}{\partial x}\right)$$

$$\qquad\qquad\text{(a)}\qquad\qquad\qquad\quad\text{(b)}\qquad\qquad\qquad\quad\text{(c)}$$

where ζ = vertical component of vorticity = $\dfrac{\partial v}{\partial x} - \dfrac{\partial u}{\partial y}$,

f = Coriolis parameter,
ω = dp/dt (equivalent of vertical velocity in pressure coordinates),
w = vertical velocity,
α = specific volume, and
p = pressure.

The first term (a) on the right-hand side of the equation is the dominant one in large-scale atmospheric motion. The second term (b) is sometimes known as the 'twisting term' and represents the transformation of vorticity from the horizontal to the vertical component. It is believed to be important within smaller-scale motions as in fronts. The third term (c) in the second form represents the direct generation of vorticity by horizontal density and pressure gradients and is usually unimportant.

vorticity theorem. The vorticity theorem is derived from the CIRCULATION THEOREM OF BJERKNES. It relates the local generation of VORTICITY to the local baroclinicity (see BAROCLINIC) of the atmosphere and may be written:

$$\mathrm{curl}\,(\rho\dot{V} - \rho C) = \nabla\rho \times \nabla(-\Phi),$$

where ρ = density,
$\dot{V}$ = acceleration of the air,
C = vector representing apparent deviating force per unit volume due to earth's rotation, and
Φ = Geopotential.

Direct application of the vorticity theorem to large-scale motions is limited because the term arising from the deviating force is usually in approximate balance with the term arising from the density gradients.

V-shaped depression. An obsolete term for a sharply defined TROUGH of low pressure, with the isobars in the form of a V.

W

warm anticyclone. See ANTICYCLONE.

warm conveyor belt. A narrow air stream conveying large quantities of heat, moisture, and westerly momentum polewards and upwards in a major confluence zone at the leading edge of a marked atmospheric trough. Warm air is drawn into the zone from the convective boundary layer in low latitudes, rises to produce cloud and precipitation and eventually produces a deck of upper-tropospheric cirrus which decays ahead of the frontal system in high latitudes. It sometimes has the characteristics of a LOW-LEVEL JET.

warm front. A FRONT whose movement is such that the warmer AIR MASS is replacing the colder.

As a warm front approaches, temperature and dew-point within the cold air gradually rise and pressure falls at an increasing rate. Precipitation usually occurs within a wide belt some 400 km in advance of the front. Passage of the front is usually marked by a steadying of the barometer, a discontinuous rise of temperature and dew-point, a veer of wind (in the northern hemisphere), and by a cessation or near cessation of precipitation. Substantial lanes of clear air separating cloud layers are found in all except a small minority of warm fronts.

The average slope of a warm-frontal surface is about 1 in 150. A warm front moves, on average, at a speed some two-thirds of the component of the geostrophic wind normal to the front and measured at it.

warm-front wave. A secondary WAVE DEPRESSION which forms on an extended WARM FRONT at a point usually a considerable distance, some 1500 km, from the parent depression; after formation, it moves quickly east or south-east away from the parent depression. This type of depression, which is not common, seldom becomes deep but is responsible for a considerable spread and intensification of the warm-front precipitation. Formation may be aided either by a frontal distortion produced by a range of hills or by movement towards a col.

warm-occlusion depression. A SECONDARY DEPRESSION which forms at the point (the 'triple point') where a cold and warm front unite to form a warm OCCLUSION. Such a secondary generally moves quickly away from the primary depression and deepens, though seldom to a marked extent, at the expense of the primary.

warm pocket. A term applied in upper-air analysis to a closed centre of high pressure on a FRONTAL CONTOUR CHART. Such a region on the chart indicates the isolation of warm air at low levels from the main body of warm air which is seen on the chart, usually at lower latitudes.

warm rain. The term 'warm rain' is sometimes applied to rain which falls from clouds whose tops do not reach the freezing level. Such rain is initiated by the coalescence process — see PRECIPITATION.

warm ridge. A pressure RIDGE (or ridge on an isobaric contour chart) in which temperature is generally higher than in adjacent areas.

warm sector. In the early stages of the life history of at least the majority of the DEPRESSIONS of temperate latitudes, and of the more important SECONDARY DEPRESSIONS, there is a surface sector of warm air, which disappears as the system deepens and the cold front catches up the warm front (see OCCLUSION). The warm sector is usually composed of tropical air, sometimes of polar maritime air.

wash-out. The removal of solid material from the air, and its deposition on the earth's surface, due to capture by falling PRECIPITATION elements. See also FALL-OUT.

water. The oxide of hydrogen of chemical formula H_2O. Its maximum DENSITY of 999.97 $kg\,m^{-3}$ occurs at 4 °C. Its THERMAL CONDUCTIVITY at 0 °C and 20 °C is respectively, 0.5527 and 0.5987 $W\,m^{-1}\,°C^{-1}$. Its physical properties are slightly modified by the small but variable amounts of impurities, due mainly to dissolved salts, which occur in natural water. It constitutes, as liquid or ice, 70.8 per cent of the earth's total surface — see EARTH.

Water plays a fundamental part in the energy balance of the earth–atmosphere system, notably because of the LATENT HEAT exchanges involved in its widespread changes of state. See also ICE, WATER VAPOUR, SALINITY.

Water (-droplet) cloud. A cloud which is composed entirely of water droplets, either in the supercooled state or at temperatures above 0 °C, as opposed to ice crystals. The CLOUD GENERA ALTOCUMULUS, STRATUS and CUMULUS are normally water clouds.

watershed. In physical geography, the line separating the head streams which are tributaries to different river systems or basins, i.e. the line enclosing a CATCHMENT AREA.

water sky. Term applied, mainly in polar regions, to the dark appearance presented by the underside of a cloud layer which lies above a water-covered region relative, in particular, to that of a cloud layer above a snow- or ice-covered region (see ICEBLINK). Such an appearance is often useful in indicating the presence of open water which is not itself then visible.

water smoke. An alternative for ARCTIC SEA SMOKE.

waterspout. A funnel-shaped TORNADO cloud which extends from the surface of a sea or inland water to the base of a cumulonimbus cloud.

A cone-like point of cloud descends from the cumulonimbus base to the agitated sea below and assumes the appearance of a column of water, the diameter of which may vary between a few tens and a few hundreds of feet. The duration of a waterspout ranges up to about half an hour, during which time the column may be appreciably bent by vertical wind shear. A circular and violent circulation of air is caused near the waterspout with an associated confused sea. The phenomenon is more common in the tropics and subtropics than in higher latitudes.

water table. The depth at which the soil is persistently saturated with water. Such depth generally varies appreciably with the wetness of the season.

water vapour. Water substance in the vapour form is, meteorologically, a very important constituent of the atmosphere and is also the most variable in space and time.

Supplied to the atmosphere by EVAPORATION and SUBLIMATION at the earth's surface, water vapour has a concentration which decreases fairly steadily with height from a mass ratio to dry air of about 1×10^{-2} near the ground to about 2×10^{-6} in the lower stratosphere. There are some recent indications of an increase of concentration with increase of height between about 20 and 35 km. Partial dissociation of water vapour by ultraviolet radiation into HYDROGEN atoms and HYDROXYL molecules may be effective above about 60 km.

The meteorological importance of water vapour derives from the part it plays in forming cloud and precipitation elements, in controlling the long-wave radiation balance of the atmosphere, in determining atmospheric stability, and in affecting the heat balance conditions of the earth–atmosphere system by the powerful absorption of heat in the course of evaporation and sublimation from liquid water and ice, and by the eventual release of the stored latent, or 'hidden', heat which is involved in the reverse processes.

The amount of water vapour held in the atmosphere is specified by various 'humidity elements' which include VAPOUR PRESSURE, HUMIDITY MIXING RATIO, RELATIVE HUMIDITY, VAPOUR CONCENTRATION, DEW-POINT, FROST-POINT and WET-BULB TEMPERATURE. Different types of HYGROMETER are commonly used to measure the humidity at different atmospheric levels.

Water vapour is by far the most strongly absorbing constituent of the atmosphere and has a wide range of absorption bands over a range of wavelengths extending from the near infra-red upwards — See Figure 1. A conspicuous feature of the absorption spectrum, of particular importance at terrestrial radiation temperatures, is the region between about 5.5 and $7\mu m$ in which water vapour is almost opaque — see also ATMOSPHERIC WINDOW.

See also SPECIFIC HEAT and LATENT HEAT for values referring to water vapour.

watt (W).   The unit of power in SI UNITS. It is the rate at which energy is transformed into heat in a lamp using 1 ampere at 1 volt.

1 watt $=$ 1 JOULE per second.

Units derived from the watt are used by meteorologists for the measurement of the intensity of radiation, e.g. 1 mW cm^{-2}.

wave clouds.   Clouds which form in the crests of MOUNTAIN WAVES.

wave depression.   A DEPRESSION which forms at the tip of a wave-like distortion of a FRONT. Most of the depressions of middle and high latitudes are of this type.

wave motion.   An oscillatory movement of the particles of a medium as the result of which 'waves' are propagated through the medium. If the particle movement is perpendicular to the direction of wave propagation, the waves are 'transverse'; if the particle movement is a rhythmic advance and retreat along the direction of wave propagation, the waves are 'longitudinal'. Energy but not, in general, matter is propagated with the waves. In ELECTROMAGNETIC RADIATION, periodic disturbances of electric and magnetic fields, but not movement of particles, are involved; such waves can be propagated through space or through a medium.

The simplest type of wave is the 'simple harmonic wave', as represented by a sine curve. The main characteristics of waves are the amplitude (a, half the distance between the extremes of the oscillations), the wavelength (λ, distance between successive maxima), the period (τ, time interval between successive crests passing the same point), and the frequency (f, number of complete oscillations per second). In a TIME SERIES the wavelength and period are identical. The speed of propagation (v) of the wave pattern (termed the PHASE VELOCITY) is related to τ, λ and f by $v = \lambda/\tau$ and $v = f\lambda$, while energy is propagated with the GROUP VELOCITY. The wave number (k) is alternatively defined as $k = 1/\lambda$ or $k = 2\pi/\lambda$) (i.e. number of waves per unit distance, or 2π times this quantity).

A large variety of types of wave motion, or of quasi-wave motion, occurs in the atmosphere. See, for example, SOUND WAVES, SHOCK WAVES, ATMOSPHERIC TIDES, GRAVITY WAVE, SHEAR WAVE, INERTIA WAVE, ROSSBY WAVE, BAROTROPIC WAVE, BAROCLINIC WAVE, LEE WAVES, WAVE DEPRESSION.

wave number. See WAVE MOTION.

wave recorder. An instrument for recording OCEAN WAVES. The record obtained from such an instrument shows the variation of the height of the sea surface above a fixed point with time.

weakening. See INTENSIFICATION.

weather. The changing atmospheric conditions, more especially as they affect man, which, in synthesis, constitute the CLIMATE of a region.
Weather in its wider sense is the study pursued in SYNOPTIC METEOROLOGY. In this branch of meteorology, however, the term itself is used in a more limited sense to denote the state of the sky and the occurrence of precipitation or of mist or fog. Codes of 'present weather' and of 'past weather' are two of the codes used in synoptic meteorology.
A concise system of notation of weather was introduced by Admiral Beaufort (see BEAUFORT NOTATION). A concise, international method of recording weather and optical phenomena by means of symbols is given in *International cloud atlas* [2].
In order to facilitate the representation and entry of METEORS in meteorological documents, symbols have been assigned to most of them. A table of these symbols can be found in the *Observer's handbook* [1, pp. 72–75].
It is possible to provide information concerning the character (intermittent or continuous) and intensity (slight, moderate or heavy) of precipitation by certain arrangements of the basic symbols. The following table, established for rain, illustrates various arrangements which may be used for this purpose.

Intensity	Character	
	intermittent	continuous
slight	•	••
moderate	⁝	⸭
heavy (dense)	⁝	⸬

Combinations of two basic symbols of meteors may be used to indicate the occurrence of mixed precipitation or the occurrence of a thunderstorm accompanied by precipitation or duststorm or sandstorm. For example, the symbol ⁎ or ⁎ denotes a mixture of falling raindrops and snowflakes; the symbol ↺ indicates thunderstorm with rain at the place of observation.
In addition to the basic symbols, several auxiliary symbols have been established to provide information concerning the showery character of precipitation and also the variation with time of various meteors and their location with respect to the station. These symbols can be found in the *International cloud atlas* [2].
Useful supplementary information about meteors can thus be given by combining the above auxiliary symbols with one, or sometimes two, basic symbols. For example, the symbol ≡| denotes fog which has become thinner during the preceding hour; the symbol ∇̇ indicates shower(s) of rain during the preceding hour, but not at the time of observation.

Weather Centre. In the Meteorological Office, an office set up to provide a service to commerce and industry and to supply meteorological information and advice to the general public through radio, television, the Press, by telephone and personal inquiry.

weather lore. Empirical weather forecasting rules, world-wide in origin, many of which are expressed in rhyme. They include rules based on the influence of the moon and tides, the appearance of plants and trees, the behaviour of animals, the weather prevailing on specified key dates and the colour and appearance of the sky. Comparison of the various rules reveals many contradictions, while tests of statistical significance lend no support except to certain of the short-period forecasting rules based on local observation of sky and wind, etc. See *Weather lore* compiled by R. Inwards [77].

weather map. A chart of a geographical area on which selected meteorological elements observed at a particular time at various points over the area are plotted in symbolic code; the positions of mean-sea-level isobars and surface fronts (also, on occasion, of other features, for example isallobars) are subsequently drawn.

The elements usually plotted on the weather map, which is also termed 'synoptic chart' or 'surface chart', are: atmospheric pressure, reduced to mean sea level; barometric characteristic and tendency; wind direction and mean speed; air temperature; dew-point; visibility; 'present weather'; 'past weather'; type, amount and height of clouds. In the case of ship observations, sea temperature, the direction and amplitude of the swell and the direction and speed of movement of the ship are also plotted. See SYNOPTIC METEOROLOGY.

weather minima. Criteria, laid down by airline operators, that govern whether a particular aircraft under the control of a particular pilot may take off from or land at a particular civil aerodrome. They comprise RUNWAY VISUAL RANGE and CRITICAL HEIGHT for landing, and RUNWAY VISUAL RANGE and CLOUD CEILING for take-off. A variety of weather minima may be simultaneously in force at the same aerodrome.

weather report. Statement of the values of meteorological elements observed at a specified place and time. The elements included depend upon the purpose for which the report is required.

It is a record of an OBSERVATION, not a FORECAST.

weather routeing. See SHIP ROUTEING.

weather window. An interlude, in a period of predominantly inclement weather, during which a weather-sensitive operation may be carried out successfully. Such operations range from towing oil rigs, and major structural engineering work in exposed situations, down to minor jobs of outdoor house-painting. Accurate forecasting of weather windows is often of great economic importance. What particular weather elements are involved depends on the type of operation.

weber (Wb). The unit of magnetic flux in SI UNITS.

wedge. In synoptic meteorology, an alternative for RIDGE (of high pressure); the term is mainly applied to a relatively fast-moving ridge on a surface synoptic chart.

weighted (moving) averages. See FILTERING, WEIGHTS.

weights. A term used for coefficients introduced when averaging estimates of the same kind to give more weight to the more reliable estimates. If $x_1, ..., x_n$ are the estimates and $w_1, ..., w_n$ the weights, then

$$\frac{w_1x_1 + w_2x_2, ..., + w_nx_n}{w_1 + w_2 + , ..., + w_n}$$

is the weighted average. In this application the weights are generally all positive, and are often adjusted for computational convenience so that their sum is unity.

wet adiabatic. An alternative for 'saturated adiabatic'. See ADIABATIC.

wet air. A term used to define the condition when objects become wet even when rain is not falling. It occurs when a warm, saturated or practically saturated air mass replaces a cold, dry air mass and is denoted by the letter 'e' in the BEAUFORT NOTATION.

wet-bulb depression. In a dry- and wet-bulb PSYCHROMETER, the amount by which the wet-bulb reading is below that of the dry-bulb.

wet-bulb potential temperature. The wet-bulb potential temperature (θ_w) at any level is obtained on an AEROLOGICAL DIAGRAM as that temperature at which the saturated ADIABATIC through the WET-BULB TEMPERATURE at the level concerned intersects the 1000 mb isobar.

θ_w is, for practical purposes, conservative for such processes as evaporation or condensation and for both dry adiabatic and saturated adiabatic temperature changes; it is therefore a useful property in AIR-MASS ANALYSIS.

wet-bulb temperature. That temperature (T_w) at which pure water must be evaporated into a given sample of air, adiabatically and at constant pressure, in order to saturate the air at temperature T_w under steady-state conditions. The temperature recorded by the wet-bulb of a psychrometer may not exactly accord with this definition. See also THERMODYNAMIC TEMPERATURES, PSYCHROMETER.

wet day. See RAIN DAY AND WET DAY.

wetness recorder. An instrument used in plant pathology to detect and record the amount of moisture deposited on foliage.

wet season. An alternative for RAINY SEASON.

whirling psychrometer. A PSYCHROMETER in which the thermometers are mounted on a frame which is rapidly rotated by hand in order to provide the required ventilation of the bulbs. It is also termed a 'sling psychrometer'.

whirlwind. A small revolving storm of wind in which the air whirls round a core of low pressure. Whirlwinds sometimes extend upwards to a height of many hundreds of metres and cause DUST WHIRLS when formed over a desert.

whistlers. A type of disturbance heard on a suitable radio receiver. It comprises a succession of whistles which progressively become fainter and take longer to fall through the audio range of frequencies.

The disturbance, strongest at about 55° geomagnetic latitude, originates in a burst of electromagnetic waves in the audio-frequency range, emitted by a lightning discharge or produced artificially. The interval between successive whistles corresponds to the time taken by the waves to travel along the lines of the earth's magnetic field to the geomagnetically conjugate point in the opposite hemisphere, and, on being reflected there, to return along the same path to the radio receiver. The arrival of the waves produces a whistle because the higher frequencies travel faster through the IONOSPHERE than do the lower. The initiating discharge is close either to the recceiver or to the conjugate point.

325

Detailed study of the dispersion of frequencies in whistlers has led to inferences concerning the geomagnetic field and state of ionization above the ionosphere. See also DAWN CHORUS, SFERICS FIX,

white-out. A term applied to that condition in which the contours and natural landmarks in a snow-covered region become indistinguishable. The associated meteorological conditions appear to be a uniform layer of relatively low cloud; under such conditions the light which reaches the surface arrives in nearly equal measure from all directions, with a resulting absence of shadows. The term is frequently misapplied, in popular usage, to BLIZZARD.

Wien's (displacement) law. See RADIATION.

willy-willy. The name given in Western Australia to a severe TROPICAL CYCLONE. It is also sometimes applied to a DUST WHIRL.

wilting point. The point at which the soil contains so little water (measured as per cent of dry soil) that it is unable to supply it at a rate sufficient to prevent permanent wilting of plants. It varies with the type and structure of the soil, being about 3 per cent for light sand and 20 per cent for heavy clay, and corresponds to a CAPILLARY POTENTIAL of about 150 metres of water (pF = 4.2). See SOIL MOISTURE.

wind. The (horizontal) movement of air relative to the rotating surface of the earth; the vertical component of air movement, generally much the smaller, is identified as such, where appropriate.

In meteorology, the specified wind direction is that, relative to true (geographic) north, from which the wind blows. The converse practice has, however, been used with certain high-level winds, e.g. those inferred from radio measurement of METEOR trails; more commonly, doubt is now removed in such cases by reference, for example, to a 'westward' wind, i.e. an east wind in the normal meteorological sense. The wind direction is generally specified as a bearing in degrees clockwise from true north; the compass point direction (8, 16 or 32 points according to the accuracy required) is also used. By international agreement (in 1956) the KNOT is the meteorological unit of wind speed. The specifications of the Beaufort forces are given under BEAUFORT SCALE, together with the dynamic pressure exerted on a flat disc. The relationships between the knot and alternative speed units are:

$$1 \text{ knot} = 0.514 \text{ m s}^{-1} = 1.152 \text{ mile h}^{-1} = 1.853 \text{ km h}^{-1} = 1.689 \text{ ft s}^{-1}.$$

Surface wind velocity is normally measured by some form of ANEMOMETER or ANEMOGRAPH, and WIND VANE. Appropriate EXPOSURE of the instrument to ensure reasonable comparability of observations at different stations is especially difficult in wind measurement; 'surface wind' in synoptic reports refers to an 'equivalent' height of 10 metres. Upper-level winds are normally measured by PILOT BALLOON or RADAR WIND techniques, or by using RADIO NAVIGATION AIDS. They may be inferred from measurements of the angular velocity of clouds, excluding certain (wave) types of cloud which do not move with the wind.

Wind velocity is intimately related to the pressure distribution in extratropical regions. In large-scale motion the relationship gives rise to BUYS BALLOT'S LAW, the GEOSTROPHIC WIND and the GRADIENT WIND, while, for example, changing pressure distribution in space or time is related to the AGEOSTROPHIC WIND. Height changes of wind velocity are generally considered in terms of the THERMAL WIND. On the more local scale the pressure–wind relationship gives rise to such winds as LAND- and SEA-BREEZES, the KATABATIC WIND and the ANABATIC WIND.

Diurnal variation of atmospheric TURBULENCE is associated with variations of surface wind velocity due to stronger vertical mixing of air at lower levels by day than by night (appreciable increase of speed and very slight veer by day). Associated with the systematic diurnal variation of pressure is a variation of wind velocity of an amplitude which is too small at the surface to be revealed except by averaging over a long period but which increases with height to such an extent as to be a prominent feature of winds at very high levels — see ATMOSPHERIC TIDES.

wind-break.   A term sometimes used in such a sense as to include both natural and artificial barriers to wind flow which provide shelter to animals, crops, etc. More usually the term is now restricted to artificial barriers (palings etc.). See also SHELTER-BELT.

The degree of shelter from wind afforded by a barrier depends, among other factors, on the height, lateral extent, and permeability of the barrier. Nevertheless, it is now generally accepted that, whatever the type of barrier employed, the shelter zone for a barrier of height h reaches to a distance of about $30h$ in the lee of the barrier, and that for significant wind-speed reductions of 20 per cent or more only distances of up to $20h$ should be considered. Strongly eddying motion is a feature of the airflow to a distance of about $15h$ downwind from a dense barrier.

Other physical effects of wind-breaks include alterations of air temperature and humidity, soil temperature and moisture, and evaporation rate over the region affected by the presence of the barrier.

wind-chill.   The ability of strong winds combined with low air temperatures to cool warm-blooded animals, heated buildings, etc. much more than would low air-temperatures alone. The idea of wind-chill is relevant only to objects that need to be maintained at a constant temperature higher than that of their surroundings.

Various empirical formulae have been devised for calculating the rate of heat loss from idealized human bodies using as independent variables wind speed, ambient air temperature, skin temperature, rate of heat production, clothing resistance, etc; the most important are those of Siple and Passel and of Steadman. A critical account with references is provided by Dixon and Prior [78]].

The formulae may be used to derive a 'wind-chill equivalent temperature' defined as the temperature of still air that would produce the same chilling effect as that of the given cold wind. The Meteorological Office began, in the mid 1980s, to quote equivalent temperatures based on the work of Steadman.

wind profiling.   An instrumental technique, using ground-based DOPPLER RADAR, for measuring the variation with height — the vertical profile — of both horizontal and vertical components of the wind field. Radars used for profiling operate on greater wavelengths (metres) than those used for detecting precipitation (centimetres), and obtain their echoes from variations in the refractive index of clear air, these variations being themselves the product of quasi-isotropic irregularities generated by small-scale turbulence which moves with the wind field. The radar normally looks simultaneously in three different directions; vertically, and at about 15° from the zenith in two orthogonal directions. Such a radar profiler may also be co-located with a multi-channel vertically pointing microwave RADIOMETER which yields profiles of temperature and humidity in the sane way as do radiometers mounted on satellites. (See REMOTE SOUNDING OF THE ATMOSPHERE.) A radar capable of making measurements throughout the mesosphere, stratosphere and troposphere is often referred to as an 'MST' radar.

wind rose.   One of a class of diagrams illustrating, for a particular place and extended period, the relative frequencies of wind direction and, in general also, of wind speed. A common form is illustrated in Figure 49; the length of each symbol is proportional to its frequency of occurrence and the radius of the circle is proportional to the frequency of calms which is also written within the circle.

A wind rose may be adapted to demonstrate the relationship between wind velocity and other meteorological variables.

Key

0   5   10   15   20
% of time

4–10 11–21 22–33 ⩾ 34
Speed in knots

FIGURE 49.   Wind rose for Gatwick Airport (51° 09′W, 0° 11′W) for January for the period 1970–79 showing annual frequency of wind direction and velocity. The number in the circle shows the percentage frequency of calm or light and variable winds. The other frequencies are measured from the circle edge according to the adjacent scale.

wind shear.   The rate of change of the vector wind (V) with distance (n) in a specified direction normal to the wind direction, or $\partial V/\partial n$. If the vector wind is horizontal then with the usual notation the 'horizontal shear' is the vector with components $\partial u/\partial y$, $\partial v/\partial x$, and the 'vertical shear' is the vector with components $\partial u/\partial z$, $\partial v/\partial z$; these reduce to $\partial V/\partial y$ and $\partial V/\partial z$ respectively when the wind vector is directed along the x-axis.

Shear is an important property of the wind field. For example, shear at a well marked frontal surface may produce SHEAR INSTABILITY, while horizontal shear is closely associated with the vertical component of VORTICITY. The integral of wind shear through a vertical interval is the vector difference between the (horizontal) winds at the top and bottom of the interval and so is identical with the THERMAL WIND if the motion is geostrophic. See also JET STREAM.

wind vane.   A device for indicating or recording the direction from which the wind is blowing. It usually consists of a horizontal arm carrying at one end a fin, either a vertical flat plate with its edge to the wind or an aerofoil, and at the other end a balance weight which also serves as a pointer; the arm is carried on a vertical spindle mounted on bearings which allow it to turn freely in the wind.

For indication only, the position of the vane can be estimated with reference to a fixed cross below the vane showing the four cardinal points. For more accurate indication, the spindle may be coupled by a suitable direction-transmitting rod to a pointer and dial situated at a convenient level below the vane. For remote indication and recording, the vane spindle is coupled to some form of electrical transmitter by means of which the movements of the vane are reproduced at a distance, and displayed as dial and pointer indications, or pen recordings on a chart.

windward.   Windward of a point signifies the 'upwind' direction from the point, e.g. westward in the case of a west wind.

winter.   See SEASONS.

328

WMO.   Abbreviation for WORLD METEOROLOGICAL ORGANIZATION.

World Meteorological Centre (WMC).   A centre of the GLOBAL DATA-PROCESSING SYSTEM, which has the primary purpose of issuing meteorological analyses and prognoses on a global scale. There are three WMCs; Washington (USA), Moscow (USSR) and Melbourne (Australia).

World Meteorological Organization.   The World Meteorological Organization (WMO) is a specialized agency of the United Nations encompassing the field of meteorology. It replaced the International Meteorological Organization in 1951. The WMO comprises over 160 States and Territories and has a permanent Secretariat in Geneva.

The purposes of the WMO are, *inter alia*, to facilitate world-wide co-operation in the establishment of networks of meteorological observation stations and to promote the development of centres charged with the provision of meteorological services; to promote the rapid exchange of weather information and the standardization of meteorological observations and their publication; to further the application of meteorology to human activities and to encourage research and training in meteorology.

World Weather Watch.   A world-wide meteorological system composed of the national facilities and services provided by individual Members, co-ordinated and in some cases supported by the WMO and other international organizations. Its primary purpose is to ensure that all Members obtain the meteorological information they require both for operational work and for research. Its essential elements are: the Global Observing System, the Global Data-Processing System, the Global Telecommunication System, and research and training programmes.

WWW.   Abbreviation for WORLD WEATHER WATCH.

X, Y, Z

xenon. One of the INERT GASES, comprising 8.0×10^{-8} and 3.6×10^{-7} parts per part of dry air by volume and weight, respectively. Its molecular weight is 131.3.

x-rays. ELECTROMAGNETIC RADIATION in the approximate band of wavelengths from 0.1 to 10 Å. The x-rays which are contained in the solar radiation incident on the high atmosphere are responsible for an appreciable part of the IONIZATION of the region.

year. The year relevant to meteorology is that describing the full cycle of the seasons, that is, the time taken by the sun to transverse the ECLIPTIC completely, or *tropical year*. It is equal to 365.242194 mean solar days.

zenith. The point of the sky in the vertical produced upwards from the observer. The word is now commonly used to denote a more-or-less extensive stretch of sky immediately overhead.

zenith distance. The zenith distance of a body is the angle between the body and the ZENITH, as observed at a particular point of observation.

zenith, magnetic. The direction indicated by the upper end of a suspended magnetic needle. In the north of the British Isles it is some $17°$, and in the south some $23°$, south-south-east of the geographical zenith.

zephyr. A westerly breeze with pleasant warm weather supposed to prevail at the summer solstice.

zero. The point of origin in the graduation of an instrument; for example, the freezing-point of water on the CELSIUS SCALE of temperature is assigned the value of '0'. An error in the positioning of the entire scale of an instrument may be regarded as an incorrect location of the zero, and the term 'zero error' is commonly applied to it.

zodiac. The series of constellations in which the sun is apparently placed in succession, on account of the revolution of the earth round the sun, are called the Signs of the Zodiac, and in older writings give their names and symbols to the months, thus:

Month	Symbol	Month	Symbol
March	Aries, the Ram	September	Libra, the Scales
April	Taurus, the Bull	October	Scorpio, the Scorpion
May	Gemini, the Heavenly Twins	November	Sagittarius, the Archer
June	Cancer, the Crab	December	Capricornus, the Goat
July	Leo, the Lion	January	Aquarius, the Watercarrier
August	Virgo, the Virgin	February	Pisces, the Fishes

Owing to precession, the position of the equator relative to the zodiacal constellations has altered a good deal since classical times. The sun now enters Aries late in April and reaches the other zodiacal constellations with the same retardation, but in textbooks of astronomy the point at which the sun crosses the equator at the spring equinox, 21 March, is still called the first point of Aries.

zodiacal band.   A very faintly luminous band, a few degrees wide, joining the apexes of the morning and evening ZODIACAL LIGHTS.

zodiacal light.   A cone of faint white light in the night sky, extending along the ZODIAC from the western horizon after evening TWILIGHT and from the eastern horizon before morning twilight.

The phenomenon is caused by the scattering of sunlight from a cloud of particles lying in the ecliptic. The composition and origin of these particles — whether of dust or molecules or electrons, solar or terrestrial — is not yet certain. Molecular emission may also play a part.

zonal flow.   West–east airflow. East–west airflow is generally reckoned as negative zonal flow.

zonal index.   A measure of the strength of the ZONAL FLOW, either at the surface or in the upper air, for a specified (large) area and period of time. Thus, for example, the mean surface pressure difference between the circles of latitude 35° and 55° is a convenient zonal index of surface airflow for mid-latitudes of the northern hemisphere.

zonal wave number.   The number of complete wave forms, usually of the contours of a particular isobaric surface, around a selected circle of latitude. Such wave numbers (and the corresponding amplitudes) may be derived by the application of HARMONIC ANALYSIS to contour-height data.

The small wave numbers refer to large-scale atmospheric motion. Wave number one is a measure of the eccentricity of the flow relative to the pole; all other wave numbers represent those parts of the total flow which are symmetric about the pole, ridge opposing ridge in even-number flow and ridge opposing trough when the wave number is odd. See WAVE MOTION.

zone of silence.   See AUDIBILITY.

zone time.   A system of local TIME classification, differing from Greenwich Mean Time in steps of 1 hour per 15° of longitude. The individual zones are distinguished by the letters A, B, C, etc. (omitting J) for areas centred on 15°E, 30°E, 45°E, etc., respectively; and by the letters N, O, P, etc. for areas centred on 15°W, 30°W, 45°W, etc., respectively. GMT is, in this system, designated Z time.

BIBLIOGRAPHY

1. METEOROLOGICAL OFFICE; 1982, Observer's handbook. London, HMSO.

2. WORLD METEOROLOGICAL ORGANIZATION; 1975, International cloud atlas, Volume 1. Manual on the observation of clouds and other meteors. Geneva, WMO-No. 407.

3. KÖPPEN, W. and GEIGER, R.; 1928, Klimakarte der Erde. (Wall map.) Gotha, Justus Perthes.

4. NATIONAL OCEANIC AND ATMOSPHERIC ADMINISTRATION; 1976, U.S. standard atmosphere, 1976. Washington D.C., NOAA, No. NOAA-S/T 76-1562.

5. INTERNATIONAL CIVIL AVIATION ORGANIZATION; 1964, Manual of the ICAO standard atmosphere, 2nd edition. Montreal, ICAO, Doc.7488/2.

6. HUMPHREYS, W.J.; 1940, Physics of the air, 3rd edition. New York, McGraw-Hill.

7. MINNAERT, M.; 1959, Light and colour in the open air. London, Bell.

8. BEST, A.C.; 1935, Transfer of heat and momentum in the lowest layers of the atmosphere. *Geophys Mem, Meteorol Off*, No. 65.

9. HOLTON, J.R.; 1979, An introduction to dynamic meteorology, 2nd edition. New York, Academic Press.

10. SPIEGEL, E.A. and VERONIS, G.; 1960, On the Boussinesq approximation for a compressible fluid. *Astrophys J*, **131**, pp. 442–447.

11. SHAW, Sir N.; 1928, Manual of meteorology. Volume II, Comparative meteorology. Cambridge University Press.

12. BEST, A.C.; 1957, Physics in meteorology. London, Pitman.

13. EMANUEL, K.A.; 1983, On assessing local conditional instability from atmospheric soundings. *Mon Weather Rev*, **111**, pp. 2016–2033.

14. PETTERSSEN, S.; 1956, Weather analysis and forecasting. Volume 1, Motion and motion systems. New York, McGraw-Hill.

15. LIST, R.J.; 1984. Smithsonian meteorological tables, 6th edition. Washington D.C., Smithsonian Institution Press.

16. CHARNEY, J.G.; 1975, Dynamics of deserts and drought in the Sahel. *Q J R Meteorol Soc*, **101**, pp. 193–202.

17. WARREN, A.; 1984, The problems of desertification. *In* CLOUDESLEY-THOMPSON, J.L. (ed), Sahara. Oxford, Pergamon Press.

18. MONTEITH, J.L.; 1957, Dew. *Q J R Meteorol Soc*, **83**, pp. 322–341.

19. DOUGLAS, H.P.; 1930, *In* Sea and swell. *Mar Obs*, **7**, pp. 106–113.

20. AIR MINISTRY; 1959, Handbook of preventive medicine. London, HMSO.

21. CHARNEY, J.G.; 1973, Planetary fluid dynamics. *In* MOREL, P. (ed), Dynamic meteorology. Dordrecht, Reidel.

22. LILLY, D.K.; 1973, Lectures in sub-synoptic scales of motion and two-dimensional turbulence. *In* MOREL, P. (ed), Dynamic meteorology. Dordrecht, Reidel.

23. HALTINER, G.J. and WILLIAMS, R.T.; 1980, Numerical prediction and dynamic meteorology, 2nd edition. New York, Wiley.

24. MATSUNO, T.; 1966, Quasi-geostrophic motion in the equatorial area. *J Meteorol Soc Jpn*, **44**, pp. 25–42.

25. TAPP, M.C. and WHITE, P.W.; 1976, A non-hydrostatic mesoscale model. *Q J R Meteorol Soc*, **102**, pp. 277–296.

26. BROOKS, C.E.P. and CARRUTHERS, N.; 1953, Handbook of statistical methods in meteorology. London, HMSO.

27. LOVELOCK, J.E.; 1979, Gaia. A new look at life on Earth. Oxford University Press.

28. WORLD METEOROLOGICAL ORGANIZATION; 1984, Technical Regulations, Volume 1: General. Geneva, WMO-No. 49.

29. HOSKINS, B.J.; 1975, The geostrophic momentum approximation and the semi-geostrophic equation. *J Atmos Sci*, **32**, pp. 233–242.

30. HOSKINS, B.J.; 1982, The mathematical theory of frontogenesis. *Annu Rev Fluid Mech*, **14**, pp. 131–152.

31. BROOKS, C.E.P. and DOUGLAS, C.K.M.; 1956, Glazed frost of January 1940. *Geophys Mem, Meteorol Off*, No. 98.

32. PALMER, T.N., SHUTTS, G.J. and SWINBANK, R.; 1986, Alleviation of a systematic westerly bias in general circulation and numerical weather prediction models through an orographic gravity wave drag parametrization. *Q J R Meteorol Soc*, **112**, pp. 1001–1039.

33. INSTITUTE OF HYDROLOGY; Hydrological data United Kingdom. Wallingford, Institute of Hydrology. (Published annually.)

34. WORLD METEOROLOGICAL ORGANIZATION; 1966, International meteorological vocabulary. Geneva, WMO-No.182. TP.91.

35. LAMB, H.H.; 1972, Climate: present, past and future. Volume 1, Fundamentals and climate now. London, Methuen.

36. LAMB, H.H.; 1950, Types and spells of weather around the year in the British Isles: Annual trends, seasonal structure of the year, singularities. *Q J R Meteorol Soc*, **76**, pp. 393–429.

37. LAMB, H.H.; 1972, British Isles weather types and a register of the daily sequences of circulation patterns 1861-1971. *Geophys Mem, Meteorol Off*, No. 116.

38. LEVICK, R.B.M.; 1949, Fifty years of English weather. *Weather*, **4**, pp. 206–211.

39. MURRAY, R. and LEWIS, R.P.W.; 1966, Some aspects of the synoptic climatology of the British Isles as measured by simple indices. *Meteorol Mag*, **95**, pp. 193–203.

40. BLOOMFIELD, P.; 1976, Fourier analysis of time series: an introduction. New York, Wiley.

41. LACOSS, R.T.; 1971, Data adaptive spectral analysis methods. *Geophys*, **36**, pp. 661–675.

42. READ, P.L.; 1982, The analysis of umkehr observations of stratospheric ozone by a 'Maximum Entropy' method. *Q J R Meteorol Soc*, **108**, pp. 719–726.